DIGITAL DESIGN FROM SCRATCH WITH VHDL IN FPGAs

Volume 2, Running with Logical Legs

DIGITAL DESIGN FROM SCRATCH WITH VHDL IN FPGAs

Volume 2, Running with Logical Legs

Blaine C. Readler

DIGITAL DESIGN FROM SCRATCH
WITH VHDL IN FPGAs
Volume 2, Running with Logical Legs

Visit us at: http://www.readler.com

The Texas Instruments *TI LCD Programmer* calculator in the back-cover photo performs logic and arithmetic in binary and hexadecimal (and octal), and in 1982 cost $50 ($150 in today's dollars). This is now a free application with laptops.

ISBN: 978-0-9992296-6-8

Printed in the United States of America

Dedicated to all those who choose learning rather than passive entertainment.

ACKNOWLEDGEMENTS

A robust thanks to all the teachers, professors, text book and technical paper authors, mentor engineers, and vendor documentation (at least the subset that was comprehensible)—in other words, thanks to those in the world of science and technology that are competent at instructing.

There are basically two types of people. People who accomplish things, and people who claim to have accomplished things. The first group is less crowded.

—Mark Twain

Contents

Introduction

I hope you know by now (assuming you bought the book), that this is a continuation of volume 1 of the series. The material that follows makes a few references to structures covered in volume 1, but for the most part, this book stands alone. It stands alone, of course, on the assumption that you already have a grasp of the theory covered in volume 1. If you don't (have a grasp), then you had better put this book aside for the moment and bone-up (e.g., take an introductory college course, or, preferably, buy volume 1). I'm cautioning this for your sake, but in general I also like to avoid people cursing me under (or over) their breath.

A word about the exercises at the end of each chapter: you may be tempted to skip these (I probably would have when I was your age), but don't. Besides the obvious (that they solidify concepts for you), I cover new material within their context. This gives you the opportunity to discover some things on your own, which I think is always fun. In any case, I provide detailed answers to the exercise problems later in the book. Also, all Notepad++ files that I present are available for download on my website (http://www.readler.com) as text files.

As I'm sure you gleaned from the title, this book focuses on using the VHDL programming language to create designs in FPGAs. One thing this book does not do is walk you through the minutia of choosing among FPGA prototype boards, and the programming (i.e., configuration) thereof. Selection of FPGA prototype boards depends on budget and your project needs, and each vendor has its own methods and quirks regarding programming.

As I warned in volume 1's introduction, when you ultimately become adept at developing FPGA designs, you will want (and need) to understand the bare basics of electricity in order to interface to the world beyond the edges of your FPGA. Many people compare VHDL development with software coding (quite rightly) but in this, the two diverge—your FPGA will need to reach out and touch (and be touched by) the real world.

Chapter 1
Simulation

Introduction to Simulation

We now introduce the powerful, indeed indispensable, tool for FPGA development.

In a real-world application, an FPGA accepts input signals—whether from traffic sensors, or indications that a button has been pushed, or anything else that can be represented as binary values—and outputs signals that result from logical operations performed on these.

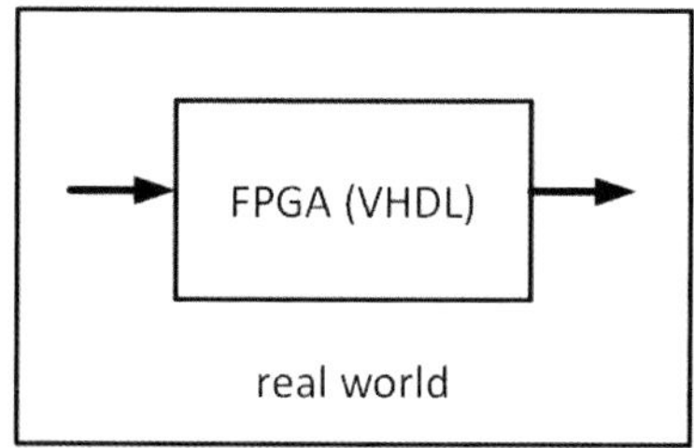

In simulation, we simply replace the real world with one that we create ourselves.

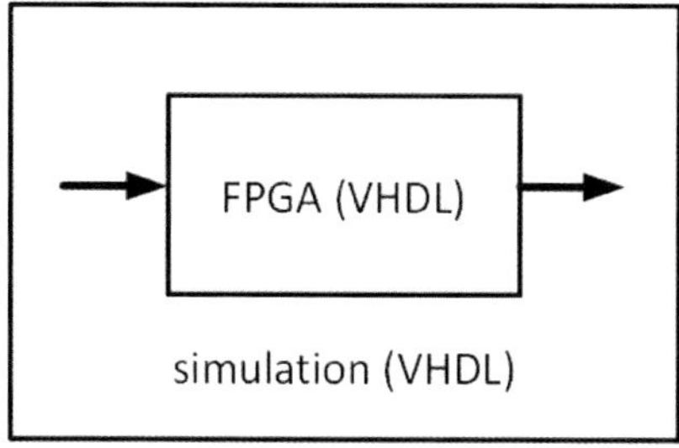

We create the simulated environment using the same coding language we use to design the logic in the FPGA, in this case, VHDL. There's no actual FPGA, just all the VHDL code that will eventually be compiled into the FPGA device.

We create the input signals to the FPGA, which we call the stimulus, and we observe the outputs—the results—to confirm that they're what we expect.

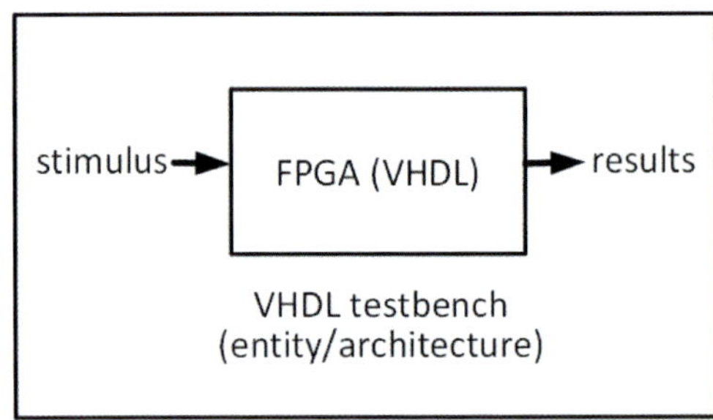

The simulation that we create is called a testbench, and consists of an entity and an architecture, the very same structure we use inside the FPGA. In our simulation, the FPGA VHDL—the top entity, if it's a modular design—is just an instantiated module inside the testbench architecture. The testbench may be just one top file, or it too may be a modular design, in which case there would be other sub-modules instantiated in the testbench top file along with that of the FPGA itself.

Whereas compilation of the FPGA logic is done using vendor-specific tools such as Xilinx's Vivado, and Altera's Quartus, simulation is generally performed using third-party simulation software.

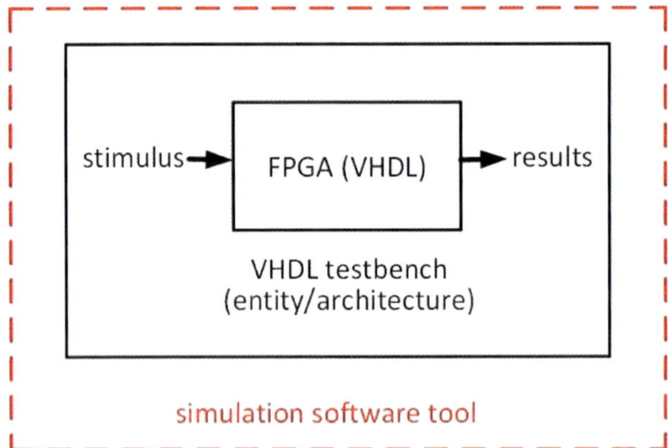

There are many full-featured simulation tool vendors, varying in price, but there are also free options available for simulating limited size designs, all of which would be candidates in this book. Also, the vendor FPGA compilation tools often come bundled with simulators, and versions of these compilation/simulation bundled software packages are generally also available for free download. These free versions can only be used with a limited set of the vendor's FPGA devices, but, again, this limitation is not an issue for the examples in this book.

I will be demonstrating with a starter version of the ModelSim simulation tool, which is available when downloading the Quartus tool from Altera—now Intel. You'll have to register with Altera/Intel to download the package (which is quite large—multiple gigabytes).

Although we do everything with VHDL in this course, I need to point out that many simulation tools allow mixed-language operation, meaning that, for example, the FPGA design may be in VHDL, while the testbench may be coded in verilog, or now, system-verilog, or vice-versa.

So, what does a simulation testbench look like?

```
entity testbench_1 is
   port (
        );
end entity testbench_1;
architecture behavioral of testbench_1 is
        } (architecture declarations and body)
end architecture behavioral;
```

Just what you're already familiar with: an entity—called "testbench_1" here—and an associated architecture.

Notice that the testbench entity port declaration list is empty. This makes sense, since the testbench represents the whole rest of the world. Every signal created and observed happens inside the testbench.

Here's the architecture.

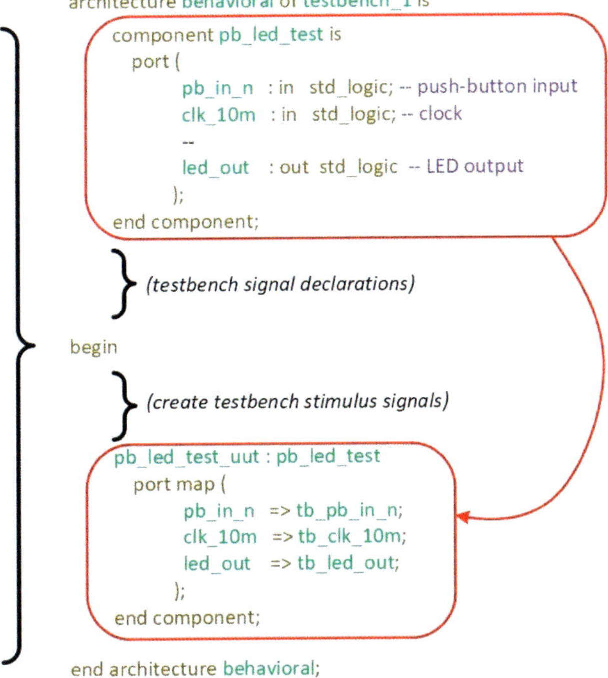
architecture behavioral of testbench_1 is
component pb_led_test is
port (
pb_in_n : in std_logic; -- push-button input
clk_10m : in std_logic; -- clock
--
led_out : out std_logic -- LED output
);
end component;
(testbench signal declarations)
begin
(create testbench stimulus signals)
pb_led_test_uut : pb_led_test
port map (
pb_in_n => tb_pb_in_n;
clk_10m => tb_clk_10m;
led_out => tb_led_out;
);
end component;
end architecture behavioral;

The design that we're simulating is declared as a sub-module, in this case the push-button LED test "pb_led_test" from volume 1, and then instantiated in the architecture body. I'm titling this instantiation as "pb_led_test_uut," where the uut stands for "unit under test." This is just the name I made up—you can use whatever you like.

The various signals that we use to stimulate and monitor the simulated VHDL "pb_led_test" design are declared as in in any other VHDL architecture, and are generated in the body.

```
architecture behavioral of testbench_1 is
    component pb_led_test is
      port (
          pb_in_n   : in  std_logic; -- push-button input
          clk_10m   : in  std_logic; -- clock
          --
          led_out   : out std_logic  -- LED output
        );
    end component;

    } (testbench signal declarations)

begin

    } (create testbench stimulus signals)

    pb_led_test_uut : pb_led_test
      port map (
          pb_in_n   => tb_pb_in_n;
          clk_10m   => tb_clk_10m;
          led_out   => tb_led_out;
        );
    end component;

end architecture behavioral;
```

This is the meat of the simulation, and we'll see a lot more about this soon.

Finally, the signals we've generated are applied to the simulated design—the "unit under test."

```
architecture behavioral of testbench_1 is
    component pb_led_test is
      port (
            pb_in_n   : in  std_logic; -- push-button input
            clk_10m   : in  std_logic; -- clock
            --
            led_out   : out std_logic  -- LED output
          );
    end component;

    } (testbench signal declarations)

begin

    } (create testbench stimulus signals)

    pb_led_test_uut : pb_led_test
      port map (
            pb_in_n  => tb_pb_in_n;
            clk_10m  => tb_clk_10m;
            led_out  => tb_led_out;
          );
    end component;

end architecture behavioral;
```

Creating Stimulus in the Testbench

We continue with just the architecture of the testbench, where, for this example, we're using the push-button test design from volume 1, "pb_led_test." Here's the FPGA design from volume 1.

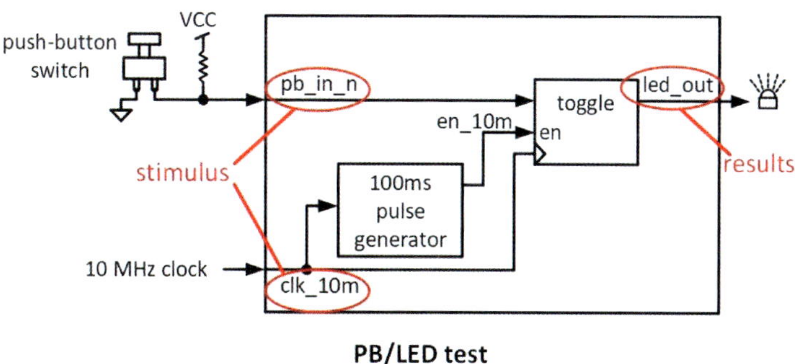

PB/LED test

The stimulus signals are "pb_in_n," and the ten MHz clock, "clk_10m." In this simple design, we have just one result, i.e., one output.

The real-world FPGA design uses a 100ms pulse generator, which samples the external push-button ten times a second. Using the provided 10MHz clock, that means that the pulse generator counts a million times between "en_10m" sampling pulses. In order to simulate this directly, we'd have to run the simulator a million clock cycles just to see two consecutive sample pulses. Although our simulator tool could do this, it would be unwieldy to observe the whole operation of the design.

Instead, for the simulation, we reduce the period of the pulse generator to just two microseconds, where now a sampling pulse is created every twenty clock cycles.

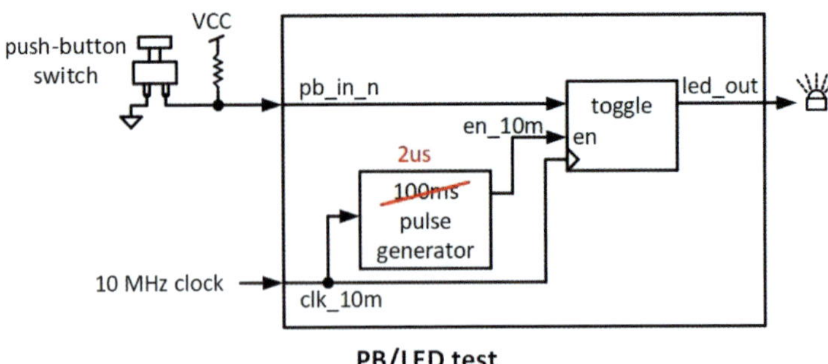

Changing the period of the pulse generator can be easily accommodated in this design, and since the pulse generator works essentially the same in both cases, other than the length of the period, we can have a good deal of confidence that what we're simulating is what will finally operate in the real word. This sort of simple modification for simulation, what we call a simulation mode, is common. If we like, once the basic design has been simulated and working, we can change the pulse generator back to its real-world 100ms operation and run one last—very long—simulation.

Here's the signal declarations.

```
architecture behavioral of testbench_1 is
    component pb_led_test is
      port (
          pb_in_n  : in  std_logic; -- push-button input
          clk_10m  : in  std_logic; -- clock
          --
          led_out  : out std_logic  -- LED output
        );
    end component;

    } (testbench signal declarations)   signal pb_in_n   : std_logic; -- push-button input
                                        signal clk_10m   : std_logic; -- clock
                                        --
begin                                   signal led_out   : std_logic;  -- LED output

    } (create testbench stimulus signals)

   pb_led_test_uut : pb_led_test
     port map (
         pb_in_n  => pb_in_n,
         clk_10m  => clk_10m,
         led_out  => led_out
       );

end architecture behavioral;
```

testbench architecture

Nothing new here. Remember that the signals that we declare are applied to the instantiated FPGA design under test.

First, we'll create the 10MHz clock, "clk_10m".

```
    signal pb_in_n        : std_logic; -- push-button input
    signal clk_10m        : std_logic; -- clock
    --
    signal led_out        : std_logic;  -- LED output

begin;

    clk_10m <= '0' after 50 ns when clk_10m = '1' else
               '1' after 50 ns;

              } (create other testbench stimulus signals)
```

We do this with one statement. This is different from everything we've seen so far. There's much that we do in simulation testbenches that we don't see in implemented FPGA designs. This is because the simulator tool is more like a software compiler than one that creates FPGA designs. After all, the simulator isn't working with actual physical connections between gates and registers, but just abstract constructs. It's sort of like playing a race-car video game, versus actually driving a car. In the video game, you can jump the gap of an opening drawbridge, survive multiple crashes, and not have to necessarily worry about fuel.

Let's see how this new statement works. The first line reads, "*Set clk_10m to zero after 50ns when clk_10m is one.*" But perhaps it makes more sense to read it in a different order: "*When clk_10m goes to one, then set it to zero after 50ns.*" Then, when clk_10m does go low, we set it back to one after another 50ns.

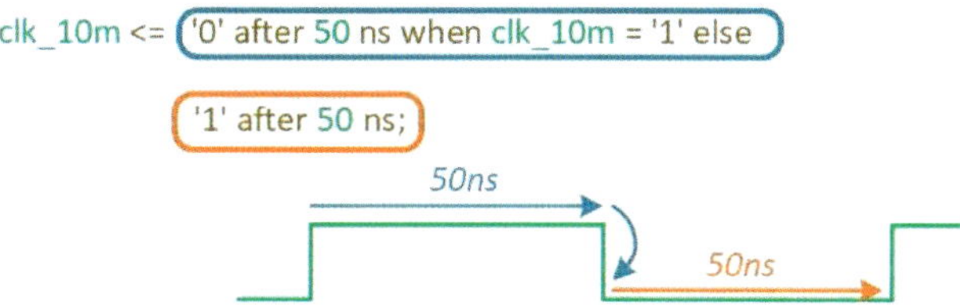

We do this over and over, creating a clock signal that is two-times-50ns, or 100ns in length—i.e., 10MHz.

Next, we generate the pb_in_n signal. Here's a timing diagram of what we expect "pb_in_not" to do, what we expect the testbench to create to drive the FPGA design, emulating what the push-button will do when pressed—the bouncing we described in volume 1:

```
    signal pb_in_n       : std_logic; -- push-button input
    signal clk_10m       : std_logic; -- clock
    --
    signal led_out       : std_logic;  -- LED output

begin;

    clk_10m <= '0' after 50 ns when clk_10m = '1' else
                  '1' after 50 ns;
```

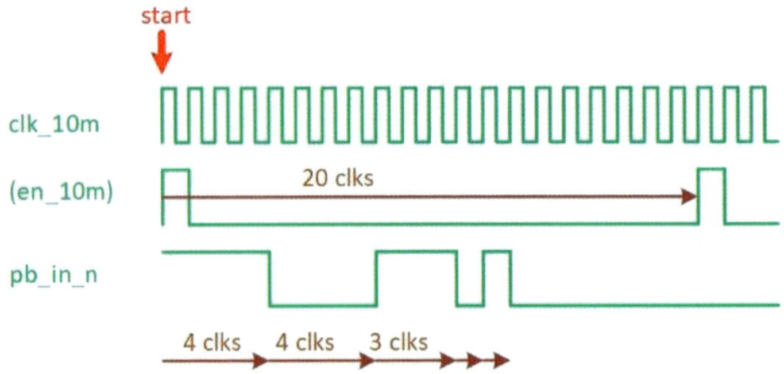

start
clk_10m
(en_10m)
20 clks
pb_in_n
4 clks
4 clks
3 clks

Note that we've made up this timing, deciding that this timing sequence is a good way to test the logic.

In the diagram, I have the button bouncing at radio-frequency rates—clearly impossible—but it would require millions of clock cycles to simulate an accurate bounce, and this vastly-shortened method adequately verifies the logic operation.

In the same way, our pulse generator generates the en_10m enable every 20 clock cycles instead of every million.

Here's one way to create that made-up "pb_in_n" signal.

```
begin;
   clk_10m <= '0' after 50 ns when clk_10m = '1' else
              '1' after 50 ns;
   stimulus : process
   begin
      pb_in_n <= '1';
      wait until rising_edge(clk_10m);
      wait until rising_edge(clk_10m);
      wait until rising_edge(clk_10m);
      wait until rising_edge(clk_10m);
      pb_in_n <= '0';
      wait until rising_edge(clk_10m);
      wait until rising_edge(clk_10m);
      wait until rising_edge(clk_10m);
      wait until rising_edge(clk_10m);
      pb_in_n <= '1';
      wait until rising_edge(clk_10m);
      wait until rising_edge(clk_10m);
      wait until rising_edge(clk_10m);
      pb_in_n <= '0';
      wait until rising_edge(clk_10m);
      pb_in_n <= '1';
      wait until rising_edge(clk_10m);
      pb_in_n <= '0';
   end process;
```

The first thing you'll notice is that we're using a process statement, and it's much different than what we're used to. Up until now we've used process statements to create clocked registers, where the sensitivity list of the process statement contained the clock signal, and we only invoked the body of the process statement when the clock went from low-to-high, the rising edge. Here, there is no sensitivity list. This is because we're not limiting the entry into—i.e., operation of—the process statement to just clock edges. In our simulation, the operation begins at the beginning of the process body (the first assignment of "pb_in_n" to '1'), and works its way through to the end. In other words, the assignment statements in a process are sequential. This is actually the way it works in our previous clocked process statements, but since we limit entry to individual points in time in those (instances associated with rising clock edges), all the signal assignments occur essentially at the same time.

The first line sets the "pb_in_n" signal to one. This is shown in the timing diagram at the "start" arrow. The next line tells the simulator to wait, to do nothing, until the next rising clock edge. This "wait" instruction is another one that you'll generally—probably always—only see in simulations, and not in FPGA compiled designs. Since there are five "wait" statements, the simulator waits that many clocks before setting "pb_in_n" to zero. Scanning down the list of statements in the process, you can see how we generate the form of "pb_in_n" shown in the timing diagram.

Since our simulation requires, at most, five clock delays at a time, it isn't too unwieldy having a separate wait line per clock. What if, however, we needed significantly more? Say, the million clocks we would need if we didn't reduce the enable period?

Here's one solution.

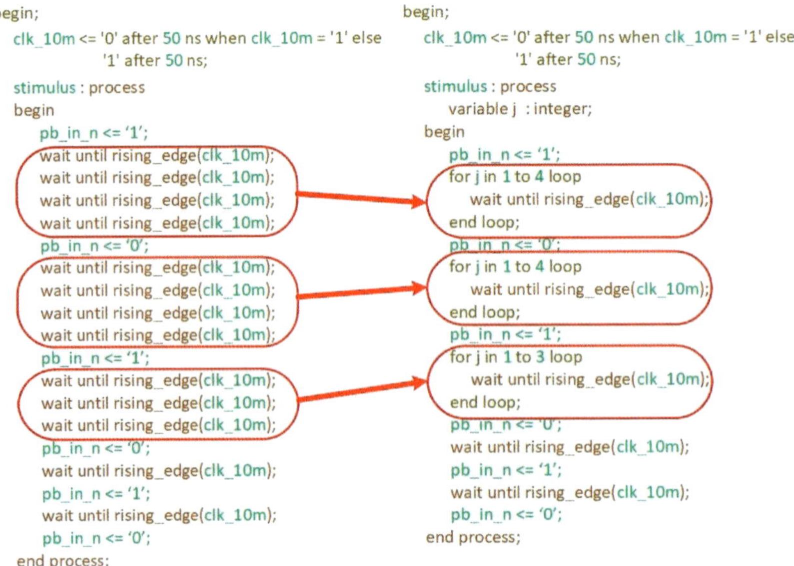

You may be familiar with "for" loops from other programming—they're ubiquitous. The loop starts with the variable "j" set to one. We then go once through the loop, in this case just waiting one clock period. The "j" variable is then incremented to two, and the loop starts over. This continues with "j" incrementing through four, and we move on to the next statement—assigning "pb_in_n" to zero. Note that we can set the loop start and stop—the one to four—to any value.

Like the wait statement, for loops are rare in compiled FPGA designs.

We've used "j" as a variable for the incrementing control of the for loop, and we can't just pull it out of the air. We have to declare it, like signals we might use. In this case, however, instead of declaring it in the declaration section of the entity architecture, we establish it within the process statement, and the declaration comes between the process label line and "begin."

```
begin;
  clk_10m <= '0' after 50 ns when clk_10m = '1' else
             '1' after 50 ns;
  stimulus : process
    variable j : integer;
  begin
    pb_in_n <= '1';
    for j in 1 to 4 loop
      wait until rising_edge(clk_10m);
    end loop;
    pb_in_n <= '0';
    for j in 1 to 4 loop
      wait until rising_edge(clk_10m);
    end loop;
    pb_in_n <= '1';
    for j in 1 to 3 loop
      wait until rising_edge(clk_10m);
    end loop;
    pb_in_n <= '0';
    wait until rising_edge(clk_10m);
    pb_in_n <= '1';
    wait until rising_edge(clk_10m);
    pb_in_n <= '0';
  end process;
```

By the way, this is the general declaration area for the process. Any variables that we declare here are local to the process statement, which means that the variable can't be used outside the process. In fact, two different process statements could use the same name for their variables since they don't exist outside the process.

Finally, if we've done all our coding correctly, the simulation should produce the "led_out" output as shown.

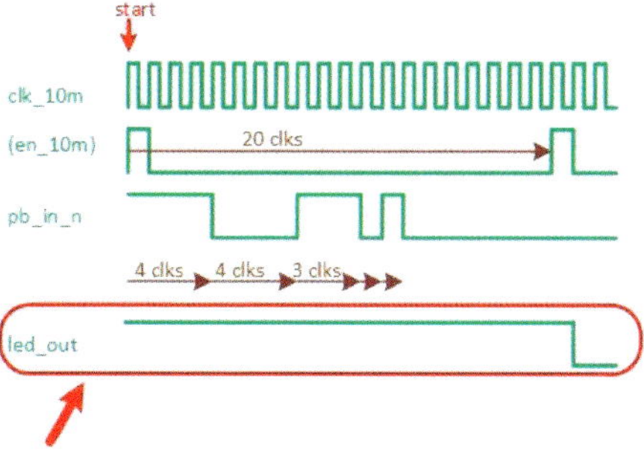

The Modelsim Simulation Tool

We now move on to actual simulations, albeit simple ones at first, and since we use Modelsim here to demonstrate simulation examples, in order to engage, you too will need the Modelsim software tool (a product of the Mentor company). There are other simulation tools that are similar to Modelsim, and although these may work for you, I encourage you to use Modelsim, particularly since it's free.

One source of free Modelsim is to download the free Web-pack version of the Altera/Intel Quartus suite. This has the advantage that you then have both simulation (Modelsim) and FPGA compilation (Quartus).

Altera/Intel Quartus

Another free source is the Modelsim vendor itself, Mentor:

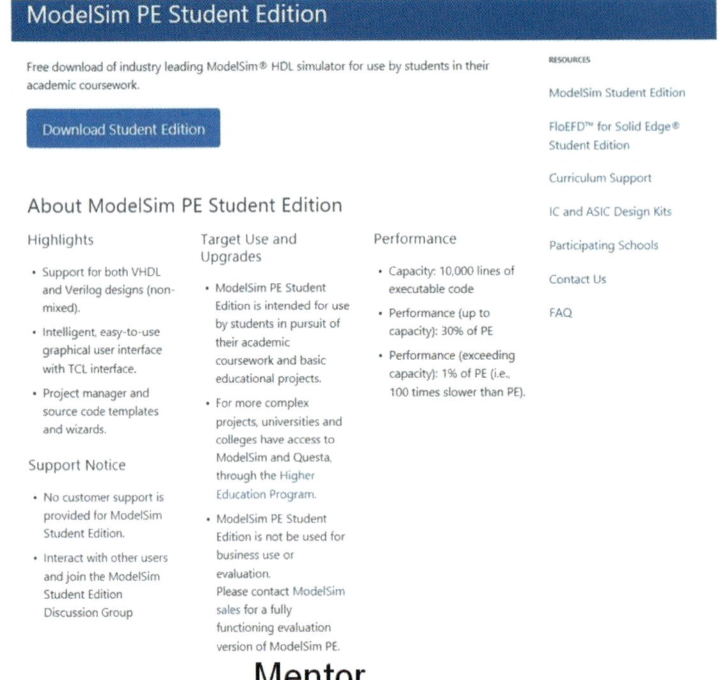

Mentor

They provide a free version for students. This student version has reduced performance, but the difference won't be noticeable for the relatively simple designs of this book. This software download is currently a bit difficult, because you must do the entire large download before registering for a free license, and if anything goes wrong on the automated process, you have de-install and download all over again.

But, again, it too is free.

Here's the testbench code as it appears in the text editor.

```
entity sim_3_7_tb is
end entity sim_3_7_tb;

architecture Behavioral of sim_3_7_tb is

    component pb_led_test is
      port (
            pb_in_n    : in    std_logic; -- push-button input
            clk_10m    : in    std_logic; -- clock
            --
            led_out     : out  std_logic   -- LED output
         );
    end component;

    constant STOP             : std_logic := '1';
    signal pb_in_n            : std_logic; -- push-button input
    signal clk_10m            : std_logic; -- clock
    --
    signal led_out            : std_logic; -- LED output

begin

      clk_10m <= '0' after 50 ns when clk_10m = '1' else
                        '1' after 50 ns;

      stimulus : process
            variable j  : integer;
      begin
            pb_in_n <= '1';
            for j in 1 to 4 loop
                wait until rising_edge(clk_10m);
            end loop;
            pb_in_n <= '0';
            for j in 1 to 4 loop
                wait until rising_edge(clk_10m);
            end loop;
            pb_in_n <= '1';
            for j in 1 to 3 loop
                wait until rising_edge(clk_10m);
            end loop;
            pb_in_n <= '0';
            wait until rising_edge(clk_10m);
            pb_in_n <= '1';
            wait until rising_edge(clk_10m);
            pb_in_n <= '0';
            wait until STOP = '0';
      end process;

     pb_led_test_uut : pb_led_test
     port map (
                pb_in_n => pb_in_n,
                clk_10m => clk_10m,
                led_out => led_out
              );

end architecture Behavioral;
```

Testbench

I'm using Notepad++, but this would be the same in any text editor.

Note that, going forward, all Notepad++ files that I present here are available for download on my website (http://www.readler.com) as text files.

The Unit Under Test is still the "pb_led_test" design from the previous volume, and this is the UUT code as entered in Notepad++:

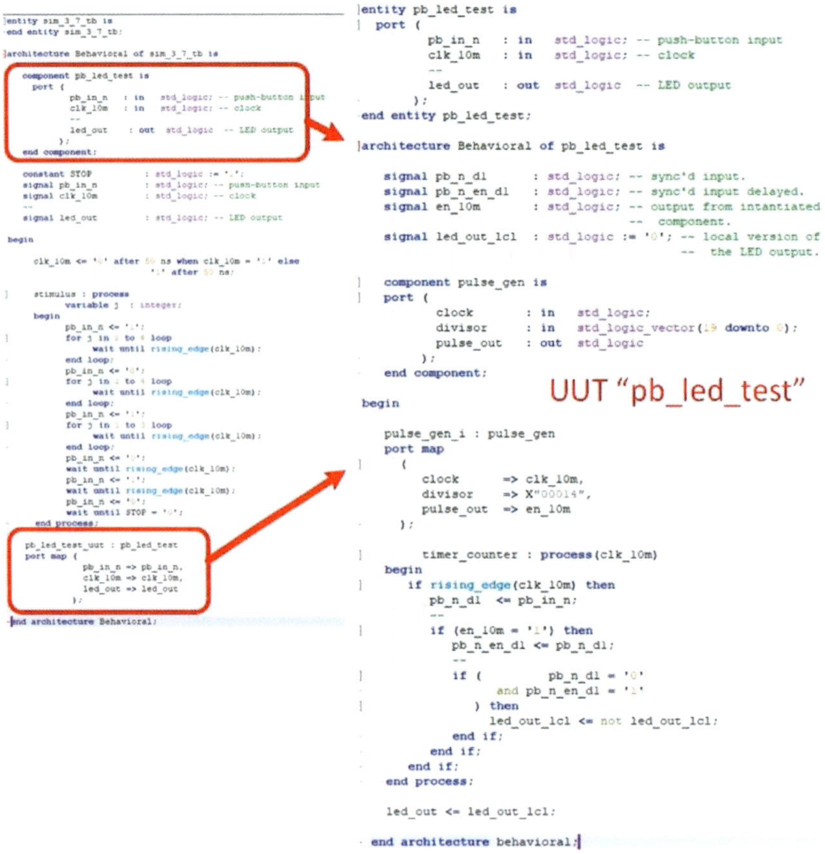

```
entity sim_3_7_tb is
end entity sim_3_7_tb;

architecture Behavioral of sim_3_7_tb is

    component pb_led_test is
      port (
            pb_in_n    : in   std_logic; -- push-button input
            clk_10m    : in   std_logic; -- clock
            --
            led_out     : out  std_logic  -- LED output
          );
    end component;

    constant STOP              : std_logic := '1';
    signal pb_in_n             : std_logic; -- push-button input
    signal clk_10m             : std_logic; -- clock
    --
    signal led_out             : std_logic; -- LED output

begin

     clk_10m <= '0' after 50 ns when clk_10m = '1' else
                          '1' after 50 ns;

     stimulus : process
           variable j  : integer;
     begin
           pb_in_n <= '1';
           for j in 1 to 4 loop
                wait until rising_edge(clk_10m);
           end loop;
           pb_in_n <= '0';
           for j in 1 to 4 loop
                wait until rising_edge(clk_10m);
           end loop;
           pb_in_n <= '1';
           for j in 1 to 3 loop
                wait until rising_edge(clk_10m);
           end loop;
           pb_in_n <= '0';
           wait until rising_edge(clk_10m);
           pb_in_n <= '1';
           wait until rising_edge(clk_10m);
           pb_in_n <= '0';
           wait until STOP = '0';
     end process;

    pb_led_test_uut : pb_led_test
    port map (
               pb_in_n => pb_in_n,
               clk_10m => clk_10m,
               led_out => led_out
              );
end architecture Behavioral;
```

UUT "pb_led_test"

```
entity pb_led_test is
  port (
           pb_in_n    : in   std_logic; -- push-button input
           clk_10m    : in   std_logic; -- clock
           --
           led_out    : out  std_logic  -- LED output
        );
end entity pb_led_test;

architecture Behavioral of pb_led_test is

    signal pb_n_dl       : std_logic; -- sync'd input.
    signal pb_n_en_dl    : std_logic; -- sync'd input delayed.
    signal en_10m        : std_logic; -- output from intantiated
                                      --  component.
    signal led_out_lcl   : std_logic := '0'; -- local version of
                                             --  the LED output.

    component pulse_gen is
    port (
            clock        : in   std_logic;
            divisor      : in   std_logic_vector(19 downto 0);
            pulse_out    : out  std_logic
         );
    end component;

begin

    pulse_gen_i : pulse_gen
    port map
      (
          clock       => clk_10m,
          divisor     => X"00014",
          pulse_out   => en_10m
      );

          timer_counter : process(clk_10m)
    begin
        if rising_edge(clk_10m) then
           pb_n_dl   <= pb_in_n;
           --
           if (en_10m = '1') then
              pb_n_en_dl <= pb_n_dl;
              --
              if (         pb_n_dl = '0'
                       and pb_n_en_dl = '1'
                 ) then
                    led_out_lcl <= not led_out_lcl;
              end if;
           end if;
        end if;
    end process;

    led_out <= led_out_lcl;

 end architecture behavioral;
```

You'll recall that this design used a sub-module called "pulse_gen" that we coded in Exercise 5-8 of volume 1, and here's the Notepad++ code for that (again, as with all upcoming Notepad++ files, available for download at http://www.readler.com).

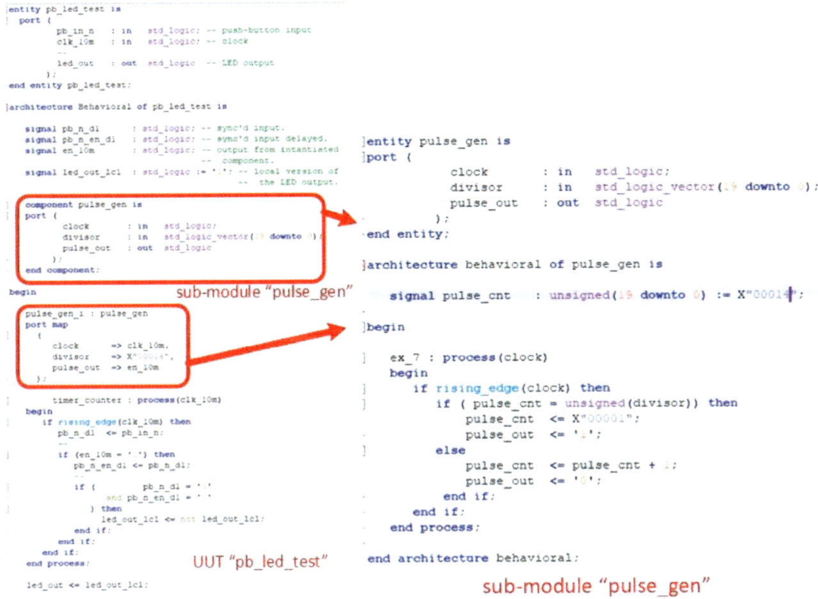

Returning to the testbench, we'll next look at how the stimulation signal "pb_in_n" is generated in the Modelsim tool. We'll look at how to use Modelsim soon, but for now, this is the correlation of each "pb_in_n" assignment as it appears in the Modelsim waveform window.

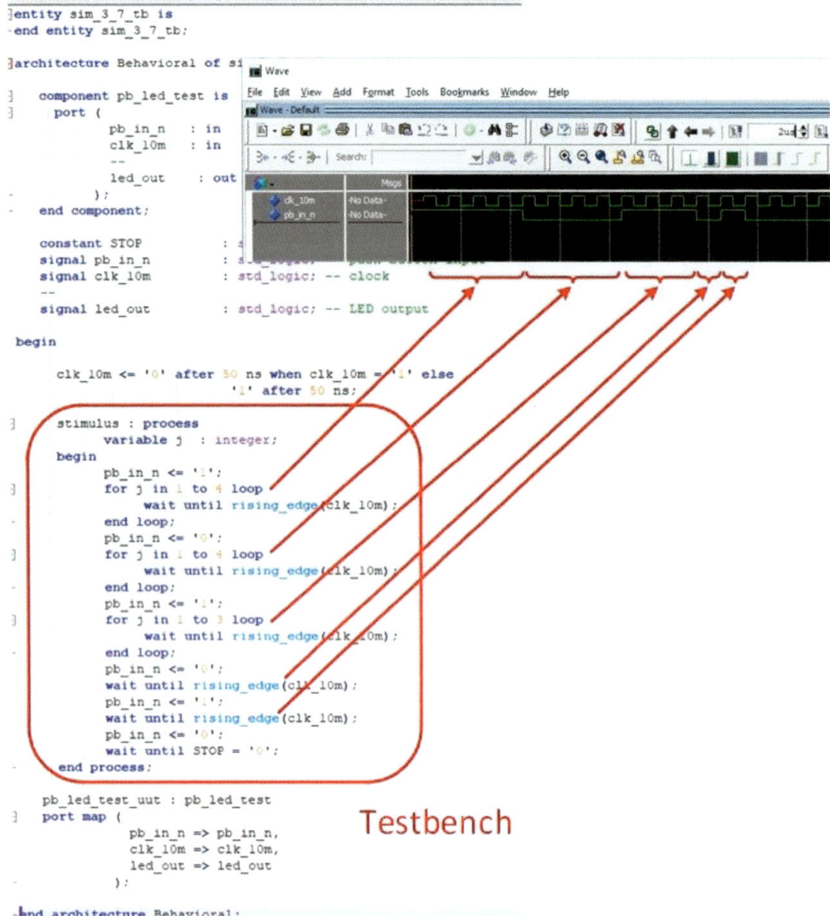

Constants and Stalling Process Statements

I included a new feature in this testbench code, a constant.

```
entity sim_3_7_tb is
end entity sim_3_7_tb;

architecture Behavioral of sim_3_7_tb is

    component pb_led_test is
      port (
              pb_in_n    : in   std_logic; -- push-button input
              clk_10m    : in   std_logic; -- clock
              --
              led_out     : out  std_logic  -- LED output
            );
    end component;

    constant STOP           : std_logic := '1';
    signal pb_in_n          : std_logic; -- push-button input
    signal clk_10m          : std_logic; -- clock
    --
    signal led_out          : std_logic; -- LED output

begin

      clk_10m <= '0' after 50 ns when clk_10m = '1' else
                         '1' after 50 ns;

      stimulus : process
            variable j  : integer;
      begin
            pb_in_n <= '1';
            for j in 1 to 4 loop
                  wait until rising_edge(clk_10m);
            end loop;
            pb_in_n <= '0';
            for j in 1 to 9 loop
                  wait until rising_edge(clk_10m);
            end loop;
            pb_in_n <= '1';
            for j in 1 to 3 loop
                  wait until rising_edge(clk_10m);
            end loop;
            pb_in_n <= '0';
            wait until rising_edge(clk_10m);
            pb_in_n <= '1';
            wait until rising_edge(clk_10m);
            pb_in_n <= '0';
            wait until STOP = '0';
      end process;

    pb_led_test_uut : pb_led_test
    port map (
                pb_in_n => pb_in_n,
                clk_10m => clk_10m,
                led_out => led_out
              );

end architecture Behavioral;
```

Testbench

If you've done any coding, you're probably already familiar with these. A constant is simply a label that refers to a fixed value, a value that's constant. Here, we define the label "stop" as an std_logic type that has a fixed value of logic '1'. We assign the value using a colon and equal sign. Anywhere that the compiler or simulator now finds the label "stop," it will replace that with a logical '1'.

We use the constant "stop" at the end of the process statement. At first glance, this statement probably looks quite strange—we're telling the simulation tool to wait until "stop" is a logical zero, but we've already defined it to be fixed as a logical one. This statement will cause the simulator to wait here forever, and this is exactly what we want. You see, the way a process statement works is that once it reaches the end, it automatically starts over again. If we don't force the process to stop here and wait forever, it will continually generate the pattern we saw in the waveform window, over and over. There are times when we may want the process to create repeating patterns, but this is not one of them.

We notice that we have a similar colon-and-equal-sign assignment in the "pb_led_test" module.

```
entity pb_led_test is
  port (
         pb_in_n    : in   std_logic; -- push-button input
         clk_10m    : in   std_logic; -- clock
         --
         led_out    : out  std_logic  -- LED output
       );
end entity pb_led_test;

architecture Behavioral of pb_led_test is

    signal pb_n_dl        : std_logic; -- sync'd input.
    signal pb_n_en_dl     : std_logic; -- sync'd input delayed.
    signal en_10m         : std_logic; -- output from intantiated
                                       -- component.
    signal led_out_lcl    : std_logic := '1'; -- local version of
                                              -- the LED output.

    component pulse_gen is
    port (
          clock       : in   std_logic;
          divisor     : in   std_logic_vector(19 downto 0);
          pulse_out   : out  std_logic
         );
    end component;

begin

    pulse_gen_i : pulse_gen
    port map
       (
        clock       => clk_10m,
        divisor     => X"00014",
        pulse_out   => en_10m
       );

        timer_counter : process(clk_10m)
    begin
       if rising_edge(clk_10m) then
          pb_n_dl   <= pb_in_n;
          --
          if (en_10m = '1') then
             pb_n_en_dl <= pb_n_dl;
             --
             if (        pb_n_dl = '0'
                     and pb_n_en_dl = '1'
                ) then
                  led_out_lcl <= not led_out_lcl;
             end if;
          end if;
       end if;
    end process;

    led_out <= led_out_lcl;

end architecture behavioral;
```

UUT "pb_led_test"

Unlike a constant assignment, when we include a value assignment in a signal's declaration, we are simply telling the tool to start off with the signal at this value, not keep it fixed forever. An FPGA complier tool generally defaults to signals being set initially to zero, but simulation tools don't assume that privilege. If we don't tell Modelsim explicitly what we want to start off with, it assumes that the signal state is unknown, and this unknown state can often then propagate through the rest of the simulation run. In this example, we have "led_out_lcl" toggling under certain conditions. If the simulator doesn't know what the initial state is, it can't know what the toggled, i.e., inverted state would be. In this example, we want "led_out_lcl" to start as a logical '1'.

Simulation Results

Finally, here's the simulation results as they compare with the expected results from earlier in the chapter. I've included some signals from the "pb_led_test" sub-module in the waveform window (under the heading label "pb_led_test").

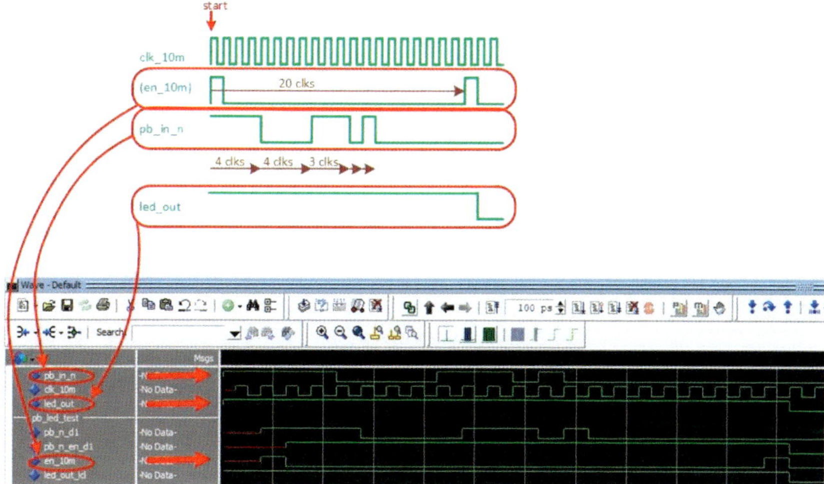

Notice that some signals (e.g., "pb_n_d1") are initially indeterminate, meaning that the simulator does not know what their state is until prior signals contributing to them become known.

We generated the input, "pb_in_n" directly in our testbench, and so it, of course, matches, and, we see that our design does indeed generate the sample enable "en_10m" correctly. And, finally and importantly, the output, "led_out," works as desired.

With a successful simulation, we now have good confidence that the logic will operate correctly when implemented in an FPGA.

Creating a Modelsim GUI Project

Now that we've wetted your appetite for simulation, let's see how to use Modelsim. We'll explore two fundamentally different methods of using Modelsim: using the tool's GUI windows, and running the tool via script files.

We'll start with the GUI approach, but first, we need to set up the folder structure that we'll use.

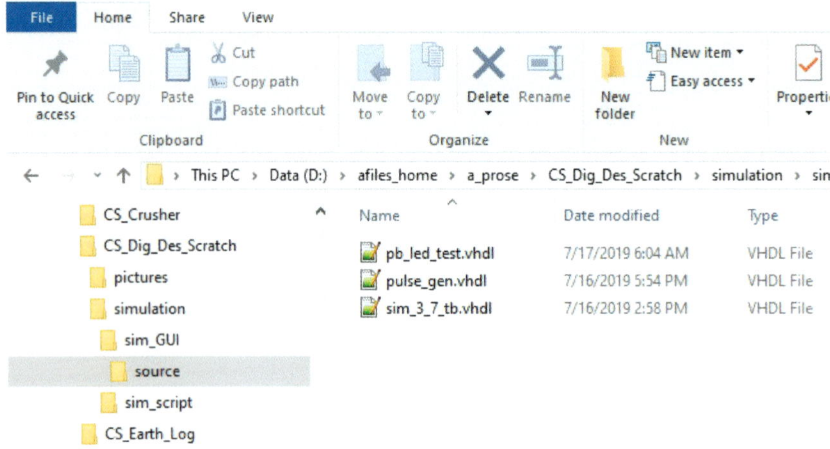

I've created a dedicated folder area called "sim_ GUI" under a general "simulation" folder for this project, and under that a folder called "source," where I've copied the three VHDL source files. There is no prescribed folder structure—this is just how I do it.

Next, we launch Modelsim, and this is how it looks initially.

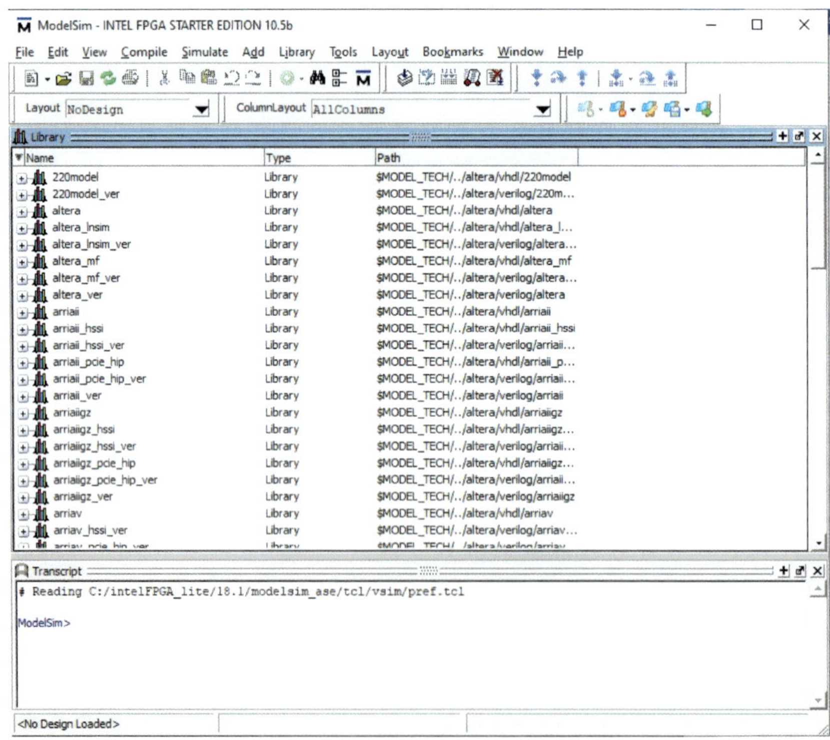

It comes with a large set of libraries, many associated with the Altera/Intel FPGA cores. You can ignore these.

The next step is setting up the project. We select File/New/and Project:

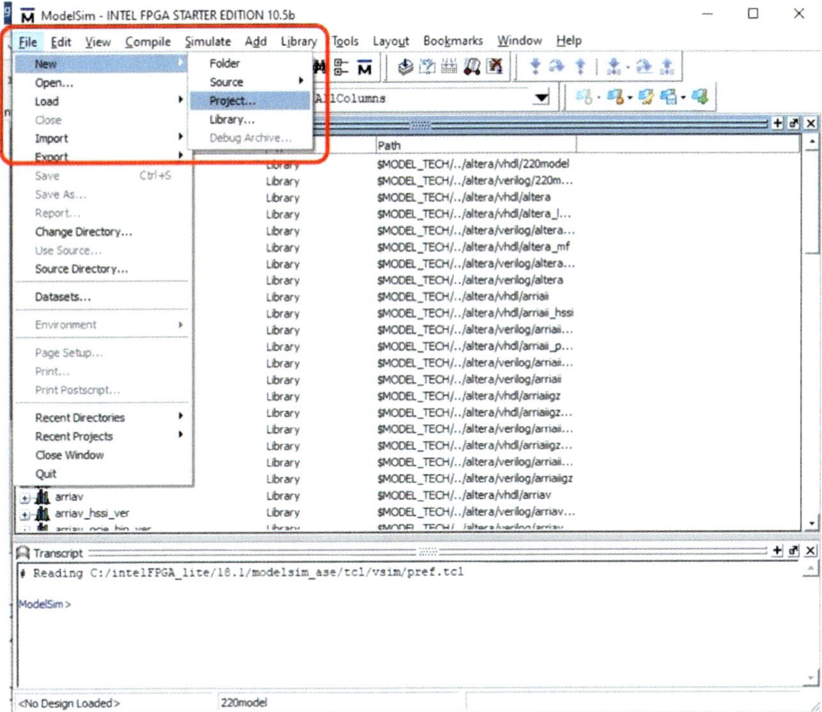

And this is the resulting GUI window:

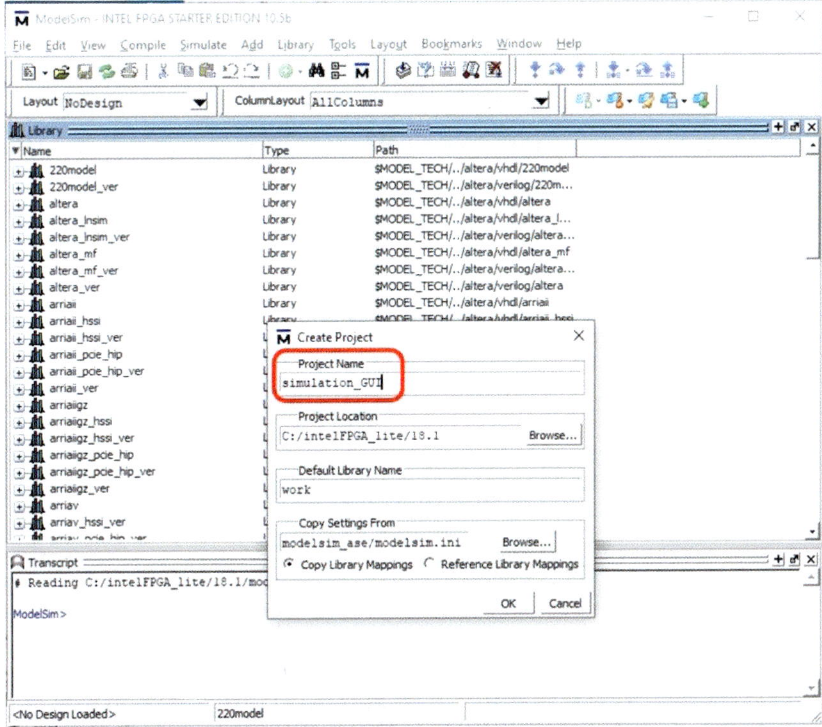

Type the name of the project, in this case, I chose "simulation_GUI"—it can be anything.

Then, browse to where you want the project to reside.

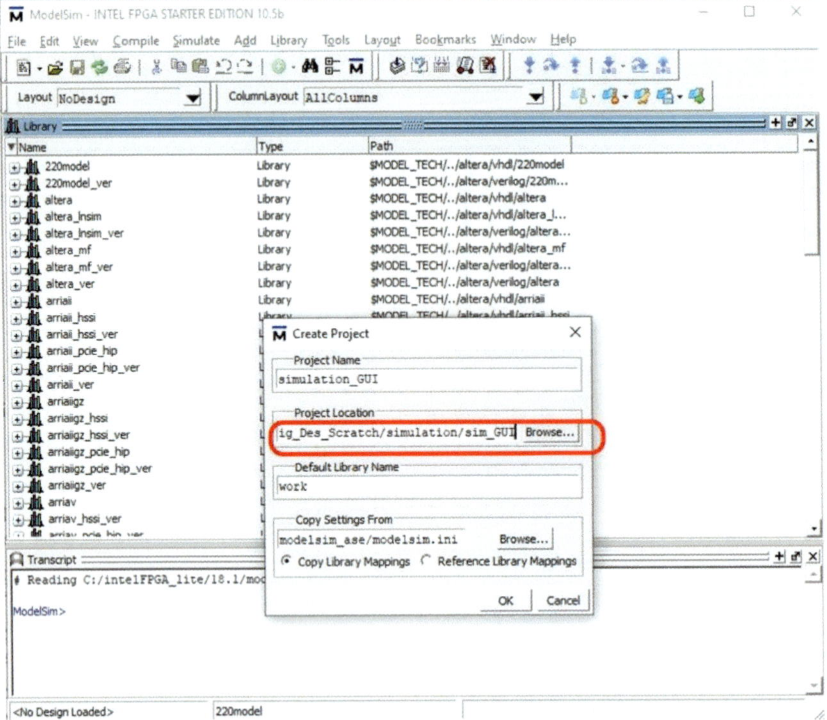

In my case, I've located it in my sim_GUI folder that I created.

When you click "OK," this window pops up, and now you select "Add Existing File":

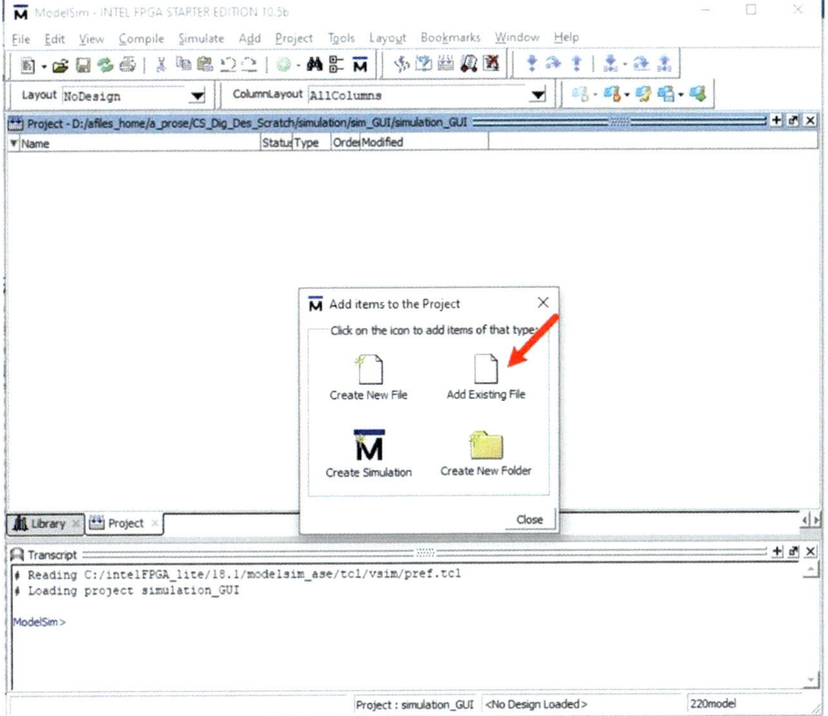

Again, you browse, but this time to where you're keeping your source files, in my case, under the "source" sub-folder.

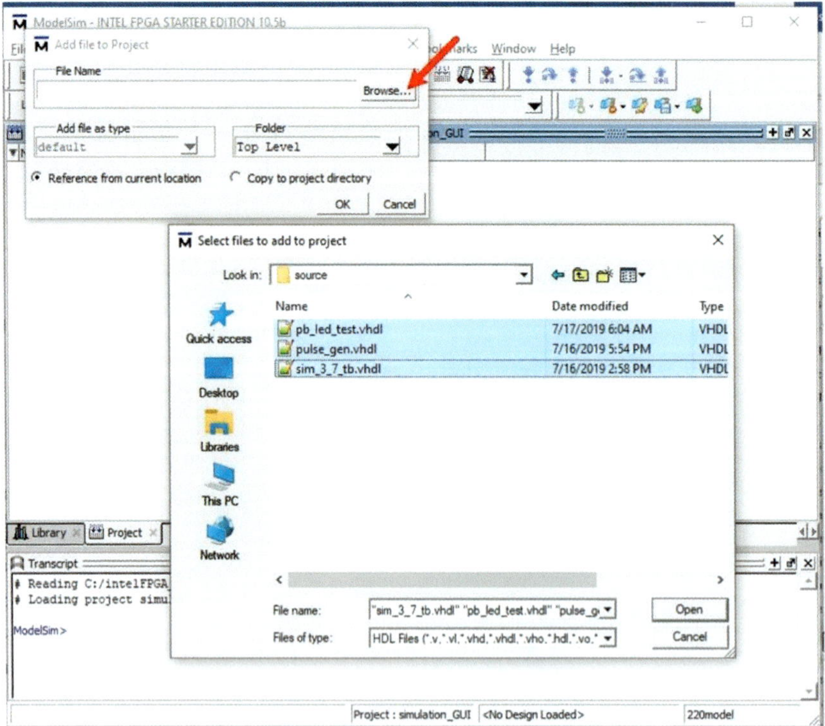

Once you select the files and click "OK," they appear in the project window. You can select them all at once, or one at a time.

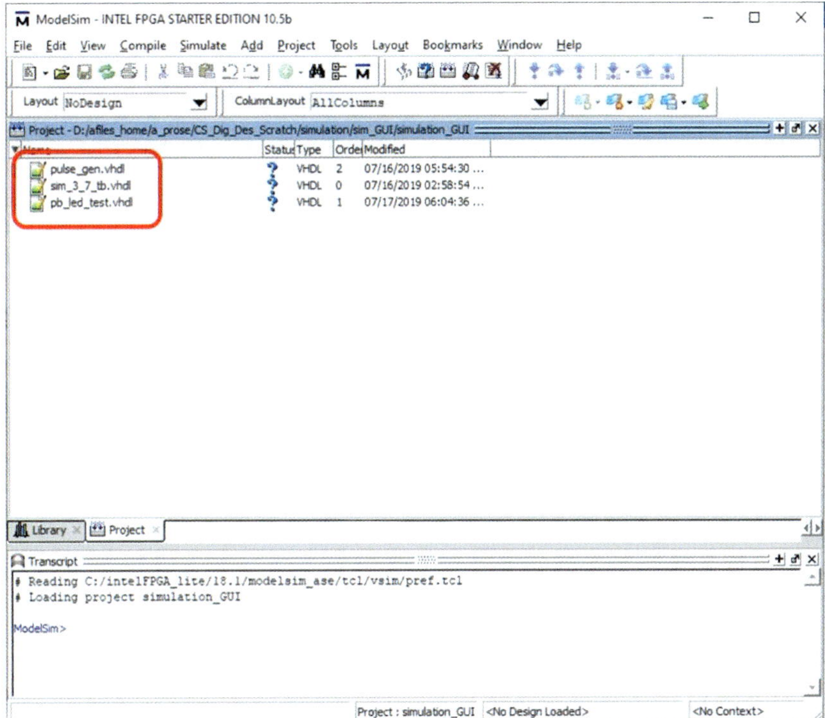

When you're done, click the "Close" button.

Next, we compile the source files. Here, Modelsim analyzes the files, makes sure there's no syntax or connection problems, and creates working files for itself that it stores in a folder it creates called "work."

We compile by selecting "Compile All" in the "Compile" menu.

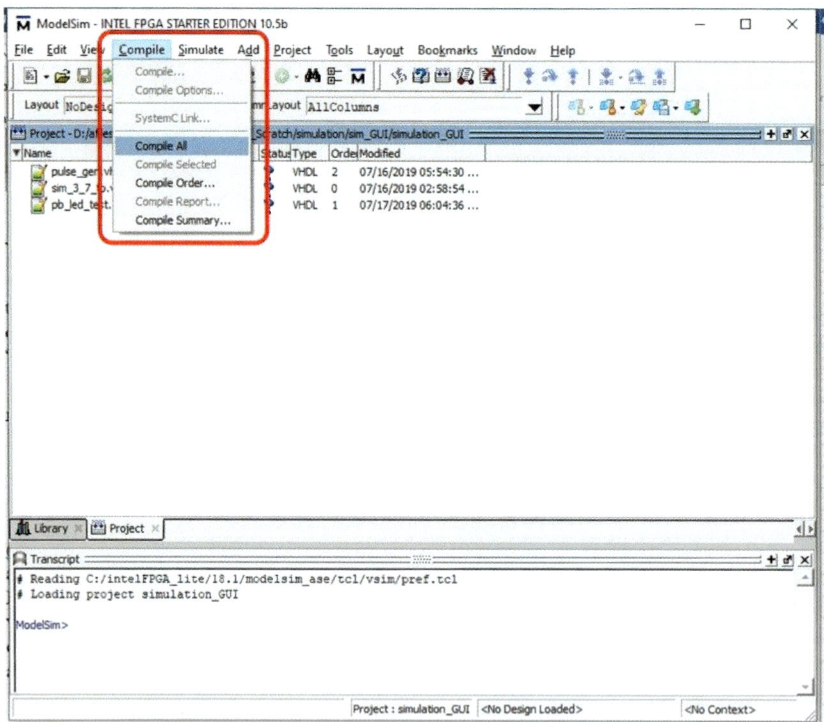

If all goes well, Modelsim indicates this with green check marks, and green messages in the transcript window. Green is good. If Modelsim finds a problem, it indicates it in red in the Transcript window, and if you double-click on the red message, it shows you details about the problem. After fixing the problem, you compile again.

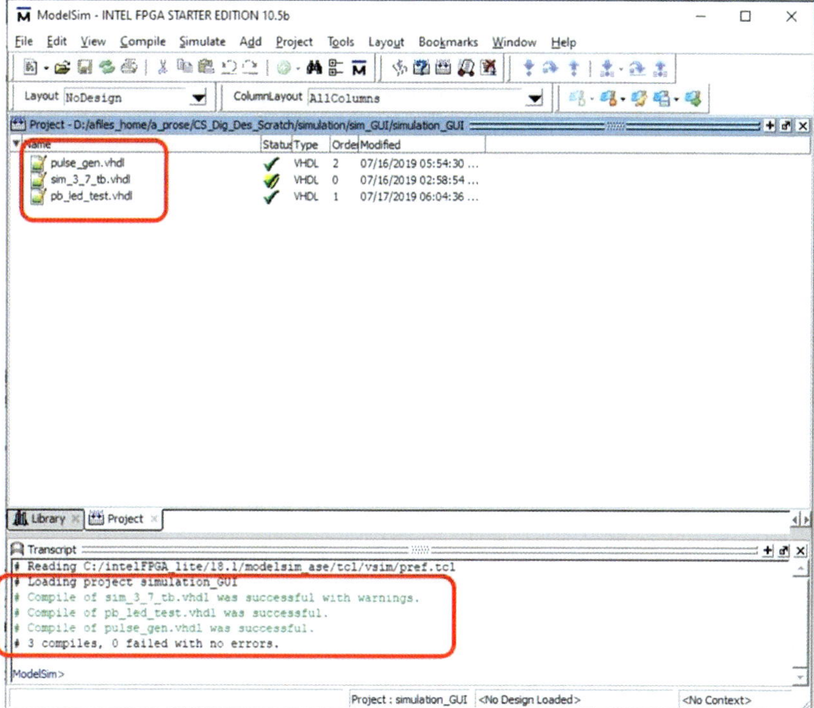

Once we get a successful compile, we're ready to simulate via the "Start Simulation" in the "Simulate" menu.

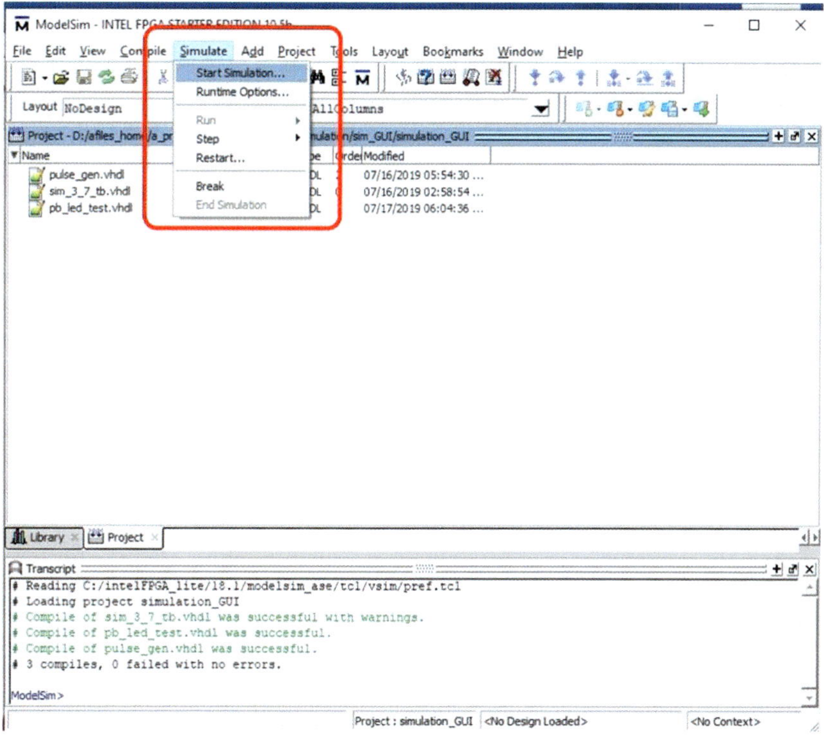

In the pop-up window, we select the testbench file, essentially telling Modelsim to simulate that file. As it happens, since all the rest of the files are included hierarchically under the testbench, they too will all be included in the simulation.

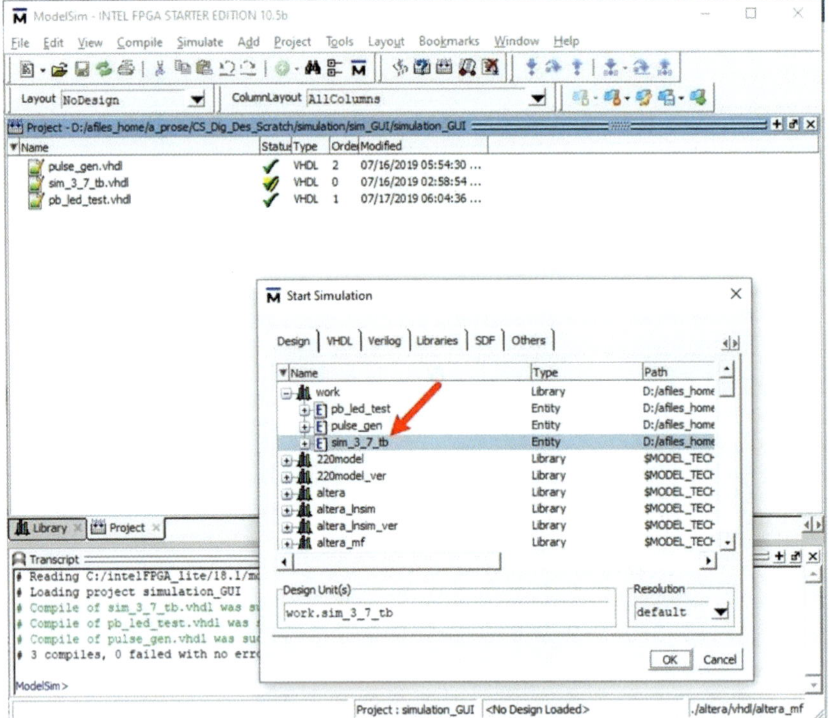

Click "OK" to simulate, and several more GUI windows appear in the overall project window. One of them is the "sim" window, showing the various instances that were compiled.

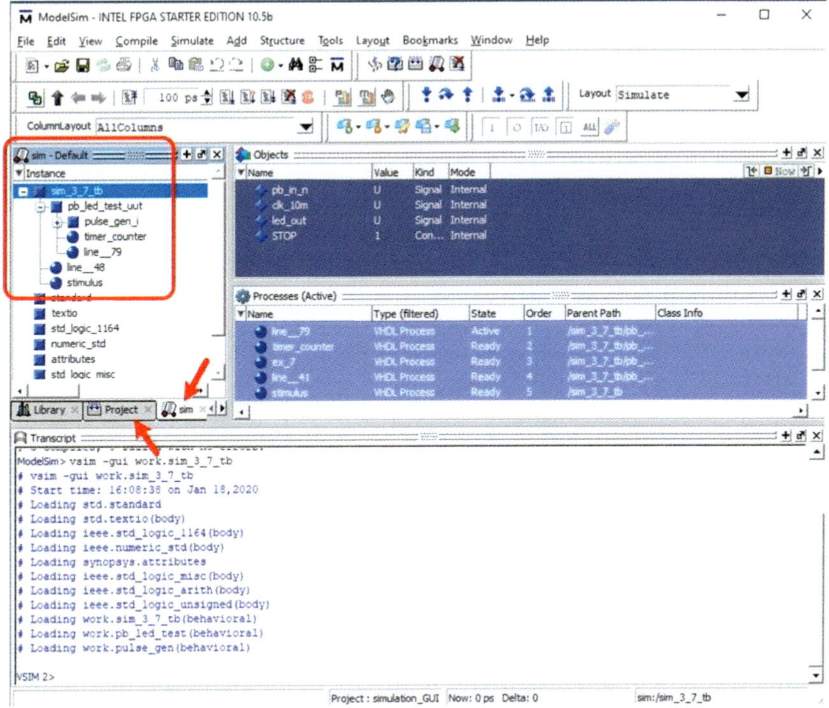

Note that Modelsim may find problems here as well—usually connection errors—and will indicate them with red entries in the "transcript" window. Note also that we haven't lost our original "Project" window listing the included files (it's underneath "sim" via a different tab).

Associated with whatever entity, i.e., source file, is selected in the "sim" window, are the various signals that you declared in that file. Here, we have the signals in the testbench.

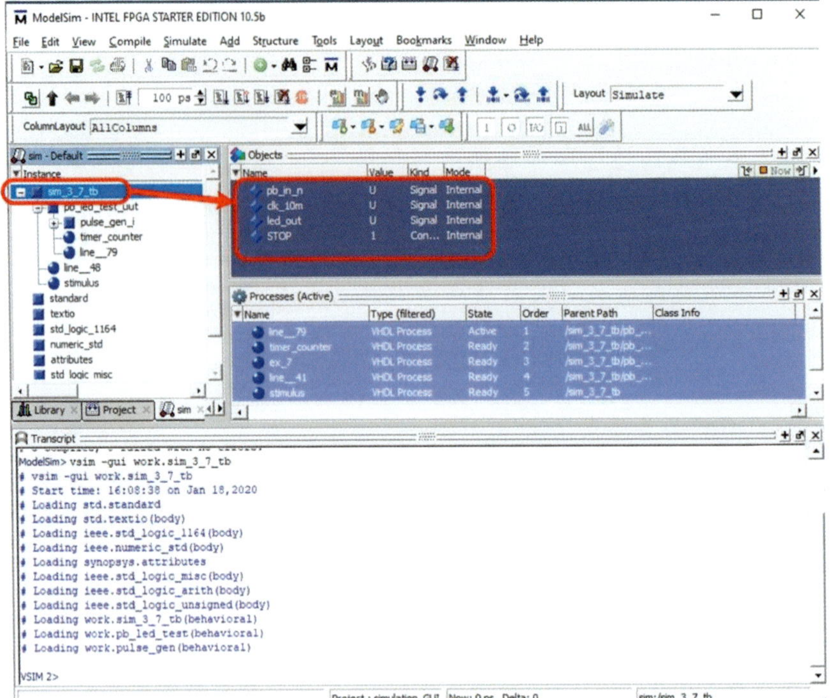

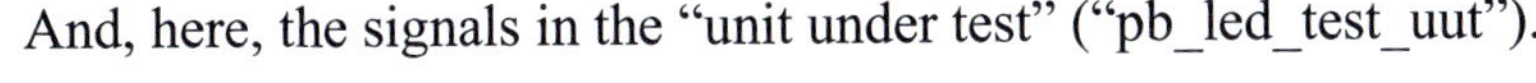

And, here, the signals in the “unit under test” (“pb_led_test_uut”).

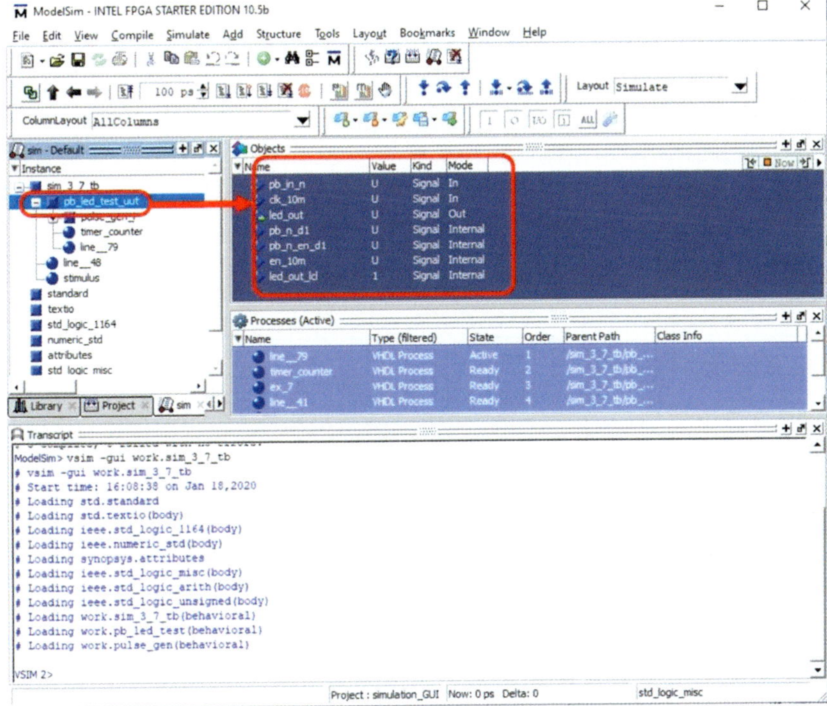

This is important to remember, since we use this window to add signals to our waveform.

Modelsim Waveforms

Next, we build our waveform. We first select the "Wave" option in the "View" menu.

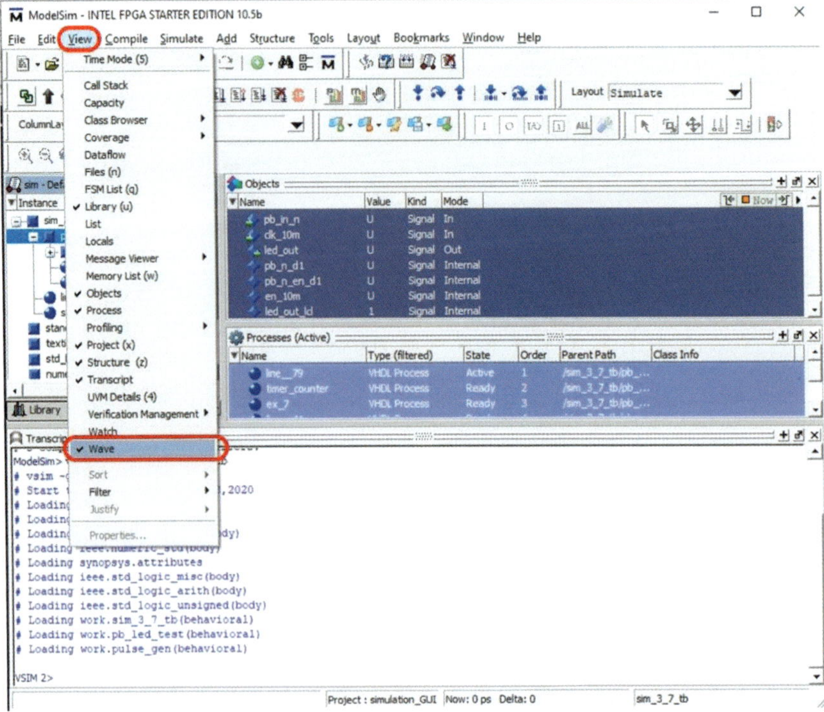

This may be automatically selected, but, if not, then select it (the wave window then opens) and start the simulation over again.

The wave window is initially empty, i.e., there are no signals selected:

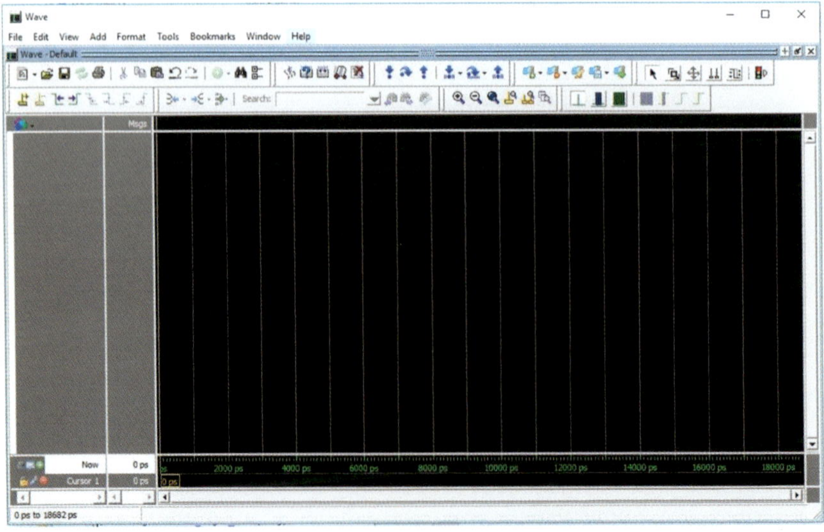

Go back to the main Modelsim window, and select the signals you want to include in the waveform. Here, I've selected three signals declared and generated in the testbench.

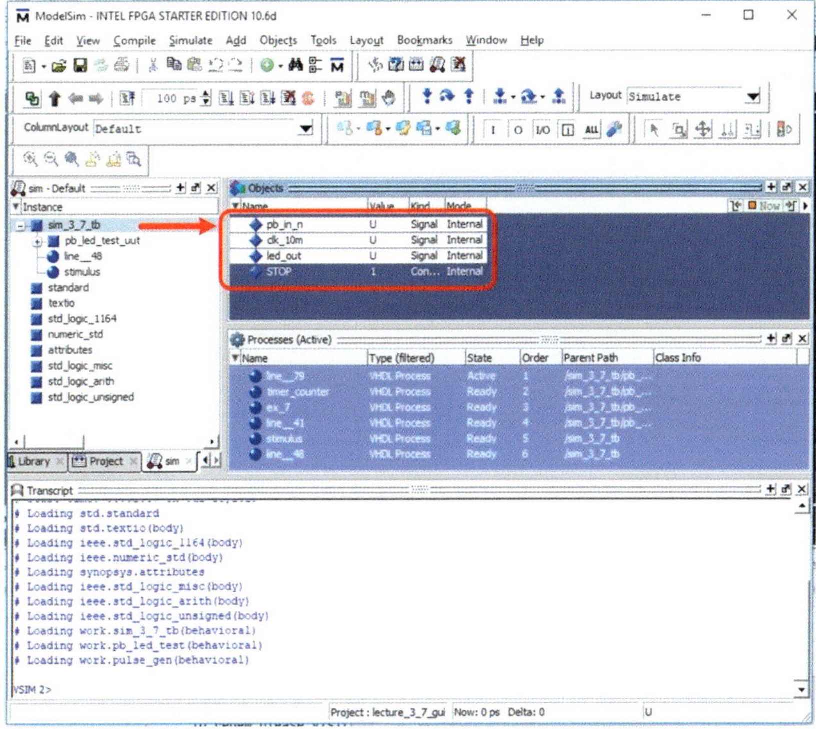

I don't need the "STOP" constant, since I know what that is. Once you've selected the signals you want, copy them (via a right-click or Ctl-C), and paste them into the wave window.

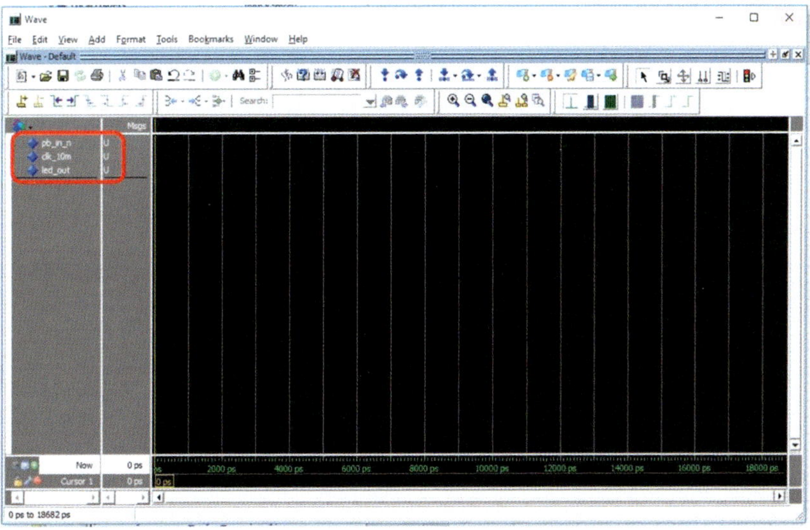

These signals are from the testbench, and we'd like to include some signals in the files that we're simulating, but first we'll create a divider by right-clicking in the signal window, and then selecting "New Divider."

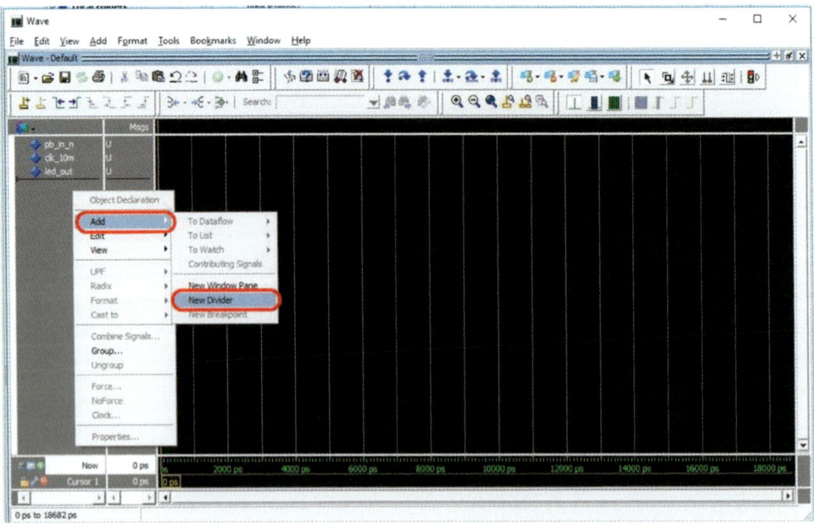

Enter the label. In this case, I've chosen the name of the source file.

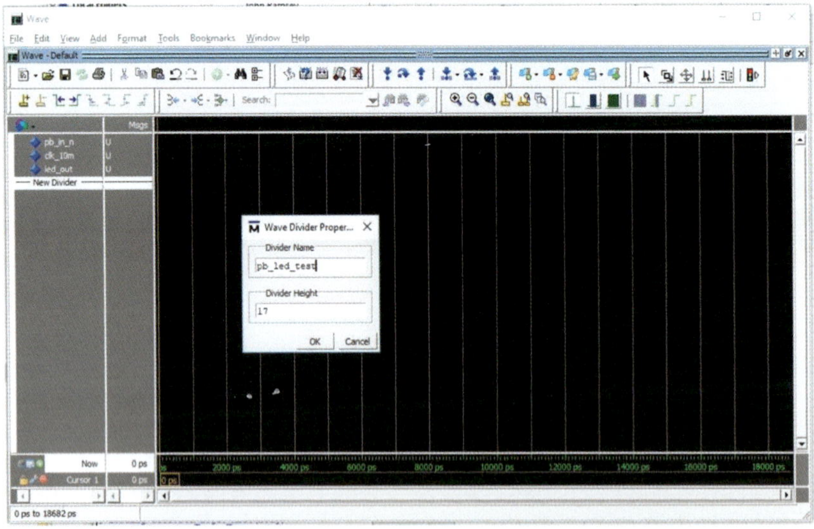

We now go back to the main Modelsim window and select the next instance in the sim window (pb_led_test_uut), and the signals declared in that file are shown.

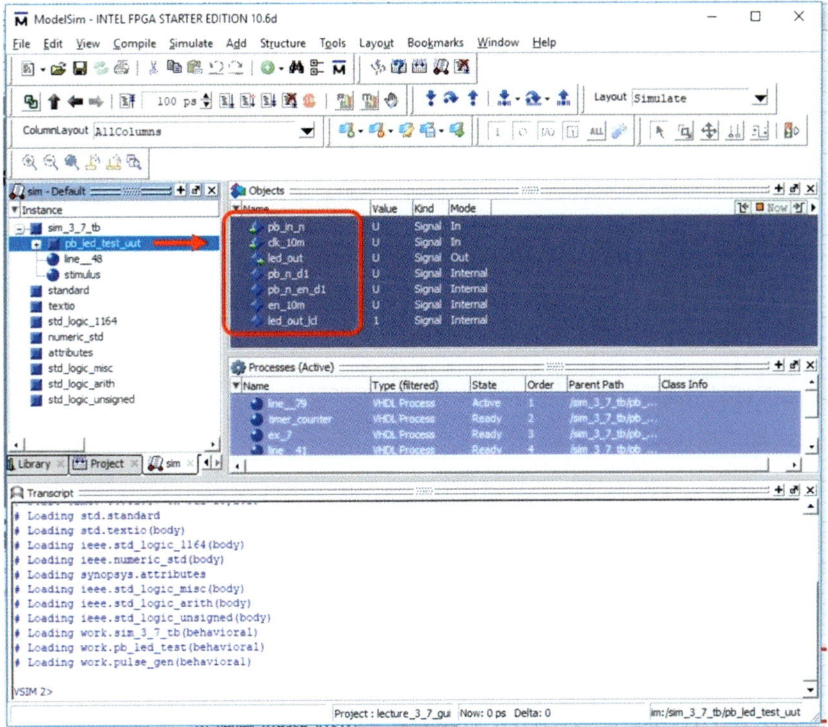

I select a subset of the signals (since the others are already included in the testbench set), and copy them,

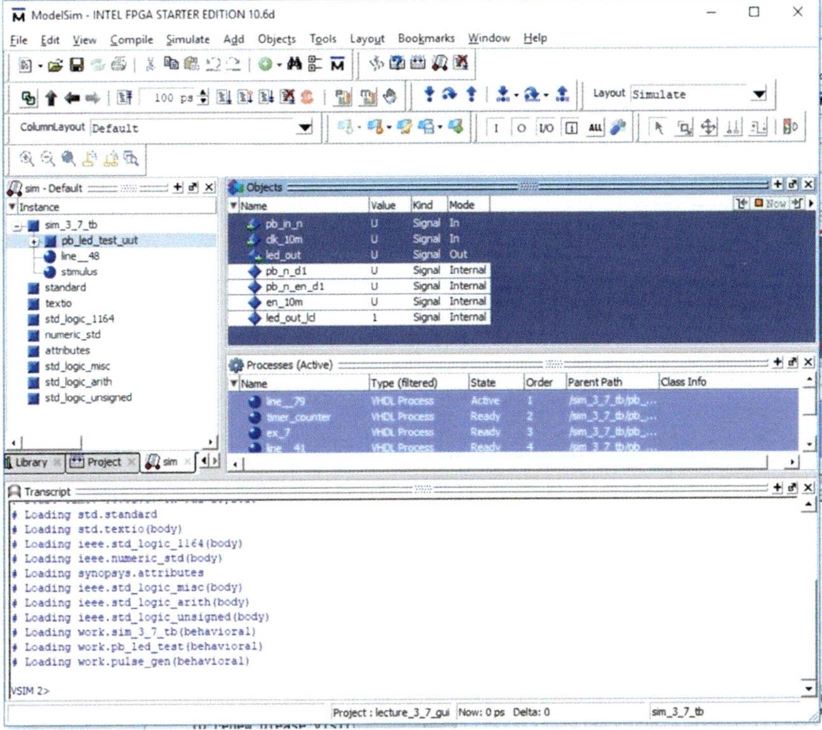

… and when I paste them into the waveform window, we see that they are inserted above the divider.

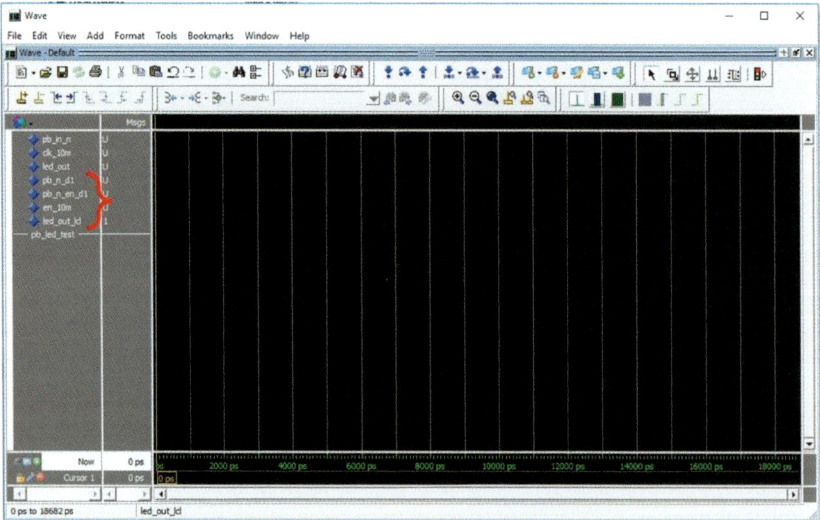

You can simply select the divider and slide it back up.

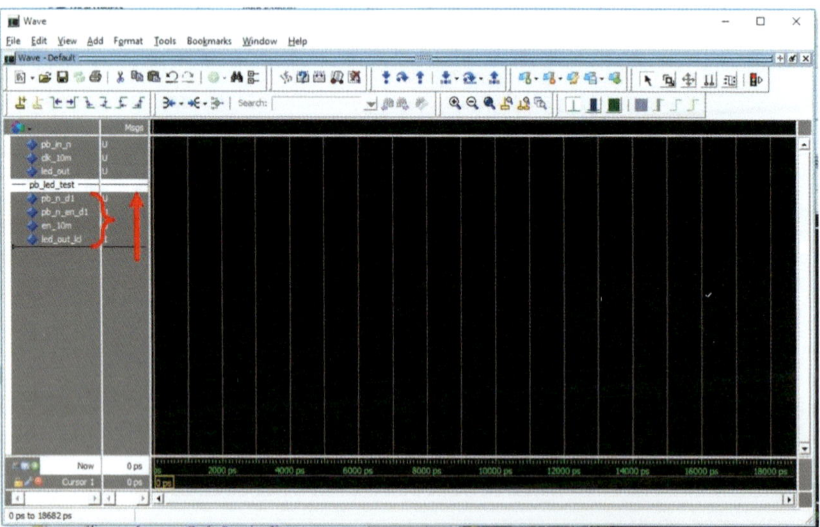

Be sure to save your waveform anytime you make changes.

Running Modelsim GUI Simulations

We're now ready to actually run a simulation—note that when we "started" the simulation earlier, we were just telling Modelsim to get ready. First, though, we tell it how long to run. We type the value into this window.

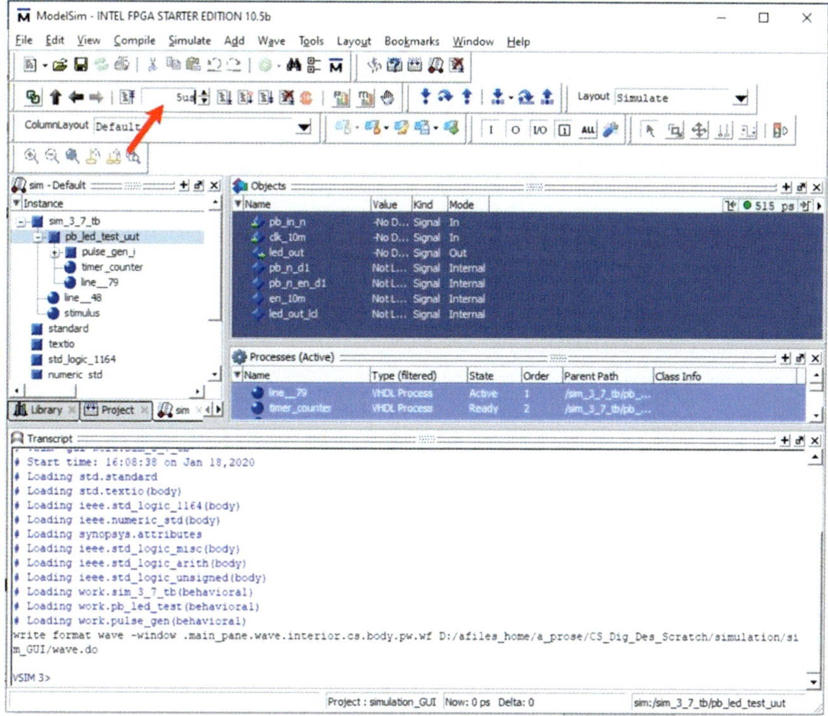

I've chosen 5 microseconds in order to match that of the earlier example. Note that Modelsim recognizes "us" as microseconds, "ms" as milliseconds, and "ns" as nanoseconds.

We kick off the simulation run by clicking this button:

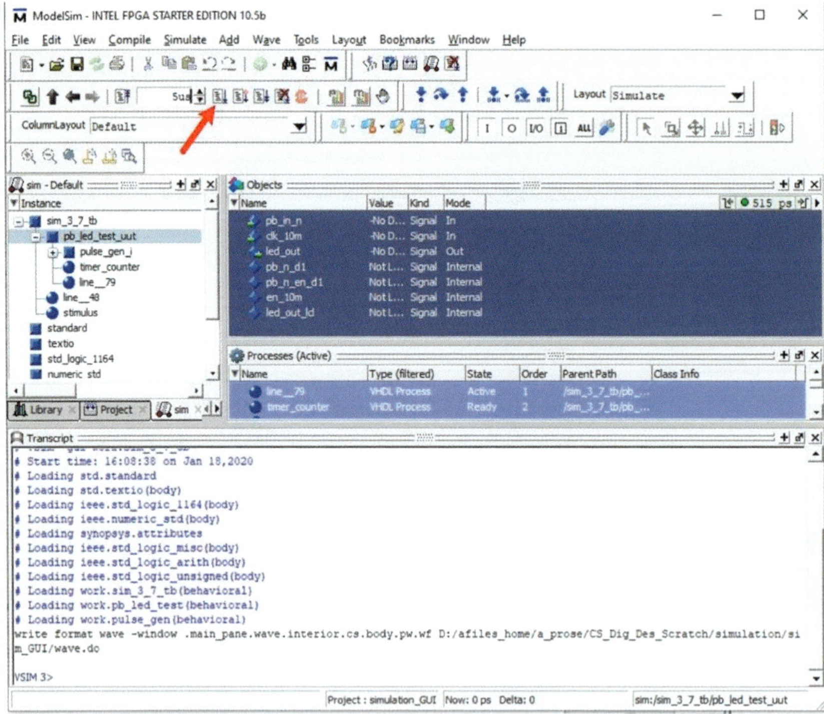

… and, the simulation results appear in the waveform window:

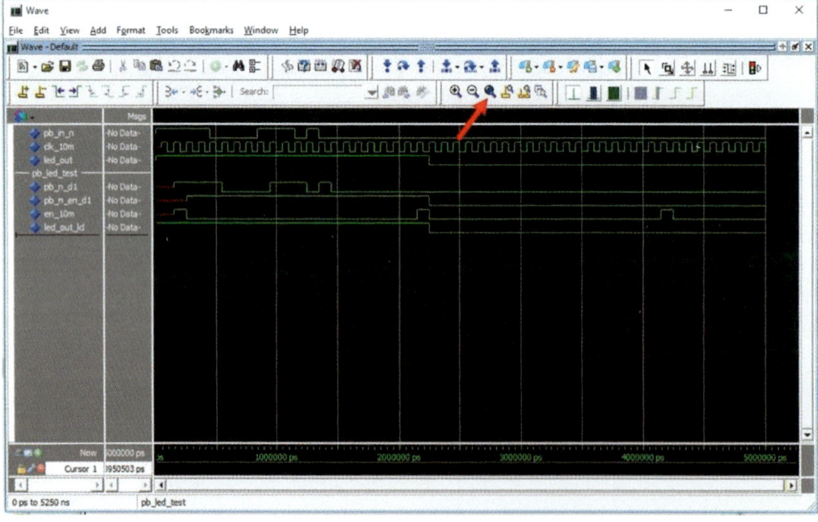

Modelsim expands the waveforms to fit the window when you click the button shown.

Creating a Modelsim Script Project

After all that window-clicking, we'll now look at an alternative way to use Modelsim, one that's much easier and quicker, particularly when running involved simulations. We generate a text file—called a script file—that contains explicit instructions for Modelsim to follow.

First, we set up our folders.

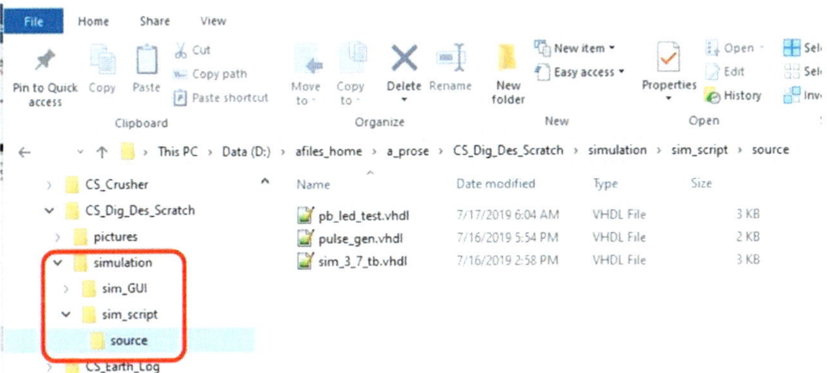

Under the general "simulation" folder already created, I've added a "sim_script" folder parallel to the GUI project we just finished, and under that, a similar source folder (containing the same VHDL source files as that of the GUI version).

I've added a "run.do" file in our main project folder.

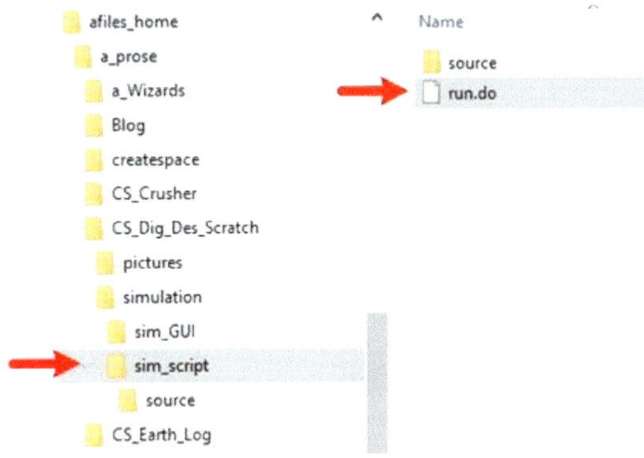

This is the script file. We'll look at this in a moment, but note that when I run this within Modelsim, the tool will create its project in this same folder.

This is the script file. I named it "run.do", but it can be any name. You can use any file extension you like (e.g., ".txt" or ".bat"), but ".do" is fairly standard for Modelsim script files.

```
1
2  # cd D:/afiles_home/a_prose/CS_Dig_Des_Scratch/simulation/sim_script
3
4
5  vlib work
6
7  vcom -work work -2002 -explicit source/pb_led_test.vhdl
8  vcom -work work -2002 -explicit source/pulse_gen.vhdl
9  vcom -work work -2002 -explicit source/sim_3_7_tb.vhdl
10
11 vsim work.sim_3_7_tb
12
13 #do wave.do
14
15 run 5 us
16
```

The first line is something I add just for convenience—note that it's commented. It's the complete path name to my main simulation project folder (yours, of course, will be different). Each time I open Modelsim, I copy and paste this (minus the pound sign) into the Modelsim transcript window. It's a shortcut method to get Modelsim where it needs to be for the scripting operation. Note the non-Microsoft forward slashes.

The next line (#5) tells Modelsim to set up its work area. With this command, Modelsim creates the "work" folder under the project folder. The three "vcom" commands (#7-9) tell Modelsim to compile each of the source files in turn, and we direct Modelsim to the "source" sub-directory where we've placed our VHDL source files. Note again the non-Microsoft forward slashes.

The "vsim" command (#11) tells Modelsim to set up the simulation, using the testbench file as the top of the hierarchy.

The "do wave.do" command (#13) would normally tell Modelsim to set up the waveform window via another script file called "wave.do", but for now it's commented, since the "wave.do" file hasn't been created yet. We could create this manually, but, as we'll see in a minute, we'll let Modelsim do that.

Finally, we tell Modelsim to run the simulation for five micro-seconds (#15).

Let's run the script through Modelsim. If you haven't opened Modelsim yet, launch it now. If you've been working with it, then close the current project (File => Close Project). You should see this empty project window:

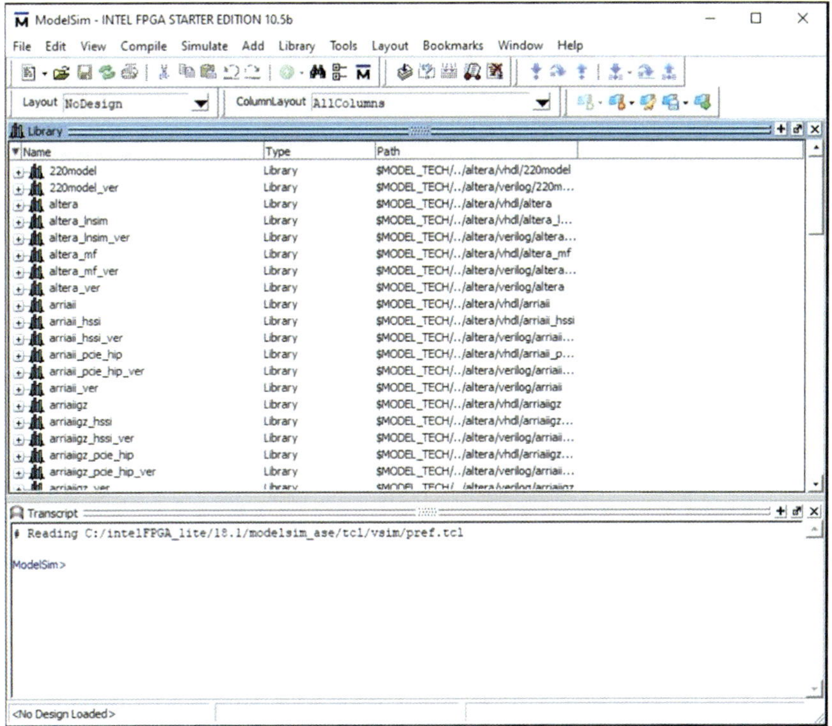

In the "transcript" window at the bottom, you can navigate to the project folder ("sim_script", using standard DOS "cd" commands), or you can copy out the commented change-directory command from your script file (after you've modified the first line) and paste it into the Modelsim transcript window. Then in the transcript window, type "do run.do". Note that the "do" is the command to Modelsim to read the "run.do" file, and perform the commands therein. Note also that you type this every time you want to compile and run a simulation.

When you hit the return key, Modelsim compiles and starts the simulation, and now we have the "sim" window on the left, and the "objects" window on the right as when we were using the GUI method.

Next, we need to launch the waveform window via the "View menu".

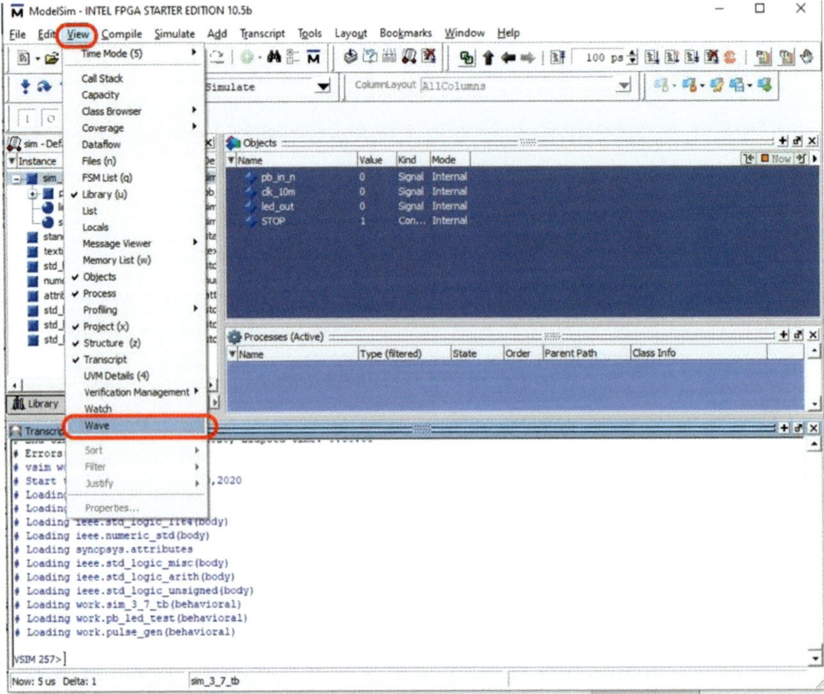

If the waveform window appears within the project window on the right side, you'll want to let it float.

From here, we need to set up the waveform window, which involves selecting the signals from the "objects" window, and pasting them into the floating waveform window—this was explained in detail back in the GUI section (don't forget to save the window after setting it up).

The last step is to uncomment the wave command in the script.

```

# cd D:/afiles_home/a_prose/CS_Dig_Des_Scratch/simulation/sim_script


vlib work

vcom -work work -2002 -explicit source/pb_led_test.vhdl
vcom -work work -2002 -explicit source/pulse_gen.vhdl
vcom -work work -2002 -explicit source/sim_3_7_tb.vhdl

vsim work.sim_3_7_tb

do wave.do

run 5 us

```

Now when we type our script command ("do run.do"), Modelsim runs a complete new simulation, and the results appear in the waveform window. As mentioned earlier, you may have to select this button to get a full-wave display.

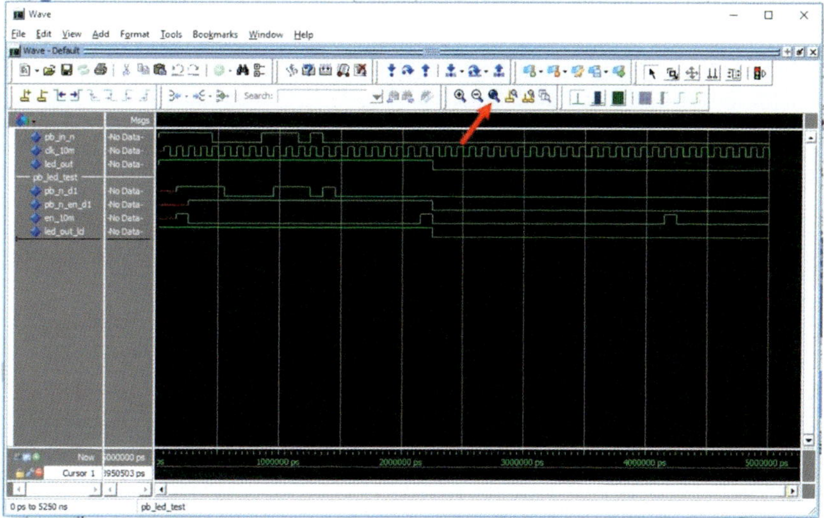

Exercises, Chapter 1

Note: these exercises use the script method of running Modelsim (the preferred method).

Also, the labeling of the files in these exercises was made for the combined, single-volume, edition. Since volume 1 ended with chapter five, the exercises of this chapter (in that context) would be chapter six.

Exercise 1-1:

Correcting coding errors in Modelsim.

1) download Modelsim and a text editor (if you haven't already);

2) set up a folder for your simulation exercises;

3) download the "sim_exercise_6_1.vhdl" testbench file into your project area;

4) download the "run_6_1.do" file into your project area, and modify the "CD" line per your folder structure. Copy and paste that modified line into Modelsim;

5) run the simulation by typing "do run_6_1.do" in Modelsim, and correct the resulting errors.

Exercise 1-2:

1) download the "sim_exercise_6_2.vhdl" testbench file into your project area. Your "sim_exercise_6_1.vhdl" file from the previous step should now look like this one (after your corrections).

2) download the "run_6_2.do" into your project area, modify the first line, and run it (type "do run_6_2.do" – note that you'll have to type "quit -sim" first to exit the previous simulation).

3) open the waveform window in Modelsim, and add the "clock" and "combo_sig" signals. Save the "wave.do" file!

4) using a text editor, remove the "#" comment before the "do wave.do".

5) re-run the simulation (run the "do run_6_2.do" script).

6) expand the "combo_sig" signal in the waveform window via this "+":

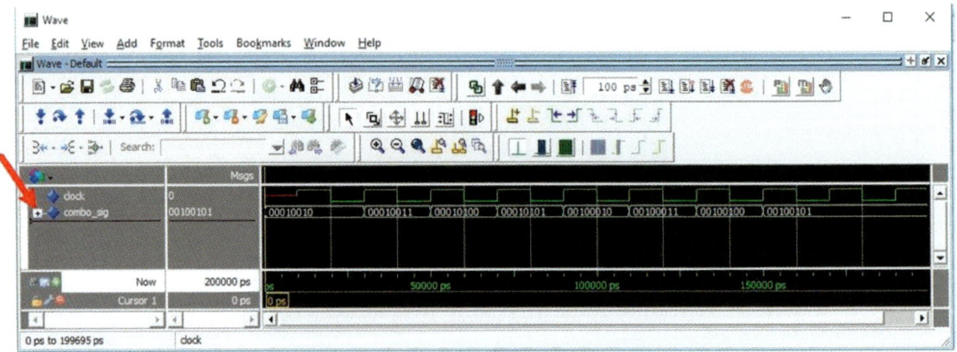

7) your waveform should look like this:

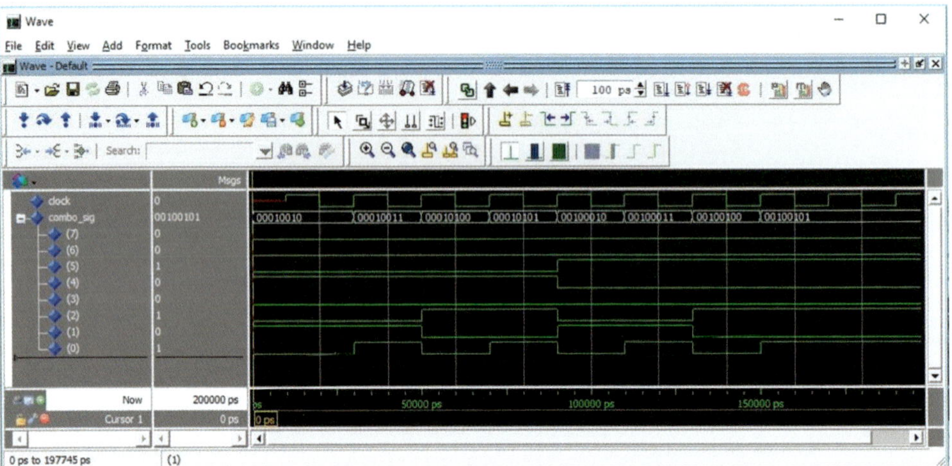

Exercise 1-3:

1) download the "pulse_gen_6_3.vhdl" file into your project area. This module creates a pulse every three clocks, but only when enabled.

2) copy the "sim_exercise_6_2.vhdl" file in your project area, and rename it "sim_exercise_6_3.vhdl".

3) copy the "do run_6_2.do" file in your project area, rename it "run_6_3.do", and modify it to include both the renamed "sim_exercise_6_3.vhdl" testbench, as well as the downloaded "pulse_gen_6_3.vhdl" file. Set it to run for 400 nanoseconds.

4) instantiate the downloaded "pulse_gen_6_3.vhdl" file in the new testbench, and modify the testbench to create an enable signal for the instantiated module that goes active for 7 clocks, and inactive for 2 clocks, repeatedly. The testbench should be able to run for at least two full cycles of the 7-clock/2-clock repetition.

5) your resulting waveform should look something like this:

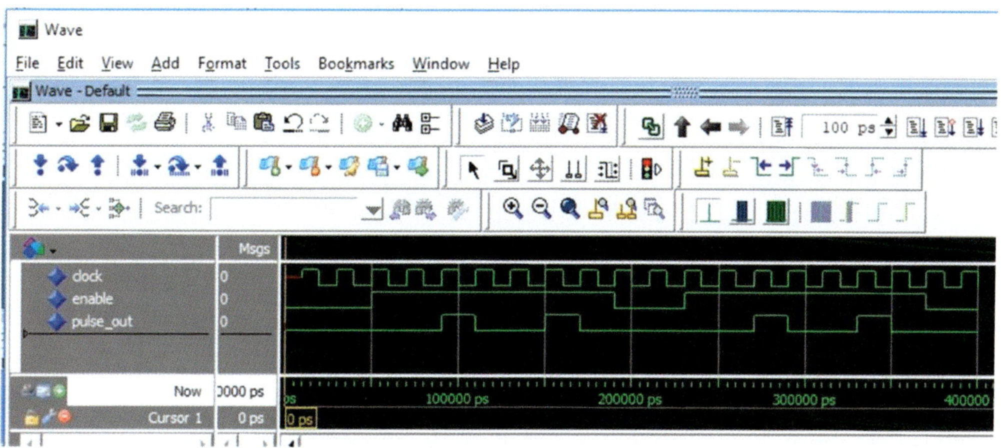

Exercise 1-4:

1) copy the "sim_exercise_6_3.vhdl" file in your project area, and rename it "sim_exercise_6_4.vhdl".

2) redesign the new "sim_exercise_6_4.vhdl" to use a clocked process statement instead of the existing one.

3) the resulting waveform should look like that from exercise 1-3.

Chapter 2

Memories

Introduction to Memories

We hear a lot about the memory in our phones, tablets, and laptops. We have a general idea of how the quantity of memory relates to the operation of our devices—how many videos we can store, etc.. In this chapter, we'll see how memory is used in FPGAs.

A memory can be thought of as a table of values that we can change, i.e., write.

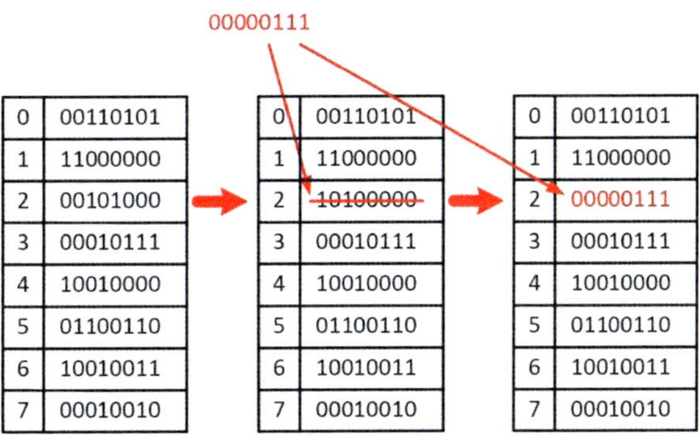

We call the location in the table the “address,” and the values themselves, the “data.”

address	data
0	00110101
1	11000000
2	00000111
3	00010111
4	10010000
5	01100110
6	10010011
7	00010010

In logic coding, the address is a vector, and, in this case, since we have eight entries in the table, we need a three-bit vector.

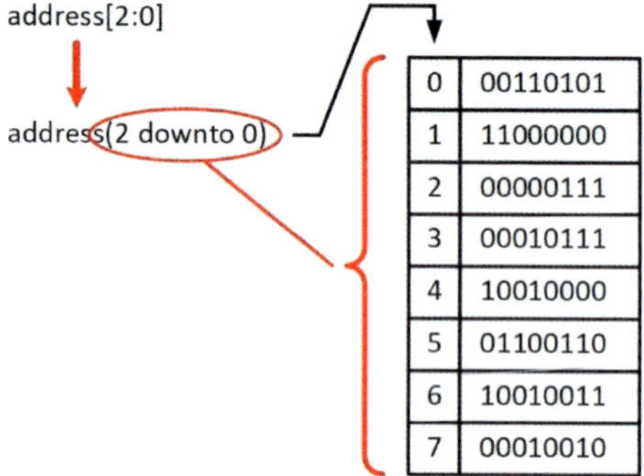

As we know, in VHDL the vector is defined with “2 downto 0”.

Since the data entries are eight-bit vectors, they are defined as “7 downto 0”.

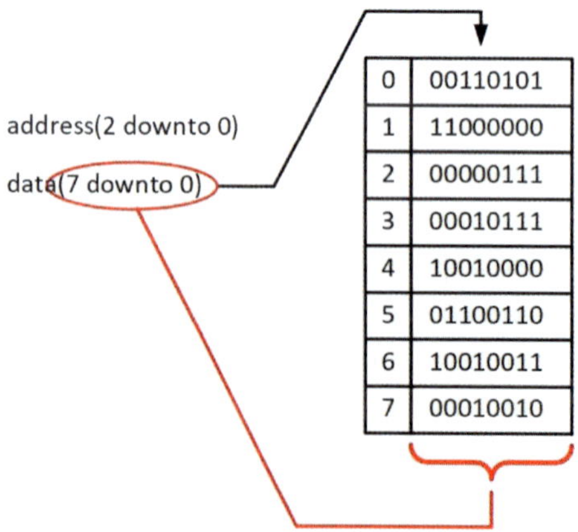

Suppose we want to write a new data value of hex 33 into address location 4, replacing the current value of hex 90. Our logic needs to know when to do the replacement, i.e., when to write the new hex 33 value into location 4. This requires an additional control signal, that we call the "write enable".

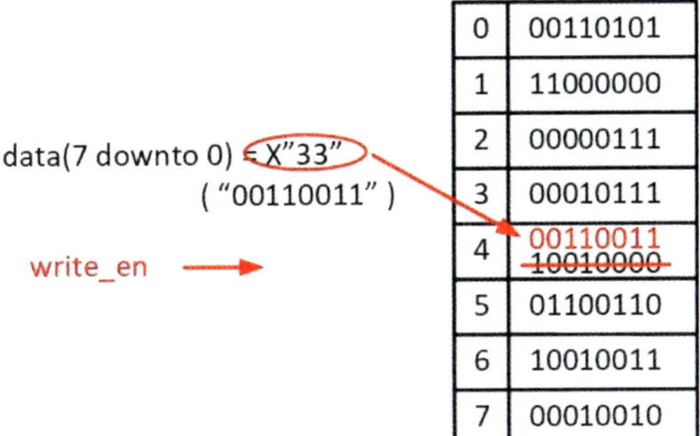

But what if we want to read the memory? All we need is the address that points to the location we want to read.

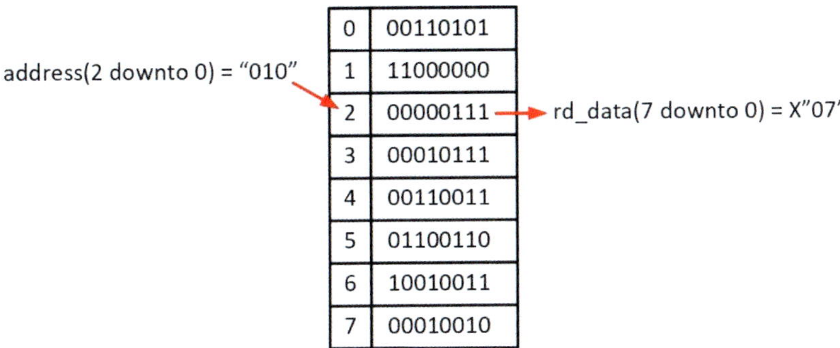

Here's the full complement of signals for our simplest memory:

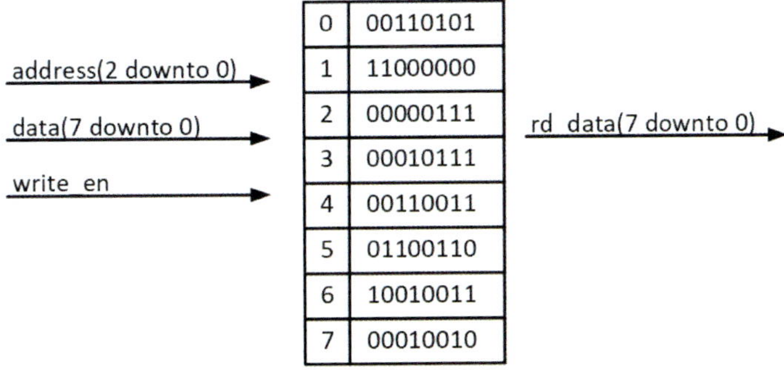

Single-port Memories

Suppose that we have a memory module that looks like this.

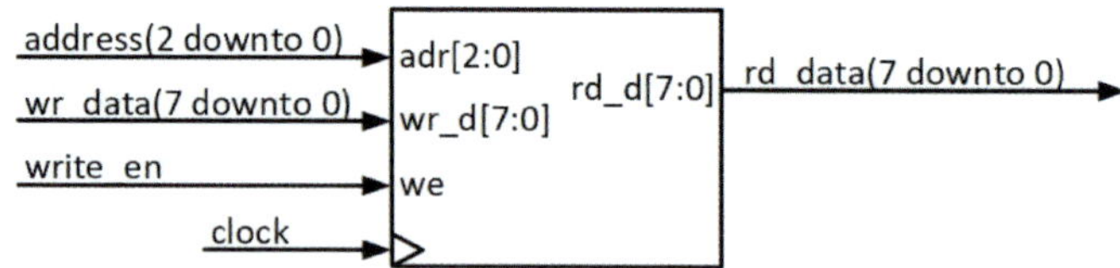

Notice that I've added a clock. There was a time when most memories were asynchronous, but you can assume that any you use now will be synchronous, meaning that they operate with a clock.

Now let's suppose that we've been applying an address of "001" for some time with the write enable inactive. That would mean that in our example memory the read data will be sitting at hex C0.

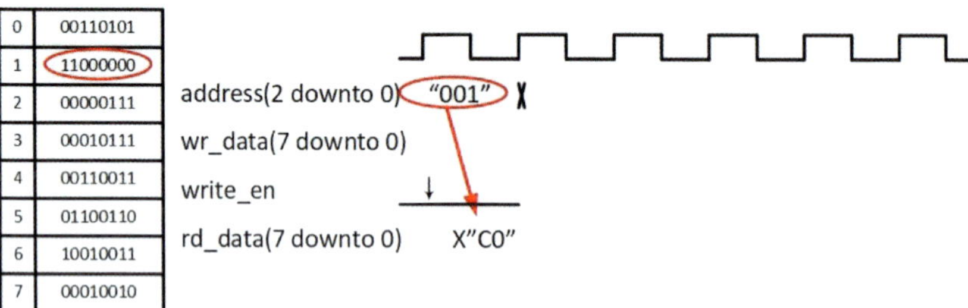

If we change our address to 5 on the next clock, what happens to the read data? It changes to the hex 66 value at location 5, but not until the next clock—it's synchronous.

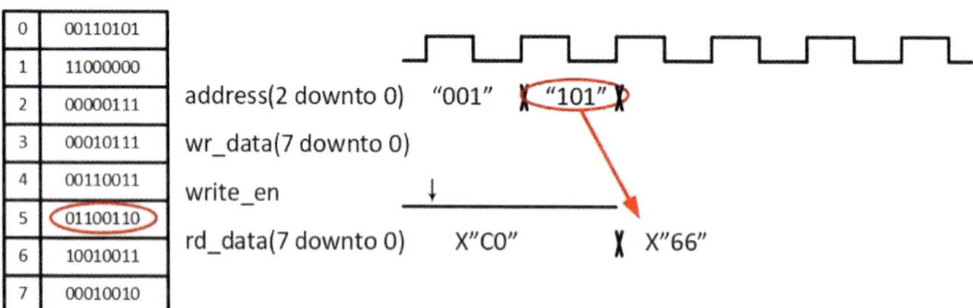

Say we want to write a new data value, and change location 5 from hex 66 to hex AA. We have to activate the write enable signal, and at the same time apply the hex value AA to the write data port. On the NEXT clock, the write actually occurs, where the value stored in the memory is replaced.

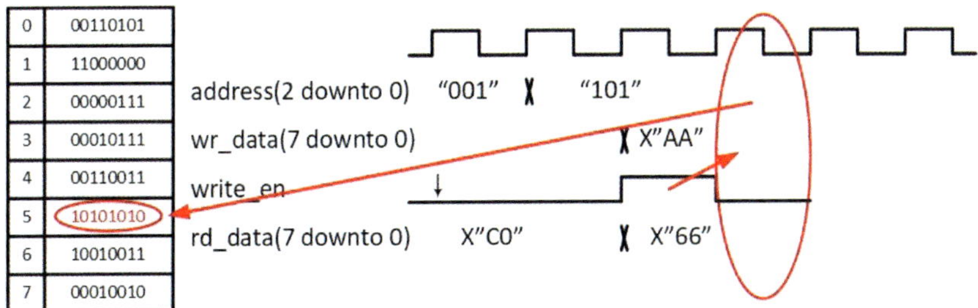

The read data then picks up the change on the next clock after that.

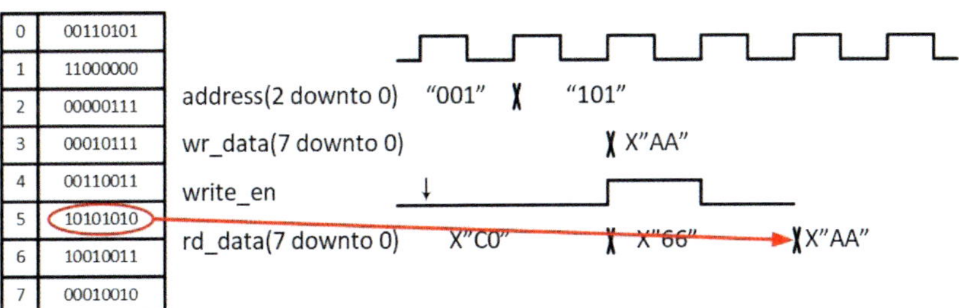

read-before-write

When the written data appears out the read port on the following clock like I've show here, the operation is called read-before-write. There's a different possibility, however. The read data may change on the same clock that the new data is written.

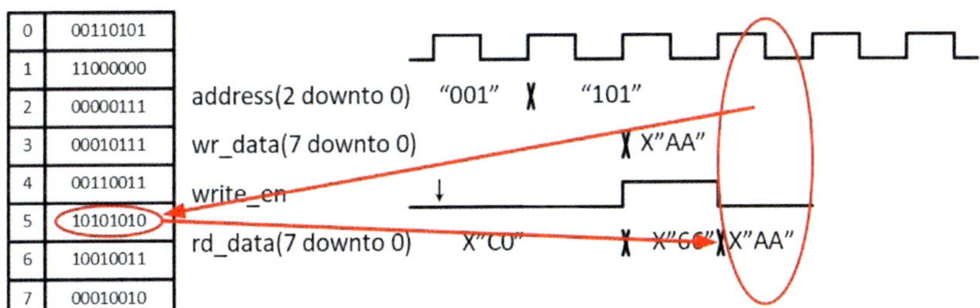

write-before-read

This is called, as you guessed, write-before-read.

The type of operation is inherent in the memory design, and may be fixed, or you may be able to configure it.

Note that these two types of memory operation apply to the very limited case where we're reading the same location that we're simultaneously writing. Normally our reads will be from a different area in memory, also remembering that practical memories are typically thousands of times larger than our small sample version here.

The memory we've been looking at is called a single-port memory, and this is evidenced by the fact that there is just one of each port bus—one address bus, and one write and one read data buses.

We usually refer to memory as RAM, which stands for "random-access memory." The name was introduced decades ago when most memory consisted of magnetic tape—those spinning reels on the computers of the movies in the seventies. If you wanted to access some particular data, you had to spin the reel forward or backward until you came to it. RAM consisting of magnetic cores or, better yet, silicon chips, was relatively expensive, but had the great advantage that you could get to any data instantly. In 1978, ten kilobytes of random-access memory was state of the art.

We can move the read data port to the left side in preparation to show you the next step in memory complexity and usefulness

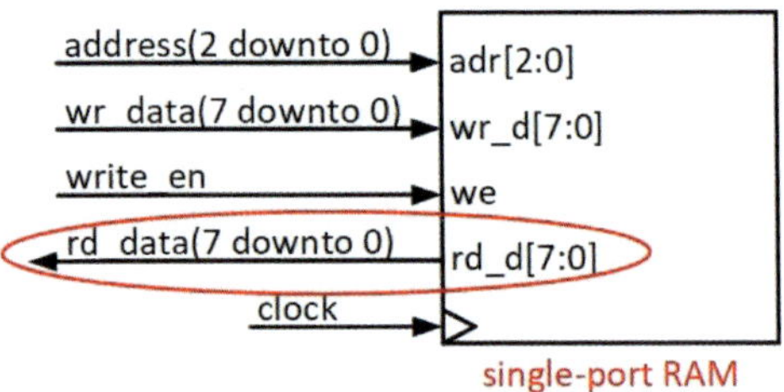

Dual-port Memories

Dual-port RAMs have two sets of access ports, often labeled A and B.

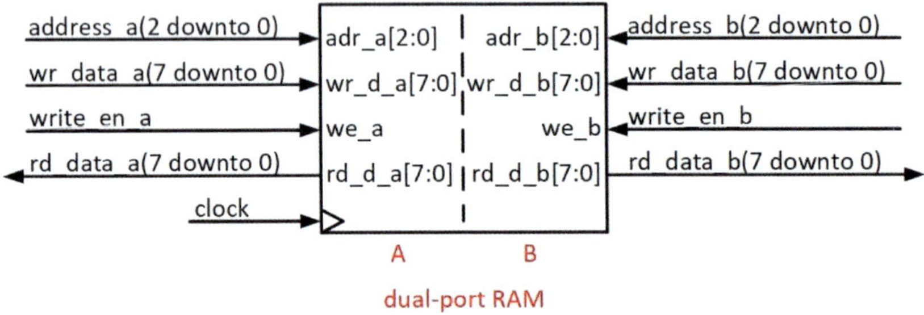

The A side (on the left) works the same as the single-port version we just covered. In fact, if you don't use the B side (on the right), the A side works identically to the original single-port memory. The B side, in turn, works exactly the same as the A side, and if you don't use the A side, then the B side works identically to the single-port memory. The point being that each side works completely independently of the other. You can have two different users—for example two different micro-processors—using the RAM at the same time.

There remains the same question of simultaneous access as that of the write and reads of the single-port memory, except that in this case, the question is what happens if accesses are attempted on both sides at the same location at the same time. Again, we choose (or accept) either write-before-read or read-before-write, but, additionally, we choose (or accept) one side's priority, meaning that if both sides try to write to the same location, the priority side wins.

There's an additional level of versatility with dual-port RAMs. The dual-port RAM above uses a single clock for both sides, but most dual-port RAMs can be configured to use different clocks on each port.

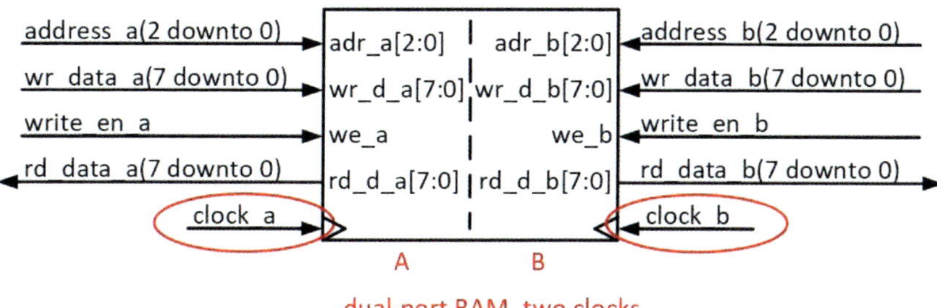

dual-port RAM, two clocks

This can be immensely useful in complex systems with multiple clock domains. Imagine the complexity required to handle the infamous same-location writes and reads.

We'll look at one last RAM configuration, pushing the complexity to the max. Instead of one address bus for both writes and reads, we can have two, each dedicated for either writes or reads.

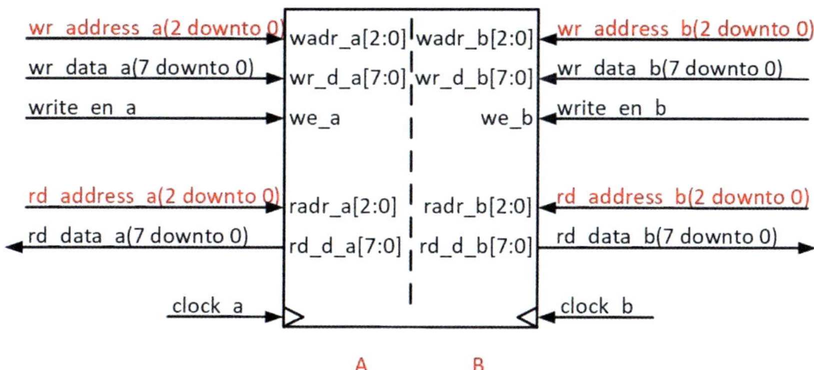

dual-port RAM, two clocks, separate write/read addresses

With this mega-RAM, we can simultaneously write and read data on both sides, using separate clocks. This is, in essence, more of a four-port RAM than a two-port.

Almost all the memory—the RAMs—that you'll use in FPGAs will be provided by the vendor via their included IP library. Their tools build the RAMs from dedicated areas of the FPGA die that implement memory, and you configure your memory blocks—different variations of what we've covered—using that IP tool.

Implementing a Simple Memory

Using the original simple single-port version, let's see how a synchronous memory might be implemented using clocked registers:

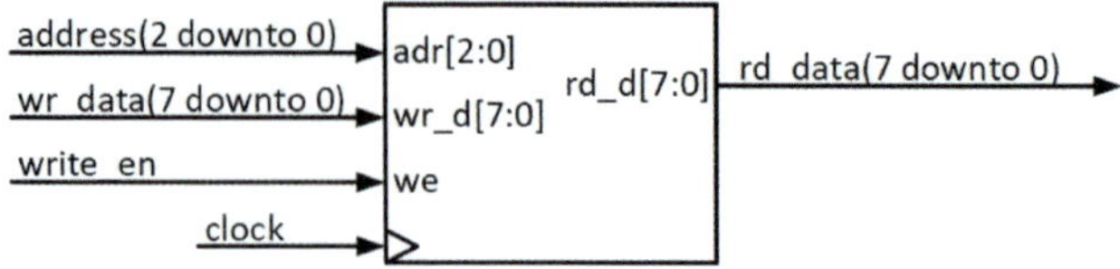

We start with eight enabled registers, where the contents change only when the enable input "en" is active.

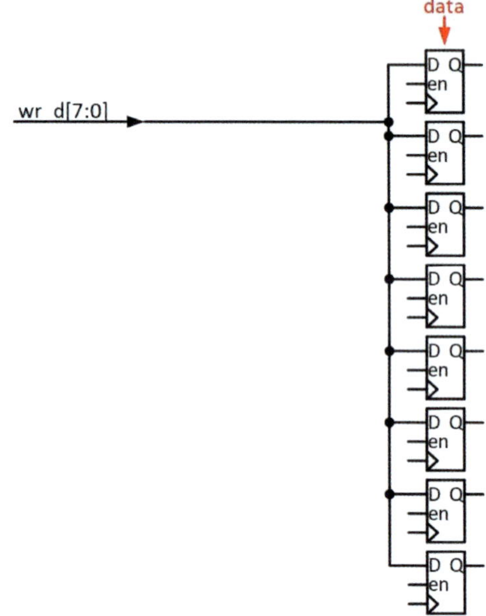

This column of registers comprises the data contents of our simple example memory, where each register represents one location in the memory. The write data bus input is fed to each register.

Note that each register shown in the column is actually 8 register d-flops, all clocked and enabled together.

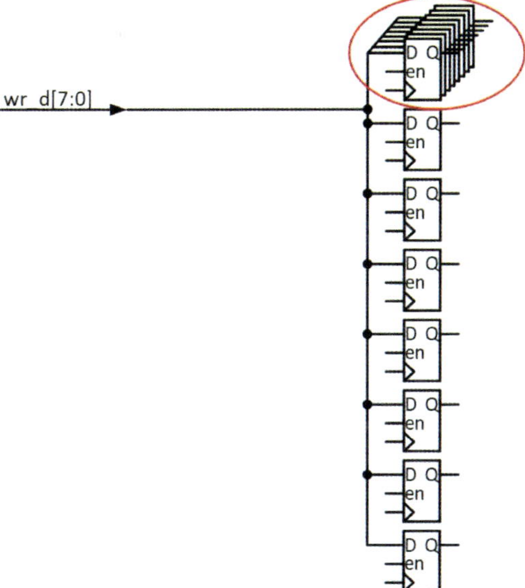

Next, we add the address, write enable, and clock inputs.

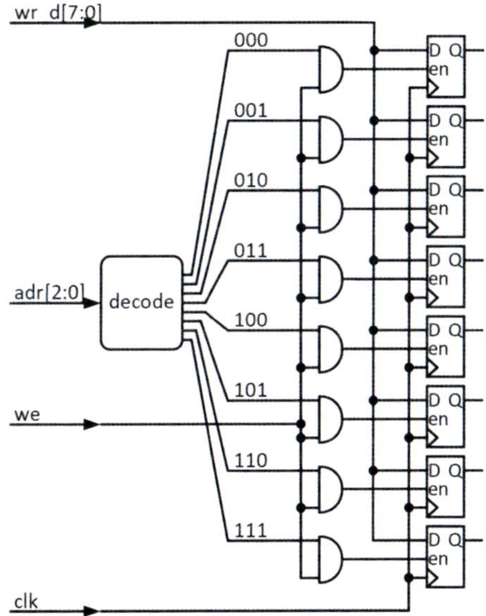

The address value ("adr") steers the write enable signal ("we") to the data register enables ("en") of one set of eight registers. In order to do this, the address must be decoded, i.e., the value must be broken out as one of eight decoded signals. The write enable signal is then ANDed with the decoded address so that when the write enable goes active, just one set of registers is enabled to receive the new data to be written.

To finish our simple example RAM, we need to provide the read data.

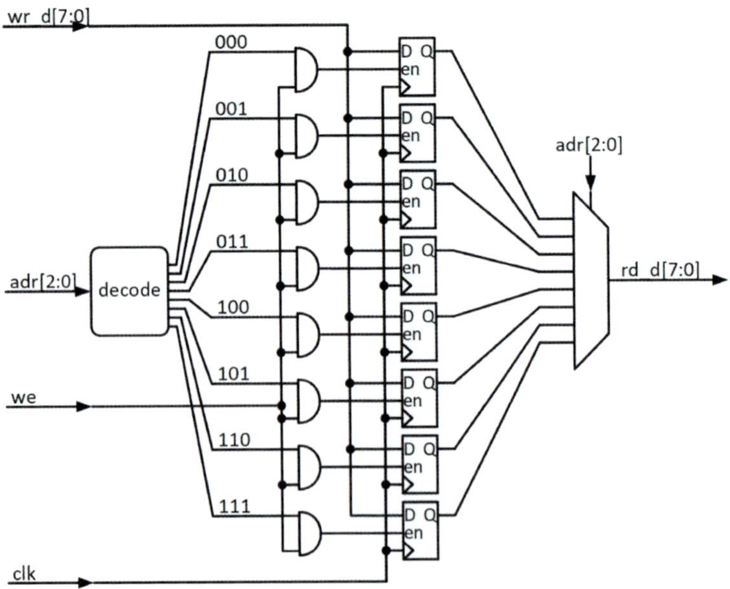

This is done with a multiplexer. A multiplexer, AKA, a mux, is simply a selector.

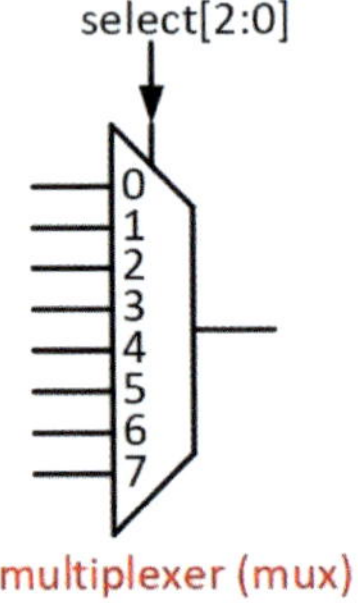

Just one of the input signals is fed to the output depending on the selection value. For example, if the select value is 010 (i.e., two), then that input is connected to the output.

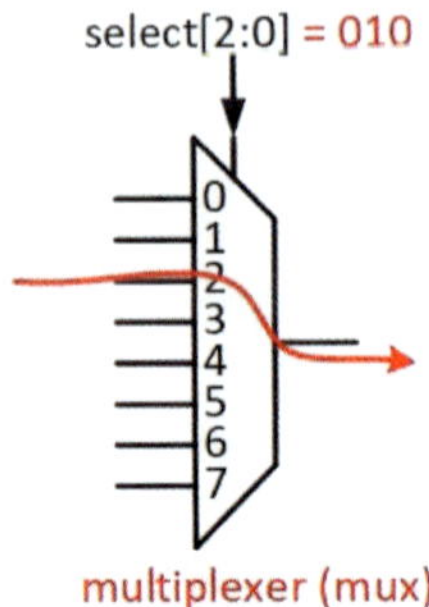

Since our address input ("adr") controls the mux selection, the memory output is the register data value stored at the addressed location.

Looking at the operation of our simple register-based memory, we can see that this performs a write-before-read version—the value that we write appears at the read port at the same clock edge that the write occurs (i.e., when the data in the addressed location is replaced). If we want to operate as a read-before-write version, we need to add a register at the output, delaying the memory change by one clock.

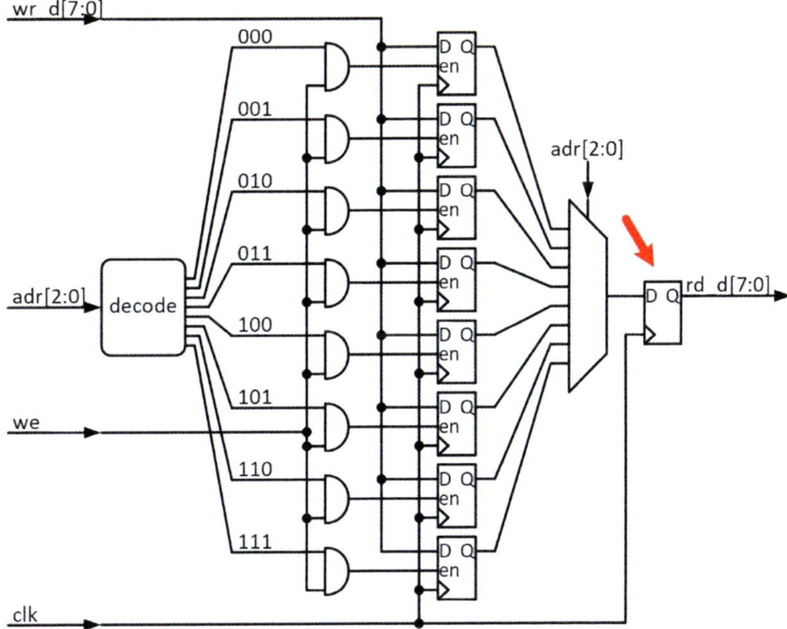

This may seem unnecessary for the operation of this memory, and for one as simple as this it is. In actual memories, where thousands, or millions, of bits need to be selected, timing may become an issue—the amount of logic required to mux so many bits may take so much time that an output register may be required to catch the data before the next clock edge.

Here's the VHDL code for this simple register-based memory, showing how the entity port declarations map to the module I/O:

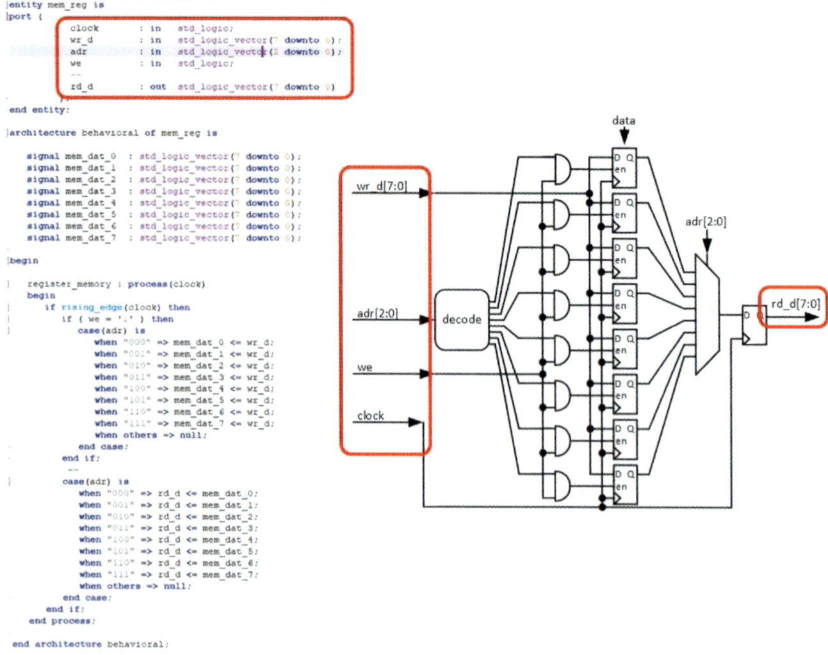

Here, we declare the signals that will hold data in registers:

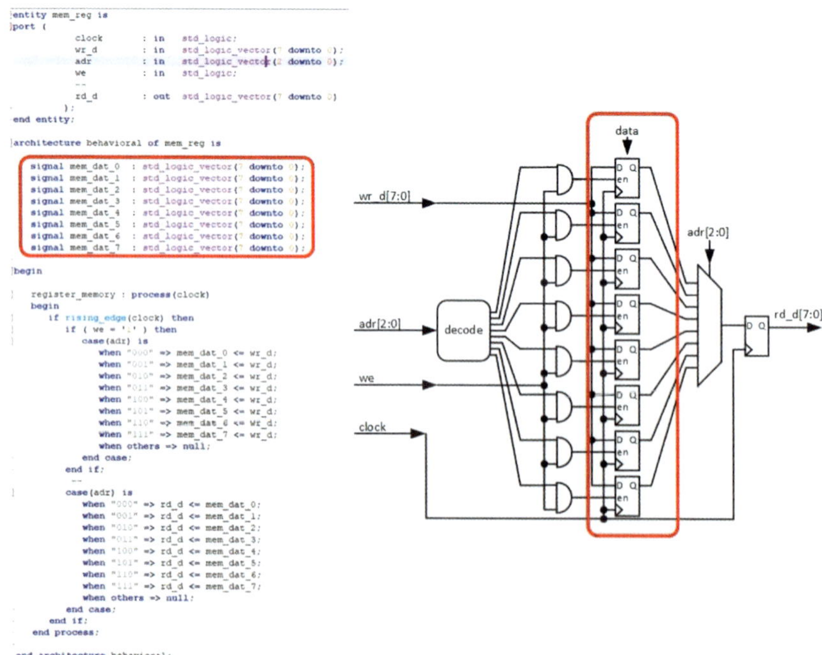

The address decode is implemented with a case statement:

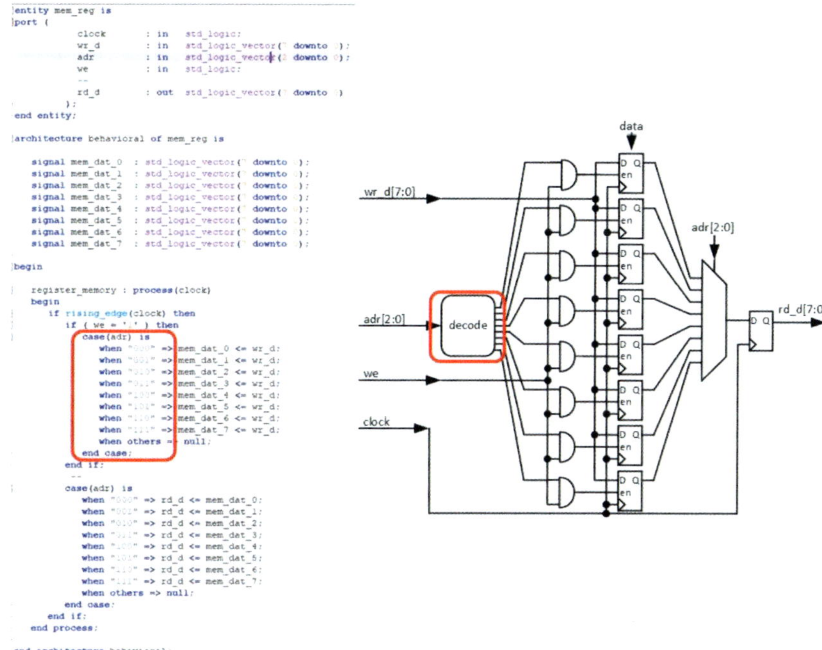

We implement the write-enable AND gates with the IF statement—the registers are loaded with new data only when the "we" write enable signal is active:

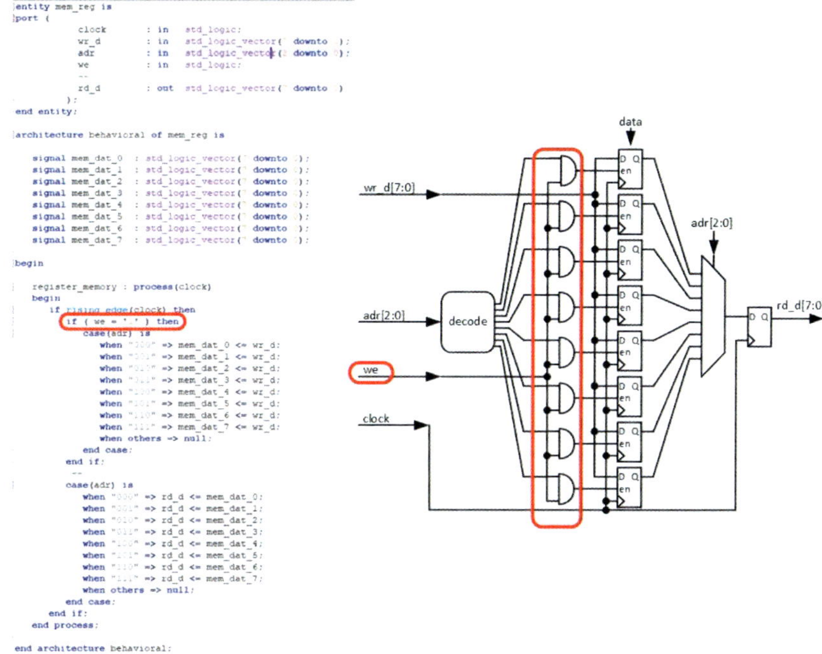

The data registers are implemented by the fact that they are assigned only on rising edges of the clock:

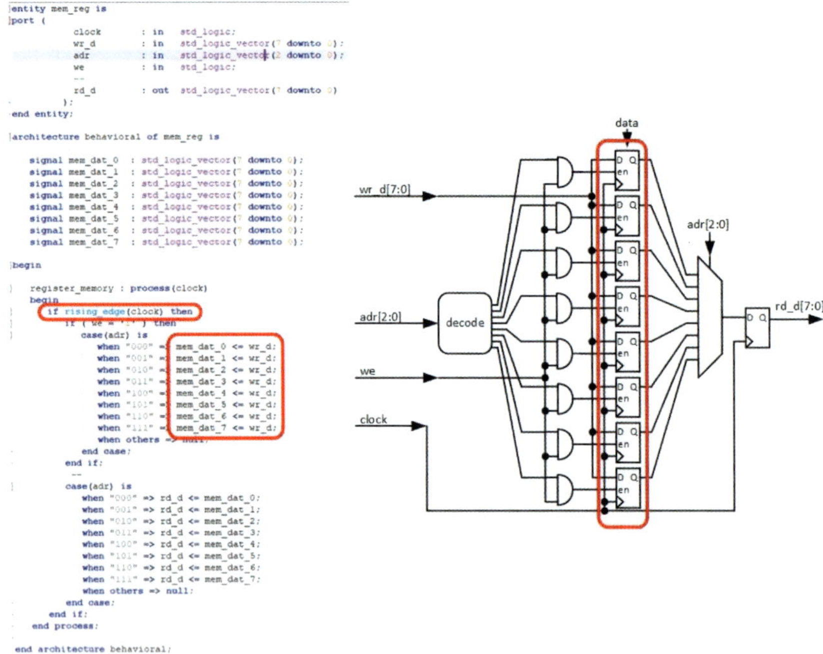

The output mux is another case statement, and this time the data registers are the source instead of the assignment destination:

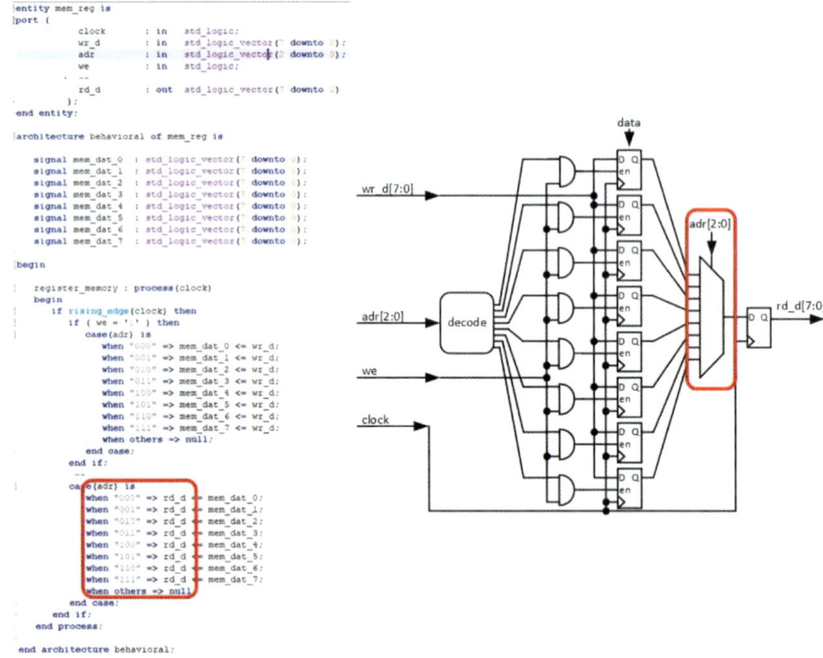

And, finally, like the data registers, the output register is implemented by assignments only with the rising edges of the clock:

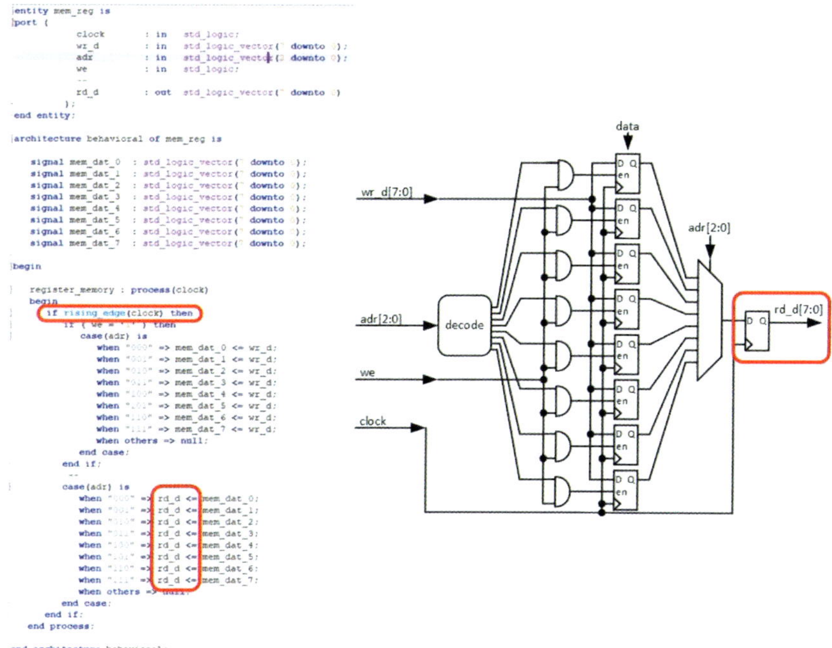

Arrays

For a simple, small memory like this, using individual VHDL signals for each memory location is reasonable, but completely impractical for real memories with thousands or millions of bits. Instead, for larger groups of registers, like a memory, we use a structure common to any programming language: an array.

We declare the array using the VHDL "type" statement—the same one we used earlier for declaring state names.

```
entity mem_reg is
port (
            clock       : in   std_logic;
            wr_d        : in   std_logic_vector(7 downto 0);
            adr         : in   std_logic_vector(2 downto 0);
            we          : in   std_logic;
            --
            rd_d        : out  std_logic_vector(7 downto 0)
        );
end entity;

architecture behavioral of mem_reg is

    type mem_array is array (7 downto 0) of std_logic_vector(7 downto 0);
    signal mem_dat   : mem_array;

begin

    register_memory : process(clock)
    begin
        if rising_edge(clock) then
            if ( we = '1' ) then
                case(adr) is
                    when "000" => mem_dat(0) <= wr_d;
                    when "001" => mem_dat(1) <= wr_d;
                    when "010" => mem_dat(2) <= wr_d;
                    when "011" => mem_dat(3) <= wr_d;
                    when "100" => mem_dat(4) <= wr_d;
                    when "101" => mem_dat(5) <= wr_d;
                    when "110" => mem_dat(6) <= wr_d;
                    when "111" => mem_dat(7) <= wr_d;
                    when others => null;
                end case;
            end if;
            --
            case(adr) is
                when "000" => rd_d <= mem_dat(0);
                when "001" => rd_d <= mem_dat(1);
                when "010" => rd_d <= mem_dat(2);
                when "011" => rd_d <= mem_dat(3);
                when "100" => rd_d <= mem_dat(4);
                when "101" => rd_d <= mem_dat(5);
                when "110" => rd_d <= mem_dat(6);
                when "111" => rd_d <= mem_dat(7);
                when others => null;
            end case;
        end if;
    end process;

end architecture behavioral;
```

```
type mem_array is array (7 downto 0) of std_logic_vector(7 downto 0);
signal mem_dat   : mem_array;
```

array

Following the keyword "type", we declare a name for this array structure (just like we did with state labels). In this case, I chose "mem_array," although it could be anything. The key words "is array" tell the compiler that this is indeed an array that we're declaring. We define the depth of the array with "(7 downto 0)". In our case, the array has eight elements. If, for example, the array had 1k elements, this would be "(1023 downto 0)". Then, "of std_logic_vector(7 downto 0)" tells the compiler that each element of the array is an 8-bit, standard logic vector value (each register location of our memory).

So far, all we've done is define a type of array, one that consists of eight 8-bit standard logic values. The next line declares the actual array that we'll use in our code. Like other VHDL signals we've used, an array is also declared as a signal, except that instead of being of an unsigned or standard logic type, this signal is defined to be the array we just created using the "type" declaration. Note that our previous state machine labels were also declared as a signal using specially created types, in that case a list of state labels.

To reiterate, "mem_array" is a general (particularly defined) type of array, whereas "mem_dat" is specifically the array that we use in our code. For example, we might have a design that includes multiple memories—e.g., "mem_dat_a", "mem_dat_b", etc.—each declared to be using the same "mem_array" type of array.

We indicate which element of the array—i.e., which register—we want to access by simply pointing to the location with an integer number. This integer is called the index.

```
entity mem_reg is
port (
            clock       : in    std_logic;
            wr_d        : in    std_logic_vector(7 downto 0);
            adr         : in    std_logic_vector(2 downto 0);
            we          : in    std_logic;
            --
            rd_d        : out   std_logic_vector(7 downto 0)
        );
end entity;

architecture behavioral of mem_reg is

    type mem_array is array (7 downto 0) of std_logic_vector(7 downto 0);
    signal mem_dat   : mem_array;

begin

    register_memory : process(clock)
    begin
        if rising_edge(clock) then
            if ( we = '1' ) then
               case(adr) is
                  when "000" => mem_dat(0) <= wr_d;
                  when "001" => mem_dat(1) <= wr_d;
                  when "010" => mem_dat(2) <= wr_d;
                  when "011" => mem_dat(3) <= wr_d;
                  when "100" => mem_dat(4) <= wr_d;
                  when "101" => mem_dat(5) <= wr_d;
                  when "110" => mem_dat(6) <= wr_d;
                  when "111" => mem_dat(7) <= wr_d;
                  when others => null;
               end case;
            end if;
            --
            case(adr) is
               when "000" => rd_d <= mem_dat(0);
               when "001" => rd_d <= mem_dat(1);
               when "010" => rd_d <= mem_dat(2);
               when "011" => rd_d <= mem_dat(3);
               when "100" => rd_d <= mem_dat(4);
               when "101" => rd_d <= mem_dat(5);
               when "110" => rd_d <= mem_dat(6);
               when "111" => rd_d <= mem_dat(7);
               when others => null;
            end case;
        end if;
    end process;

end architecture behavioral;
```

Comparing the two methods, the individual register method uses 8 different signals:

```
architecture behavioral of mem_reg is

    signal mem_dat_0    : std_logic_vector(7 downto 0);
    signal mem_dat_1    : std_logic_vector(7 downto 0);
    signal mem_dat_2    : std_logic_vector(7 downto 0);
    signal mem_dat_3    : std_logic_vector(7 downto 0);
    signal mem_dat_4    : std_logic_vector(7 downto 0);
    signal mem_dat_5    : std_logic_vector(7 downto 0);
    signal mem_dat_6    : std_logic_vector(7 downto 0);
    signal mem_dat_7    : std_logic_vector(7 downto 0);

begin

    register_memory : process(clock)
    begin
        if rising_edge(clock) then
            if ( we = '1' ) then
                case(adr) is
                    when "000" => mem_dat_0 <= wr_d;
                    when "001" => mem_dat_1 <= wr_d;
                    when "010" => mem_dat_2 <= wr_d;
                    when "011" => mem_dat_3 <= wr_d;
                    when "100" => mem_dat_4 <= wr_d;
                    when "101" => mem_dat_5 <= wr_d;
                    when "110" => mem_dat_6 <= wr_d;
                    when "111" => mem_dat_7 <= wr_d;
                    when others => null;
                end case;
            end if;
            --
            case(adr) is
                when "000" => rd_d <= mem_dat_0;
                when "001" => rd_d <= mem_dat_1;
                when "010" => rd_d <= mem_dat_2;
                when "011" => rd_d <= mem_dat_3;
                when "100" => rd_d <= mem_dat_4;
                when "101" => rd_d <= mem_dat_5;
                when "110" => rd_d <= mem_dat_6;
                when "111" => rd_d <= mem_dat_7;
                when others => null;
            end case;
        end if;
    end process;

end architecture behavioral;
```

individual signals

… while the array method, just one—the 8-element array:

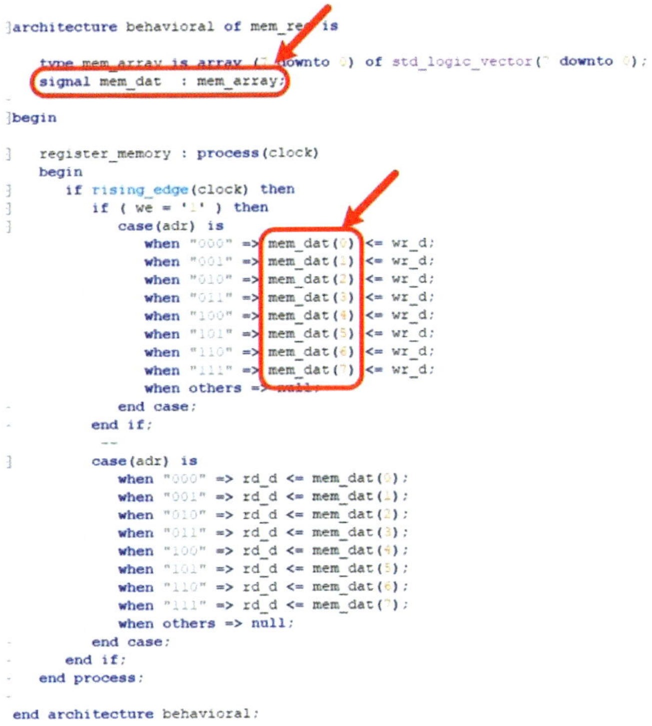

```
architecture behavioral of mem_reg is

   type mem_array is array (7 downto 0) of std_logic_vector(7 downto 0);
   signal mem_dat   : mem_array;

begin

   register_memory : process(clock)
   begin
      if rising_edge(clock) then
         if ( we = '1' ) then
            case(adr) is
               when "000" => mem_dat(0) <= wr_d;
               when "001" => mem_dat(1) <= wr_d;
               when "010" => mem_dat(2) <= wr_d;
               when "011" => mem_dat(3) <= wr_d;
               when "100" => mem_dat(4) <= wr_d;
               when "101" => mem_dat(5) <= wr_d;
               when "110" => mem_dat(6) <= wr_d;
               when "111" => mem_dat(7) <= wr_d;
               when others => null;
            end case;
         end if;
         --
         case(adr) is
            when "000" => rd_d <= mem_dat(0);
            when "001" => rd_d <= mem_dat(1);
            when "010" => rd_d <= mem_dat(2);
            when "011" => rd_d <= mem_dat(3);
            when "100" => rd_d <= mem_dat(4);
            when "101" => rd_d <= mem_dat(5);
            when "110" => rd_d <= mem_dat(6);
            when "111" => rd_d <= mem_dat(7);
            when others => null;
         end case;
      end if;
   end process;

end architecture behavioral;
```

an array

There's potential for confusion with array indexing. In the example of the next figure, we say that "mem_dat(3)" points to the fourth 8-bit element of the array, but this seems that it could also be indicating bit three of "mem_dat".

```
architecture behavioral of mem_reg is

   type mem_array is array (7 downto 0) of std_logic_vector(7 downto 0);
   signal mem_dat   : mem_array;

begin

   register_memory : process(clock)
   begin
      if rising_edge(clock) then
         if ( we = '1' ) then
            case(adr) is
               when "000" => mem_dat(0) <= wr_d;
               when "001" => mem_dat(1) <= wr_d;
               when "010" => mem_dat(2) <= wr_d;
               when "011" => mem_dat(3) <= wr_d;
               when "100" => mem_dat(4) <= wr_d;
               when "101" => mem_dat(5) <= wr_d;
               when "110" => mem_dat(6) <= wr_d;
               when "111" => mem_dat(7) <= wr_d;
               when others => null;
            end case;
         end if;
         --
         case(adr) is
            when "000" => rd_d <= mem_dat(0);
            when "001" => rd_d <= mem_dat(1);
            when "010" => rd_d <= mem_dat(2);
            when "011" => rd_d <= mem_dat(3);
            when "100" => rd_d <= mem_dat(4);
            when "101" => rd_d <= mem_dat(5);
            when "110" => rd_d <= mem_dat(6);
            when "111" => rd_d <= mem_dat(7);
            when others => null;
         end case;
      end if;
   end process;

end architecture behavioral;
```

The compiler, however, knows that "mem_dat" is an array, and understands that we're indicating an array element, not a bit location. If we wanted to point to an individual bit in the array, say bit 7 at the fourth location, we would follow the array index with the bit location, like so:

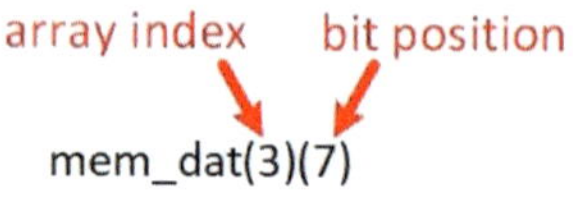

We could also indicate a range of vector bits, for example the MS nibble ([7:5]) at the sixth array location:

array index   bit range

mem_dat(5)(7 downto 5)

Coding memories using arrays is useful for small structures, but, as noted earlier, larger memories, particularly dual-port versions, are generally built using the vendor's IP and associated configuration tools.

While Loops

Here's a testbench I've created for these two simple memory designs. Note that the same testbench works with both designs.

```
entity sim_7a_tb is
end entity sim_7a_tb;

architecture Behavioral of sim_7a_tb is

   component mem_reg is
   port ( clock        : in   std_logic;
          wr_d         : in   std_logic_vector(7 downto 0);
          adr          : in   std_logic_vector(2 downto 0);
          we           : in   std_logic;
          rd_d         : out  std_logic_vector(7 downto 0)
        );
   end component;

   signal clk               : std_logic;
   signal we                : std_logic;
   signal wr_d              : std_logic_vector(7 downto 0);
   signal rd_d              : std_logic_vector(7 downto 0);
   signal adr               : std_logic_vector(2 downto 0);
   signal adr_cnt           : unsigned(2 downto 0);
   signal clock             : std_logic;

begin

   clk <= '0' after 50 ns when clk = '1' else
          '1' after 50 ns;

   stimulus : process
   begin
      adr_cnt <= "000";
      we      <= '0';
      wait until rising_edge(clk);
      while (adr_cnt < "010") loop
         we <= '1';
         wait until rising_edge(clk);
         adr_cnt <= adr_cnt + 1;
         we <= '0';
         wait until rising_edge(clk);
      end loop;
      --
      wait until rising_edge(clk);
      adr_cnt <= "000";
      wait until rising_edge(clk);
      while (adr_cnt < "011") loop
         adr_cnt <= adr_cnt + 1;
         wait until rising_edge(clk);
      end loop;
   end process;

   adr     <= std_logic_vector(adr_cnt);
   wr_d    <= adr & "00" & (adr XOR "111");
   clock   <= clk;

   mem_reg_1 : mem_reg
   port map
      ( clock        => clock,
        wr_d         => wr_d,
        adr          => adr,
        we           => we,
        rd_d         => rd_d
      );
end architecture Behavioral;
```


```
while (adr_cnt < "010") loop
   we <= '1';
   wait until rising_edge(clk);
   adr_cnt <= adr_cnt + 1;
   we <= '0';
   wait until rising_edge(clk);
end loop;
```


I've introduced a new type of statement—the "while" loop. Like arrays, "while" loops are common structures in any programming language. The loop keeps cycling as long as the "while" entry condition is true, in this case, as long as the address counter we've created is less than two. Each pass through the loop consists of two clocks, and each pass we activate the write enable, and then de-activate it on the next clock, incrementing the count in the process.

Once this loop is done, when the count reaches two, we clear the address counter, and the testbench moves on to the next loop, which simply cycles through addresses zero through two to read the written values back.

```
entity sim_7a_tb is
end entity sim_7a_tb;

architecture Behavioral of sim_7a_tb is

    component mem_reg is
    port ( clock        : in    std_logic;
           wr_d         : in    std_logic_vector(7 downto 0);
           adr          : in    std_logic_vector(2 downto 0);
           we           : in    std_logic;
           rd_d         : out   std_logic_vector(7 downto 0)
         );
    end component;

    signal clk              : std_logic;
    signal we               : std_logic;
    signal wr_d             : std_logic_vector(7 downto 0);
    signal rd_d             : std_logic_vector(7 downto 0);
    signal adr              : std_logic_vector(2 downto 0);
    signal adr_cnt          : unsigned(2 downto 0);
    signal clock            : std_logic;

  begin

    clk <= '0' after 50 ns when clk = '1' else
           '1' after 50 ns;

    stimulus : process
    begin
       adr_cnt <= "000";
       we      <= '0';
       wait until rising_edge(clk);
       while (adr_cnt < "010") loop
          we <= '1';
          wait until rising_edge(clk);
          adr_cnt <= adr_cnt + 1;
          we <= '0';
          wait until rising_edge(clk);
       end loop;
       --
       wait until rising_edge(clk);
       adr_cnt <= "000";
       wait until rising_edge(clk);
       while (adr_cnt < "011") loop
          adr_cnt <= adr_cnt + 1;
          wait until rising_edge(clk)
       end loop;
    end process;

    adr     <= std_logic_vector(adr_cnt);
    wr_d    <= adr & "00" & (adr XOR "111");
    clock   <= clk;

    mem_reg_1 : mem_reg
    port map
       ( clock      => clock,
         wr_d       => wr_d,
         adr        => adr,
         we         => we,
         rd_d       => rd_d
       );
end architecture Behavioral;
```


```
while (adr_cnt < "011") loop
    adr_cnt <= adr_cnt + 1;
    wait until rising_edge(clk);
end loop;
```


Note that the address increments to three before leaving the loop.

We need an 8-bit value to write into the memory.

```
entity sim_7a_tb is
end entity sim_7a_tb;

architecture Behavioral of sim_7a_tb is

   component mem_reg is
   port ( clock        : in   std_logic;
          wr_d         : in   std_logic_vector(7 downto 0);
          adr          : in   std_logic_vector(2 downto 0);
          we           : in   std_logic;
          rd_d         : out  std_logic_vector(7 downto 0)
        );
   end component;

   signal clk              : std_logic;
   signal we               : std_logic;
   signal wr_d             : std_logic_vector(7 downto 0);
   signal rd_d             : std_logic_vector(7 downto 0);
   signal adr              : std_logic_vector(2 downto 0);
   signal adr_cnt          : unsigned(2 downto 0);
   signal clock            : std_logic;

begin

   clk <= '0' after 50 ns when clk = '1' else
          '1' after 50 ns;

   stimulus : process
   begin
      adr_cnt <= "000";
      we      <= '0';
      wait until rising_edge(clk);
      while (adr_cnt < "010") loop
         we <= '1';
         wait until rising_edge(clk);
         adr_cnt <= adr_cnt + 1;
         we <= '0';
         wait until rising_edge(clk);
      end loop;
      --
      wait until rising_edge(clk);
      adr_cnt <= "000";
      wait until rising_edge(clk);
      while (adr_cnt < "011") loop
         adr_cnt <= adr_cnt + 1;
         wait until rising_edge(clk);
      end loop;
   end process;

   adr     <= std_logic_vector(adr_cnt);
   wr_d    <= adr & "00" & (adr XOR "111");
   clock   <= clk;

   mem_reg_1 : mem_reg
   port map
      ( clock      => clock,
        wr_d       => wr_d,
        adr        => adr,
        we         => we,
        rd_d       => rd_d
      );
end architecture Behavioral;
```

wr_d   <= adr & "00" & (adr XOR "111");

For the data assignment, I'm using the address as a convenient value that changes each write cycle, and since the address comprises only six bits (two times three bits), I'm filling the rest of the eight bits with two zeros. Remember that an ampersand is a concatenation operator. For the LS three bits, I'm tacking on an inverted version of the address for fun.

Note that all three of these assignments ("adr", "wr_d", and "clock") occur continuously, since they're located outside the process statement.

Here's the resulting waveform:

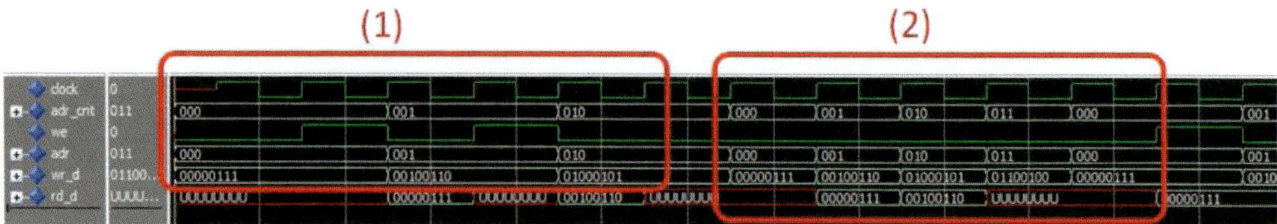

(1) this is the first while loop—the write loop,
(2) and this is the second while loop—the read loop.

In this next screenshot,

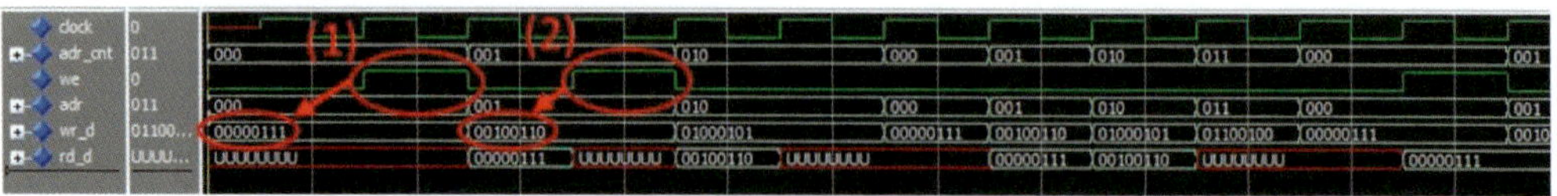

(1) the first write enable writes "00000111", to an address of all zeros, based on an address of all zeros (note the inverted LS 3 bits), and
(2) the second write enable writes "00100110, an address of "001".

Here, we read the first three addresses (zero through two).

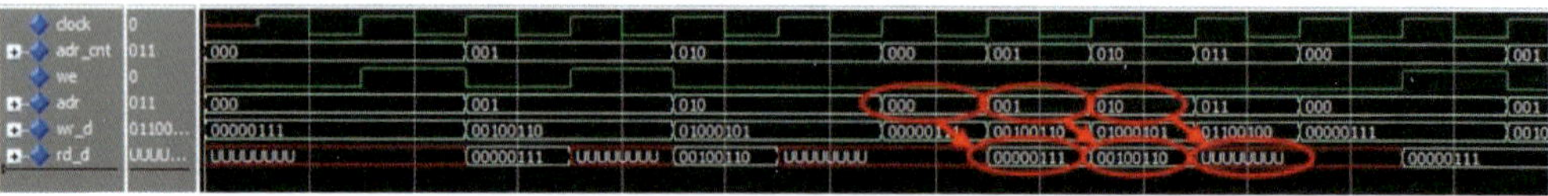

Note that the value in address two is undefined, since we did not write into that address:

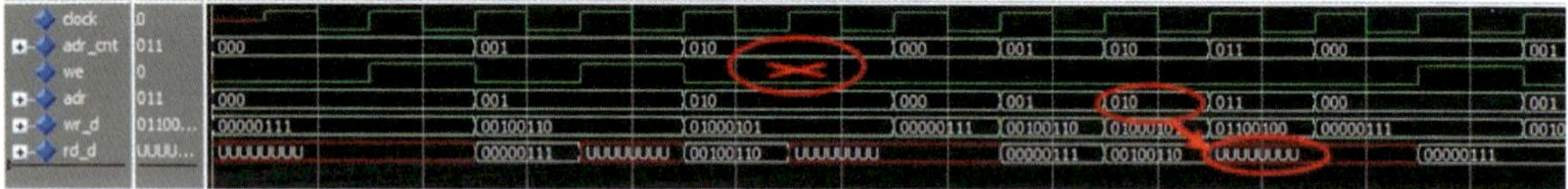

Since we did not stall the process statement, the two while loop sequences start over here:

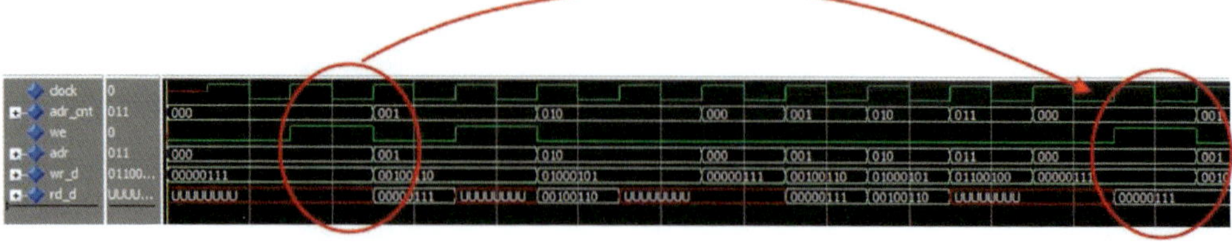

One point about the waveform GUI—the waveform defaults to binary values, but we can change them to hex by right-clicking on the signal, and selecting “radix” and “Hexadecimal”.

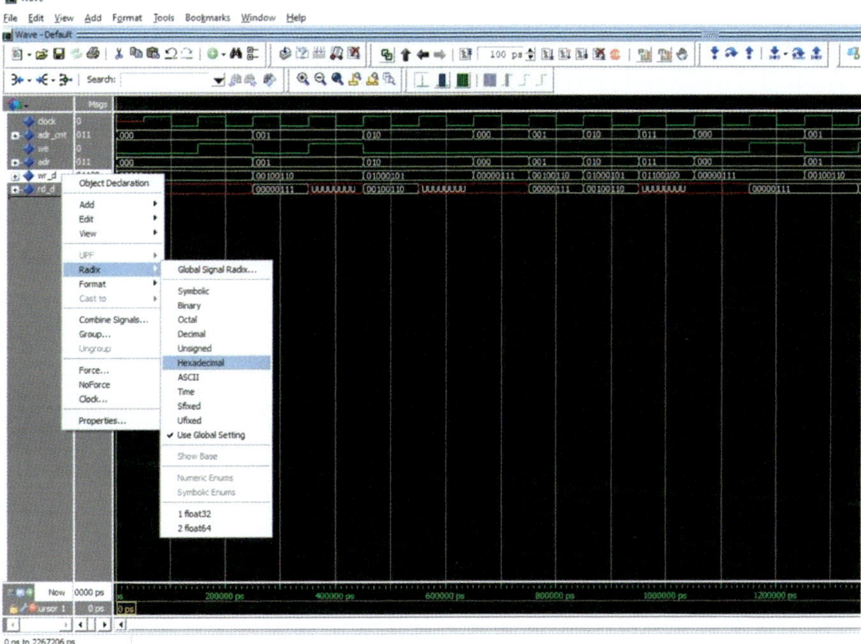

Here, I’ve converted write and read data signals to hex—much easier to read:

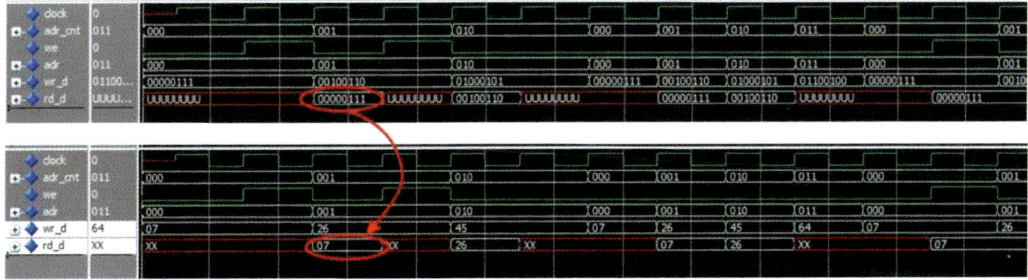

As always, don’t forget to save the waveform.

Inferred Memory

Notice how, in our simple memory design, the addresses in both the write and read case statements are the same values as those of the array indexes.

```
architecture behavioral of mem_reg is

    type mem_array is array (7 downto 0) of std_logic_vector(7 downto 0);
    signal mem_dat   : mem_array;

begin

    register_memory : process(clock)
    begin
        if rising_edge(clock) then
            if ( we = '1' ) then
                case(adr) is
                    when "000" => mem_dat(0) <= wr_d;
                    when "001" => mem_dat(1) <= wr_d;
                    when "010" => mem_dat(2) <= wr_d;
                    when "011" => mem_dat(3) <= wr_d;
                    when "100" => mem_dat(4) <= wr_d;
                    when "101" => mem_dat(5) <= wr_d;
                    when "110" => mem_dat(6) <= wr_d;
                    when "111" => mem_dat(7) <= wr_d;
                    when others => null;
                end case;
            end if;
            --
            case(adr) is
                when "000" => rd_d <= mem_dat(0);
                when "001" => rd_d <= mem_dat(1);
                when "010" => rd_d <= mem_dat(2);
                when "011" => rd_d <= mem_dat(3);
                when "100" => rd_d <= mem_dat(4);
                when "101" => rd_d <= mem_dat(5);
                when "110" => rd_d <= mem_dat(6);
                when "111" => rd_d <= mem_dat(7);
                when others => null;
            end case;
        end if;
    end process;

end architecture behavioral;
```

This means that we could substitute the addresses for the index values:

```
entity mem_reg is
port (
            clock       : in   std_logic;
            wr_d        : in   std_logic_vector(7 downto 0);
            adr         : in   std_logic_vector(2 downto 0);
            we          : in   std_logic;
            --
            rd_d        : out  std_logic_vector(7 downto 0)
        );
end entity;

architecture behavioral of mem_reg is

    type mem_array is array (7 downto 0) of std_logic_vector(7 downto 0);
    signal mem_dat  : mem_array;

begin

    register_memory : process(clock)
    begin
        if rising_edge(clock) then
            if ( we = '1' ) then
                case(adr) is
                    when "000" => mem_dat(0) <= wr_d;
                    when "001" => mem_dat(1) <= wr_d;
                    when "010" => mem_dat(2) <= wr_d;
                    when "011" => mem_dat(3) <= wr_d;
                    when "100" => mem_dat(4) <= wr_d;
                    when "101" => mem_dat(5) <= wr_d;
                    when "110" => mem_dat(6) <= wr_d;
                    when "111" => mem_dat(7) <= wr_d;
                    when others => null;
                end case;
            end if;
            --
            case(adr) is
                when "000" => rd_d <= mem_dat(0);
                when "001" => rd_d <= mem_dat(1);
                when "010" => rd_d <= mem_dat(2);
                when "011" => rd_d <= mem_dat(3);
                when "100" => rd_d <= mem_dat(4);
                when "101" => rd_d <= mem_dat(5);
                when "110" => rd_d <= mem_dat(6);
                when "111" => rd_d <= mem_dat(7);
                when others => null;
            end case;
        end if;
    end process;

end architecture behavioral;
```

```
entity mem_single_port is
port (
            clock       : in   std_logic;
            wr_d        : in   std_logic_vector(7 downto 0);
            adr         : in   std_logic_vector(2 downto 0);
            we          : in   std_logic;
            --
            rd_d        : out  std_logic_vector(7 downto 0)
        );
end entity;

architecture behavioral of mem_single_port is

    type array_8x8 is array (7 downto 0) of std_logic_vector(7 downto 0);
    signal mem_dat  : array_8x8;
    signal adr_int  : integer;

begin

    adr_int <= to_integer(unsigned(adr));

    register_memory : process(clock)
    begin
        if rising_edge(clock) then
            if ( we = '1' ) then
                mem_dat(adr_int) <= wr_d;
            end if;
```

Notice that I'm starting a new VHDL file called "mem_single_port". Note also that we need to convert the standard-logic address to an integer via the unsigned intermediary, "adr_int" (we covered this in Exercise 5.8 of volume 1).

Finally, we pull in the read assignments as well using addresses as array indexes.

```
entity mem_reg is
port (
            clock       : in   std_logic;
            wr_d        : in   std_logic_vector(7 downto 0);
            adr         : in   std_logic_vector(2 downto 0);
            we          : in   std_logic;
            --
            rd_d        : out  std_logic_vector(7 downto 0)
          );
end entity;

architecture behavioral of mem_reg is

    type mem_array is array (7 downto 0) of std_logic_vector(7 downto 0);
    signal mem_dat   : mem_array;

begin

    register_memory : process(clock)
    begin
       if rising_edge(clock) then
          if ( we = '1' ) then
             case(adr) is
                when "000" => mem_dat(0) <= wr_d;
                when "001" => mem_dat(1) <= wr_d;
                when "010" => mem_dat(2) <= wr_d;
                when "011" => mem_dat(3) <= wr_d;
                when "100" => mem_dat(4) <= wr_d;
                when "101" => mem_dat(5) <= wr_d;
                when "110" => mem_dat(6) <= wr_d;
                when "111" => mem_dat(7) <= wr_d;
                when others => null;
             end case;
          end if;
          --
          case(adr) is
             when "000" => rd_d <= mem_dat(0);
             when "001" => rd_d <= mem_dat(1);
             when "010" => rd_d <= mem_dat(2);
             when "011" => rd_d <= mem_dat(3);
             when "100" => rd_d <= mem_dat(4);
             when "101" => rd_d <= mem_dat(5);
             when "110" => rd_d <= mem_dat(6);
             when "111" => rd_d <= mem_dat(7);
             when others => null;
          end case;
       end if;
    end process;

 end architecture behavioral;
```

```
entity mem_single_port is
port (
            clock       : in   std_logic;
            wr_d        : in   std_logic_vector(7 downto 0);
            adr         : in   std_logic_vector(2 downto 0);
            we          : in   std_logic;
            --
            rd_d        : out  std_logic_vector(7 downto 0)
          );
end entity;

architecture behavioral of mem_single_port is

    type array_8x8 is array (7 downto 0) of std_logic_vector(7 downto 0);
    signal mem_dat   : array_8x8;
    signal adr_int   : integer;

begin

    adr_int <= to_integer(unsigned(adr));

    register_memory : process(clock)
    begin
       if rising_edge(clock) then
          if ( we = '1' ) then
             mem_dat(adr_int) <= wr_d;
          end if;
          --
          rd_d <= mem_dat(adr_int);
       end if;
    end process;

end architecture behavioral;
```

The resulting process statement structure can have a special meaning to FPGA compilers.

```
inferred_memory : process(clock)
begin
   if rising_edge(clock) then
      if ( we = '1' ) then
         mem_dat(adr_int) <= wr_d;
      end if;
      --
      rd_d <= mem_dat(adr_int);
   end if;
end process;
```

Inferred memory

Most vendor compilers will understand that this is describing a memory and may automatically use the fixed memory blocks we saw earlier in the chapter. We say that the compiler has inferred a memory.

Coded Dual-port Memory

Using our new VHDL memory construction techniques, let's see how the simplest single-clock dual-port RAM from earlier in the chapter would be implemented in VHDL.

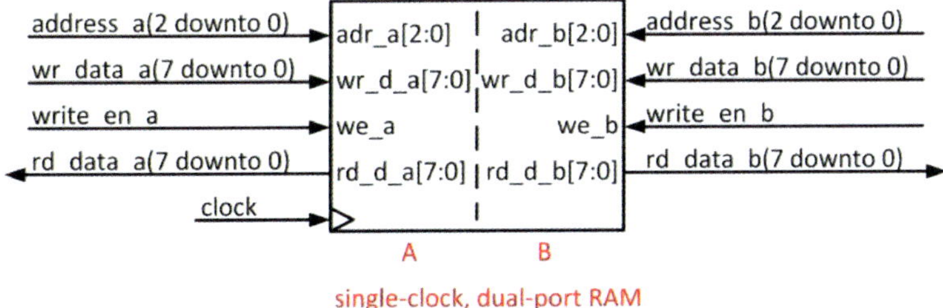

single-clock, dual-port RAM

Here's the code.

```
entity mem_dual_port is
port (
            clock          : in   std_logic;
            --
            wr_d_a         : in   std_logic_vector(7 downto 0);
            adr_a          : in   std_logic_vector(2 downto 0);
            we_a           : in   std_logic;
            rd_d_a         : out  std_logic_vector(7 downto 0);
            --
            wr_d_b         : in   std_logic_vector(7 downto 0);
            adr_b          : in   std_logic_vector(2 downto 0);
            we_b           : in   std_logic;
            rd_d_b         : out  std_logic_vector(7 downto 0)

          );
end entity;

architecture behavioral of mem_dual_port is

   type array_8x8 is array (7 downto 0) of std_logic_vector(7 downto 0);
   signal mem_dat  : array_8x8;

begin

   inferred_dp_memory : process(clock)
   begin
      if rising_edge(clock) then
         -- writes ---
         if ( we_a = '1' ) then
            mem_dat(to_integer(unsigned(adr_a))) <= wr_d_a;
         elsif ( we_b = '1' ) then
            mem_dat(to_integer(unsigned(adr_b))) <= wr_d_b;
         end if;
         --
         -- reads ---
         rd_d_a <= mem_dat(to_integer(unsigned(adr_a)));
         rd_d_b <= mem_dat(to_integer(unsigned(adr_b)));
      end if;

   end process;

end architecture behavioral;
```

The A port is essentially the same as the single-port memory:

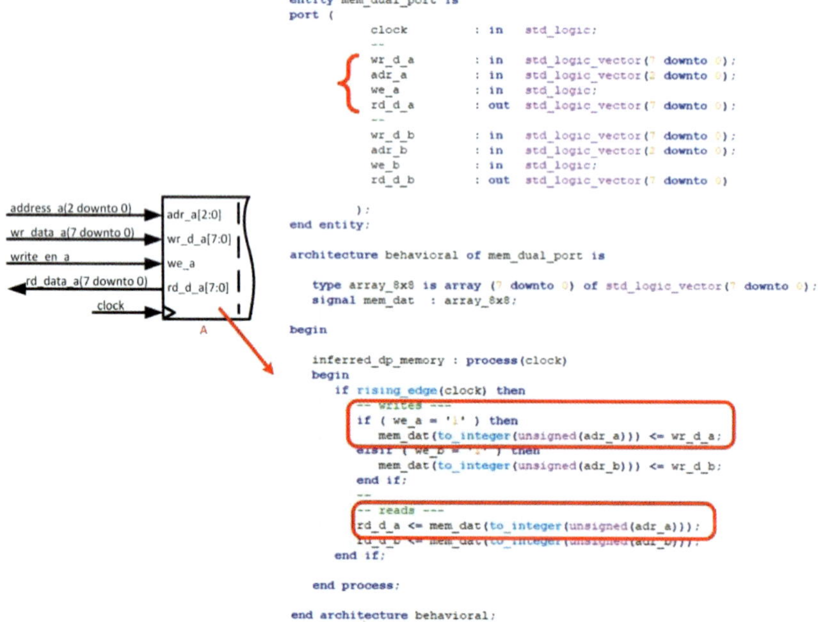

and the code for the B side is virtually a duplicate of that:

```
entity mem_dual_port is
port (
            clock          : in   std_logic;
            --
            wr_d_a         : in   std_logic_vector(7 downto 0);
            adr_a          : in   std_logic_vector(2 downto 0);
            we_a           : in   std_logic;
            rd_d_a         : out  std_logic_vector(7 downto 0);
            --
            wr_d_b         : in   std_logic_vector(7 downto 0);
            adr_b          : in   std_logic_vector(2 downto 0);
            we_b           : in   std_logic;
            rd_d_b         : out  std_logic_vector(7 downto 0)

        );
end entity;

architecture behavioral of mem_dual_port is

   type array_8x8 is array (7 downto 0) of std_logic_vector(7 downto 0);
   signal mem_dat  : array_8x8;

begin

   inferred_dp_memory : process(clock)
   begin
      if rising_edge(clock) then
         -- writes ---
         if ( we_a = '1' ) then
            mem_dat(to_integer(unsigned(adr_a))) <= wr_d_a;
         elsif ( we_b = '1' ) then
            mem_dat(to_integer(unsigned(adr_b))) <= wr_d_b;
         end if;
         --
         -- reads ---
         rd_d_a <= mem_dat(to_integer(unsigned(adr_a)));
         rd_d_b <= mem_dat(to_integer(unsigned(adr_b)));
      end if;

   end process;

end architecture behavioral;
```

adr_b[2:0]
address_b(2 downto 0)
wr_d_b[7:0]
wr_data_b(7 downto 0)
we_b
write_en_b
rd_d_b[7:0]
rd_data_b(7 downto 0)
B

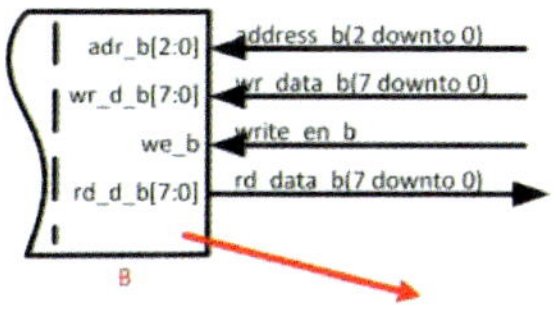

Both sides use the same memory array ("mem_dat"), of course, since they share the same memory locations, i.e., the same memory registers. Only one side can write during any particular clock period, and with the "elsif" we give the A side the priority.

The reads don't get in the way of each other—we say they are non-blocking—and occur independently.

```
begin

   inferred_dp_memory : process(clock)
   begin
      if rising_edge(clock) then
         -- writes ---
         if ( we_a = '1' ) then
            mem_dat(to_integer(unsigned(adr_a))) <= wr_d_a;
         elsif ( we_b = '1' ) then
            mem_dat(to_integer(unsigned(adr_b))) <= wr_d_b;
         end if;
         --
         -- reads ---
         rd_d_a <= mem_dat(to_integer(unsigned(adr_a)));
         rd_d_b <= mem_dat(to_integer(unsigned(adr_b)));
      end if;

   end process;

end architecture behavioral;
```

Note that I've carried the std_logic-to-integer conversion down into the assignment statements.

```
begin

   inferred_dp_memory : process(clock)
   begin
      if rising_edge(clock) then
         -- writes ---
         if ( we_a = '1' ) then
            mem_dat(to_integer(unsigned(adr_a))) <= wr_d_a;
         elsif ( we_b = '1' ) then
            mem_dat(to_integer(unsigned(adr_b))) <= wr_d_b;
         end if;
         --
         -- reads ---
         rd_d_a <= mem_dat(to_integer(unsigned(adr_a)));
         rd_d_b <= mem_dat(to_integer(unsigned(adr_b)));
      end if;

   end process;

end architecture behavioral;
```

Single-port, Shared Bus RAM

We now turn our attention back in the other direction, back to the simple single-port RAM first introduced, which, as with all the memories we've looked at so far, has both a write data bus, and a read data bus. However, we don't necessarily need to have separate buses. Here, we combine the functions of the write and read buses into a single, shared bus ("data"):

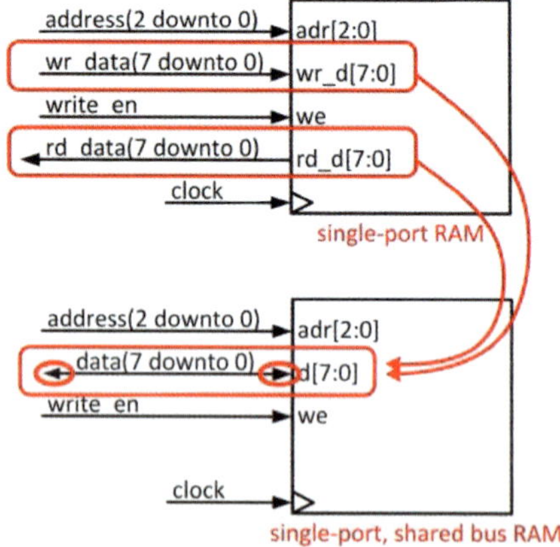

This shared bus now must operate in both directions, what we call bi-directional operation. For writes, the bus points into the memory, and for reads, the bus points out.

In order to share the same bus, however, we need a new control signal—a read enable:

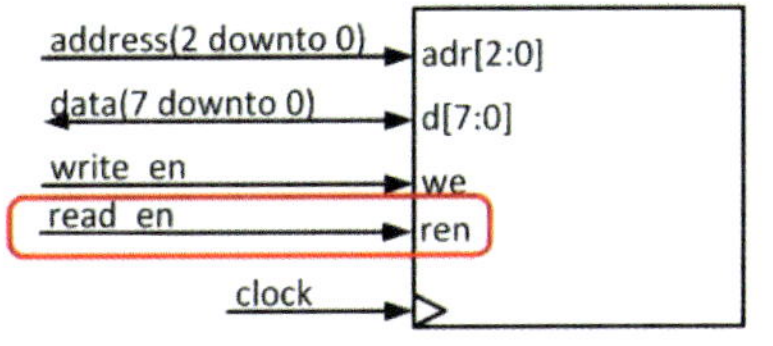

single-port, shared bus RAM

This new read enable signal tells the memory when to drive out onto the bus. When the write enable signal is active, the memory knows that data is being driven into it:

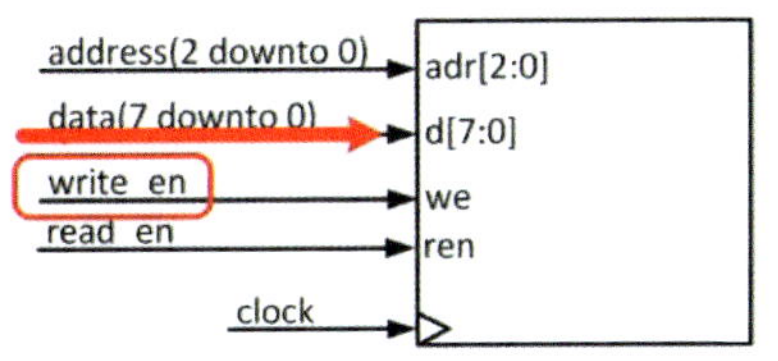

single-port, shared bus RAM

and when the read enable is active, the memory knows that it must drive data out:

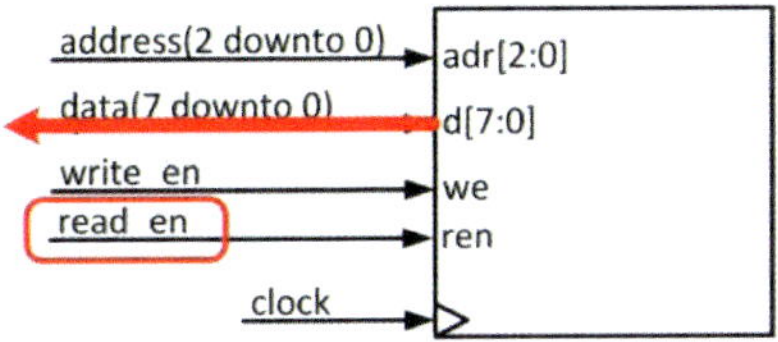

single-port, shared bus RAM

Tri-state Buffers

In order to both receive and drive data on the same bus, we need to introduce a new concept, a tri-state buffer:

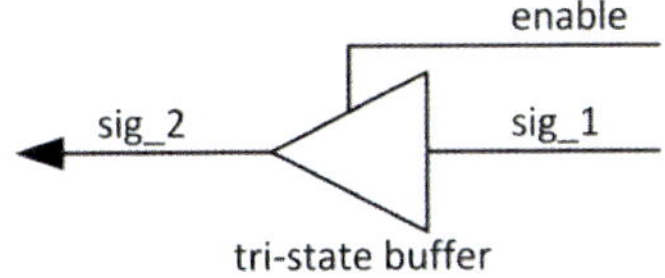

tri-state buffer

When the enable signal is active (high) the signal "sig_1" is passed directly through to sig_2:

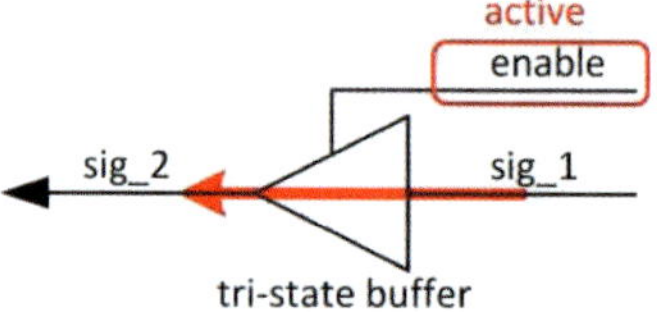

And when enable signal is inactive(low) the output of the buffer is essentially disconnected. The sig_2 signal is now effectively floating.

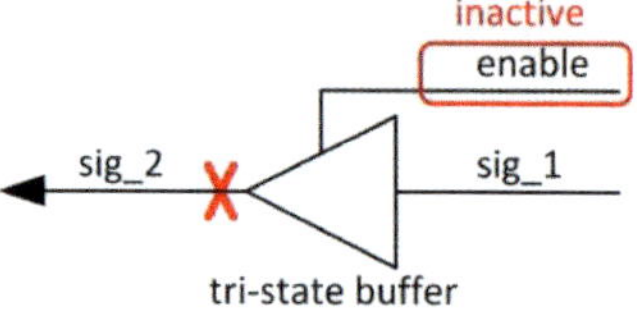

This is what we call a high-impedance state, and any other source (presumably another tr-state buffer) can now drive the signal. We call these buffers tri-state, i.e., *three*-states, since their outputs can be high or low (when the enable is active) or high-impedance (when the enable is inactive). That's three states instead of just the normal high/low two.

Here's a timing diagram of how it works.

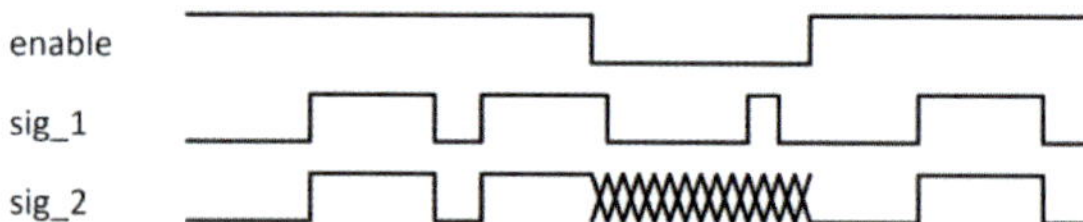

When the enable signal is high, sig_2 follows sig_1, and when the enable signal is low, the output floats, the high-impedance state. We don't know what the signal will do, and we indicate it with X'es.

We often say that when the enable signal is inactive, the output of the buffer is "tri-stated," i.e., placed into its third, indeterminate state.

Let's see how the tri-state buffer is used to implement a shared-bus memory.

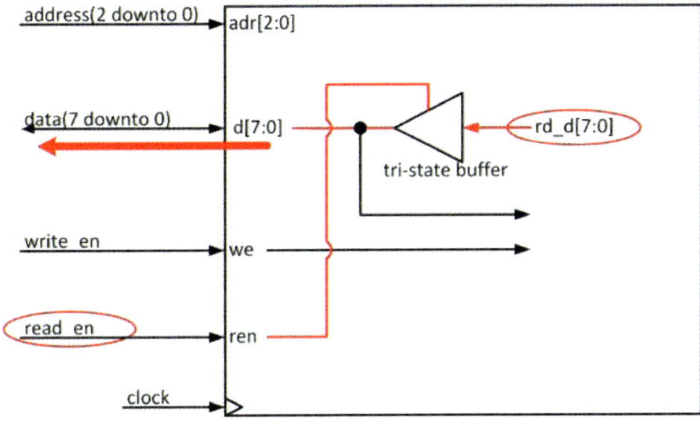

single-port, shared bus RAM

When the read enable signal is active, the tri-state buffer is enabled, and the internal read data—what comprises the read bus on the non-shared memories—passes through, and is driven out the memory,

When the read enable signal is in-active:

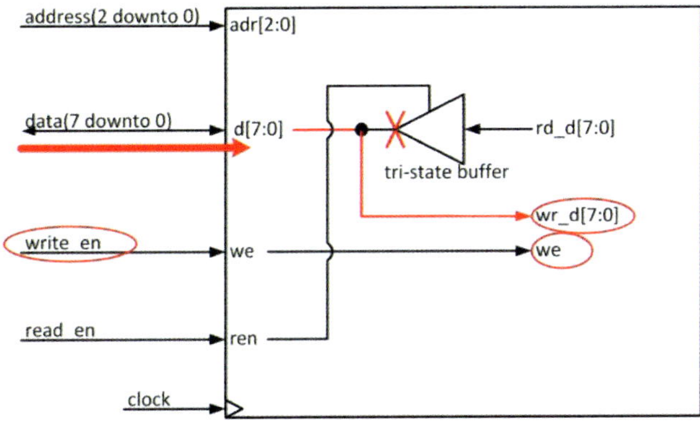

single-port, shared bus RAM

the tri-state buffer is disconnected, and the external data can be driven into the memory, and becomes what comprises the write bus on the non-shared memories. If the write-enable signal is active, then that data is written internally.

Exercises, Chapter 2

Exercise 2-1:

Given this read-before-write dual-port memory, fill in the rd_data_a and rd_data_b signals in the timing diagram.

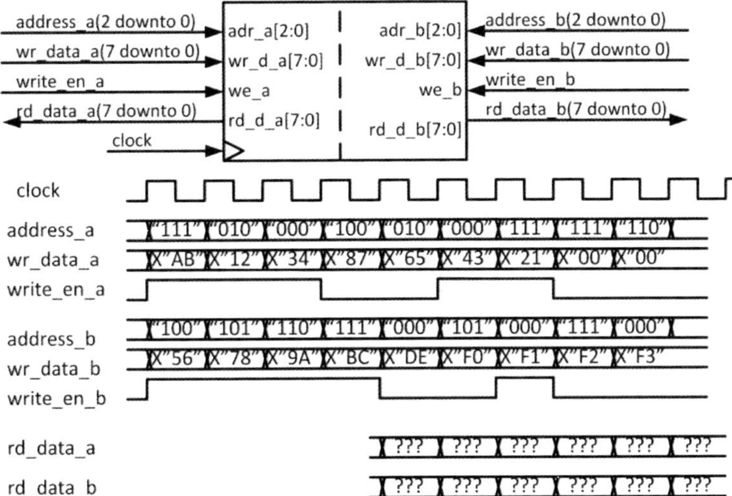

Exercise 2-2:

Implement these two registered muxes in VHDL code.

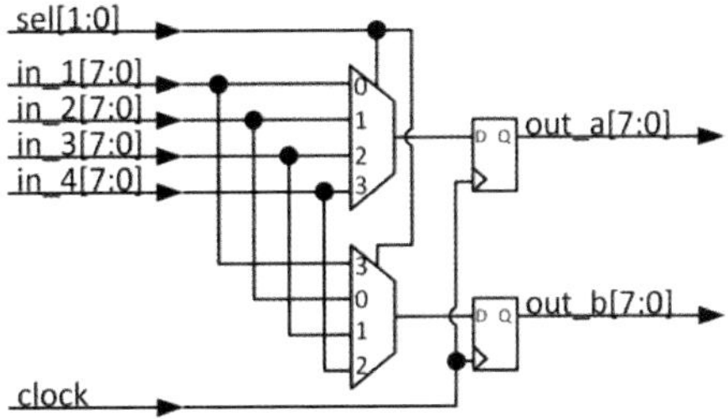

Exercise 2-3:
Implement these registers as an array in VHDL code.

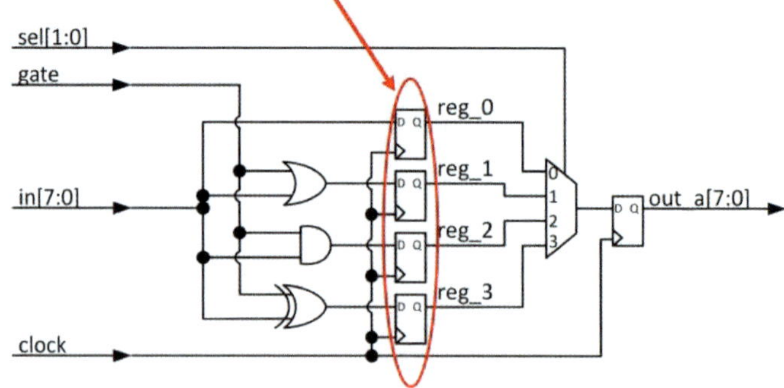

Exercise 2-4:
Code a dual-port memory with separate read and write clocks.

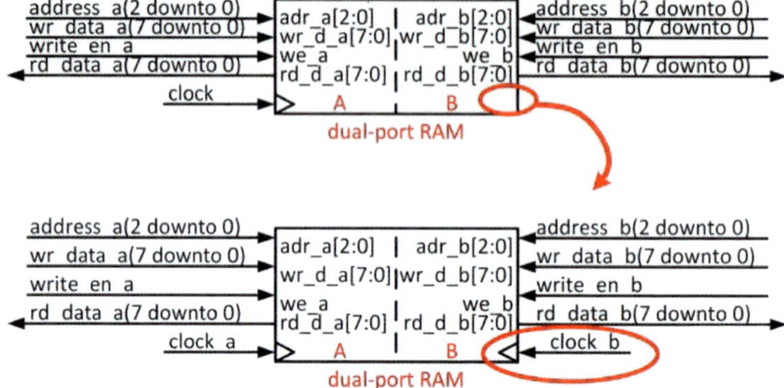

Chapter 3
FIFOs

FIFO Concepts

Now that we have a good handle on implementing memories, we'll look at a specialized version that's nearly ubiquitous.

We start with a common application that most are familiar with—audio communicated via the internet. Here's a representation of, for example, the waveform of a portion of some spoken word as we saw way back in the first chapter.

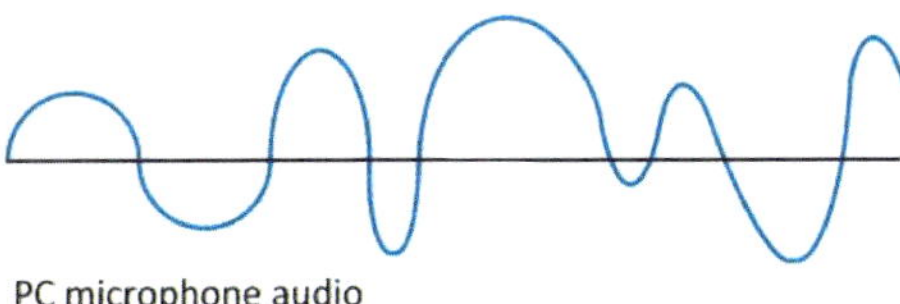

As we did then, the first step is to digitize the waveform, i.e., to record the amplitude of the audio signal at periodic sampling times.

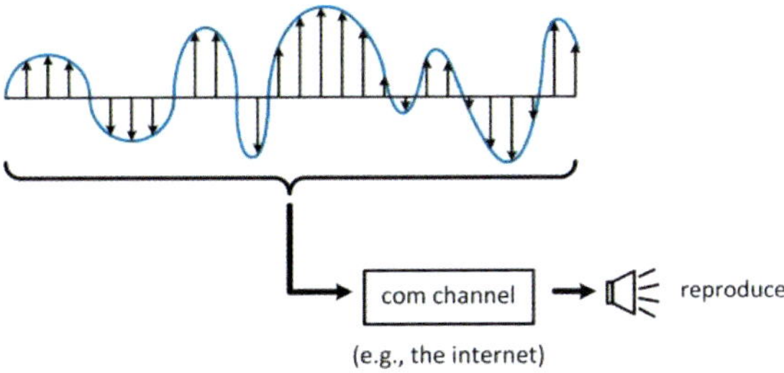

These few dozen samples will be carried across the internet for reproduction via a speaker at the far end of the internet connection.

In order to carry them across the internet, however, the samples must be divided into groups and collected together in transmission units called packets.

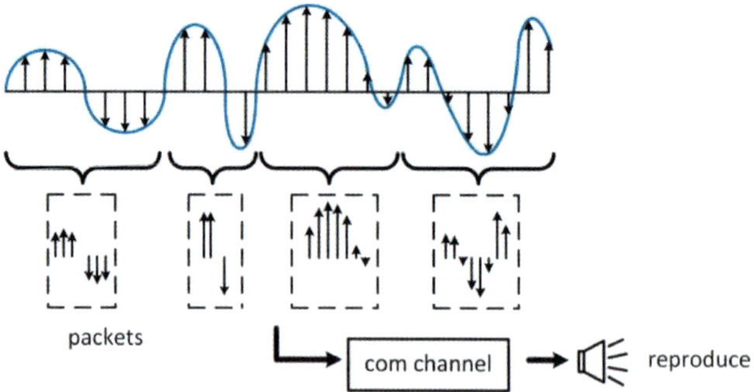

Typically, hundreds of samples would be carried in each packet, but we'll assume just a few each here for illustration. These packets are sent out onto the internet in the order that they occur, and, since the packets arrive at the far end spaced in time, if the arriving samples were simply applied as they arrive, the resulting waveform would look something like this:

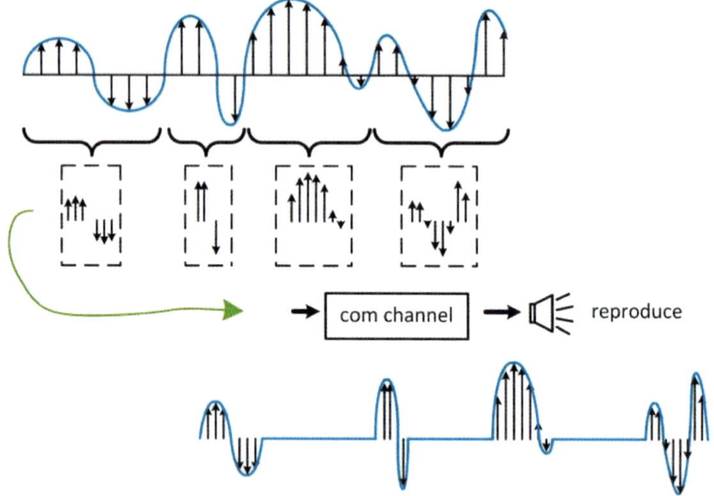

… and would sound pretty nasty.

Clearly, the arriving samples need to spread out in time, re-creating the same sample period spans as when they were originally sampled.

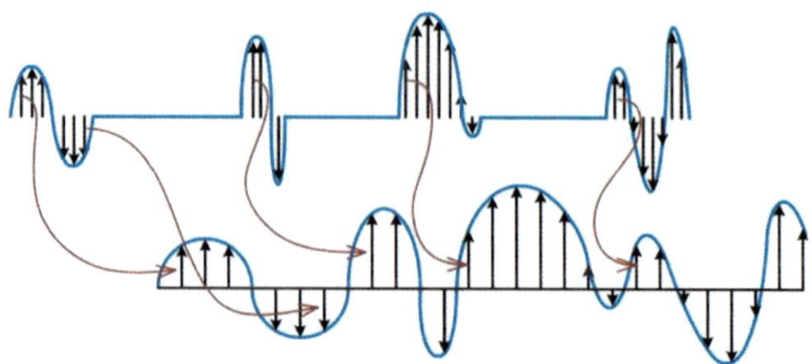

Let's take a look at how this works with one of the packets:

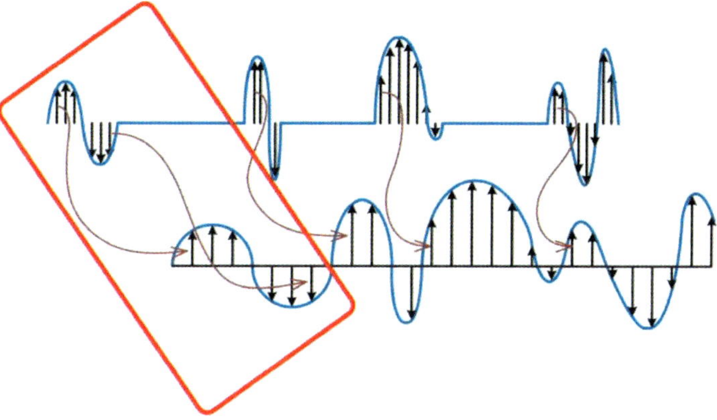

We'll label the original and destination samples …

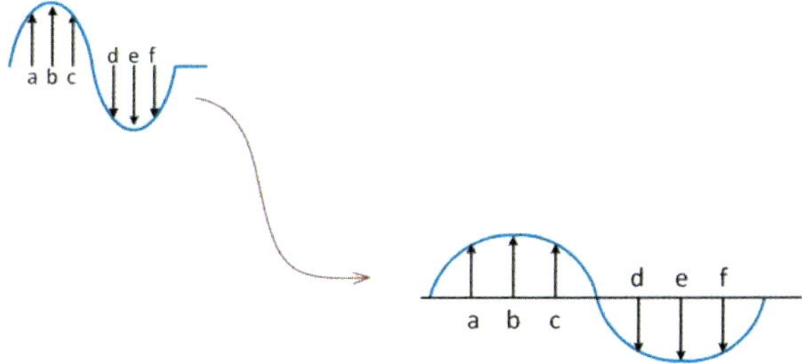

and, we start by storing the first few samples of the packet in a memory:

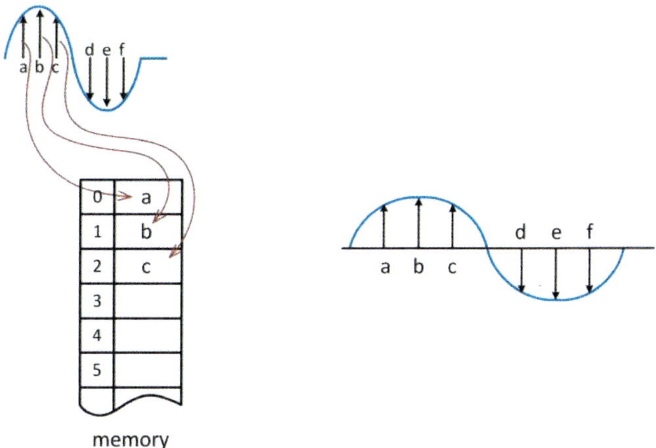

After the rest of the samples for this packet are stored, we read them in turn for reproduction, at the same time intervals as when they were originally sampled.

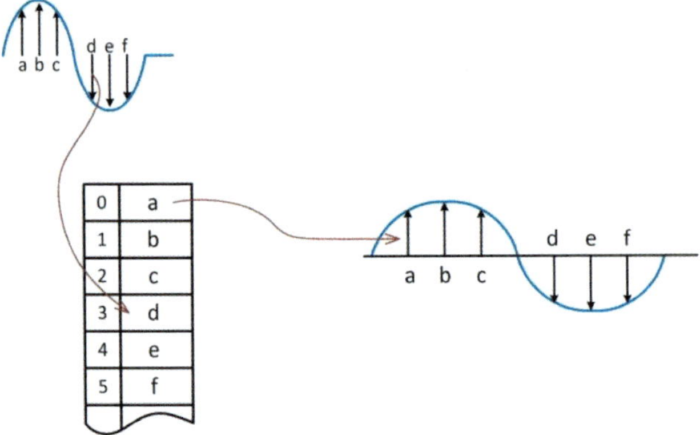

Note that after we read each sample from memory, that memory location becomes free—we're done with that sample, and there's no need to keep it around any longer.

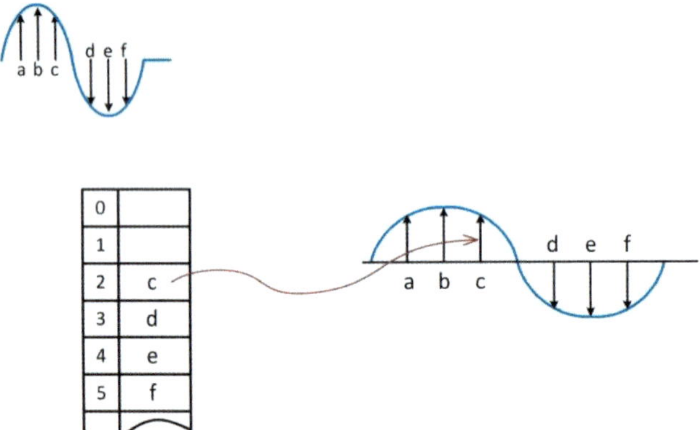

Here, we see, however, that the next packet arrives before we've finished re-creating the samples of the first packet.

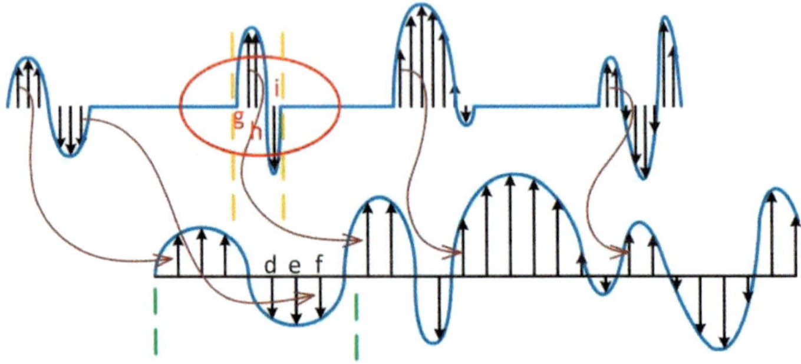

The second packet's "g", "h", and "i" samples begin arriving before we've used the "d", "e", and "f" samples from the first packet.

However, we're done with the first packet's "a", "b", and "c" locations in the memory:

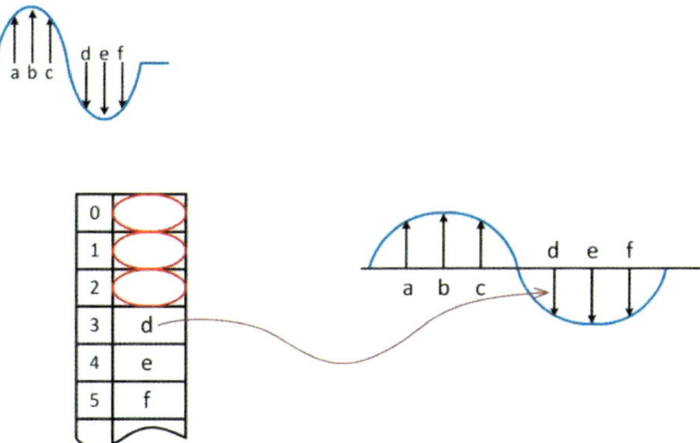

and we can use them for the newly arrived "g, h, and i" samples.

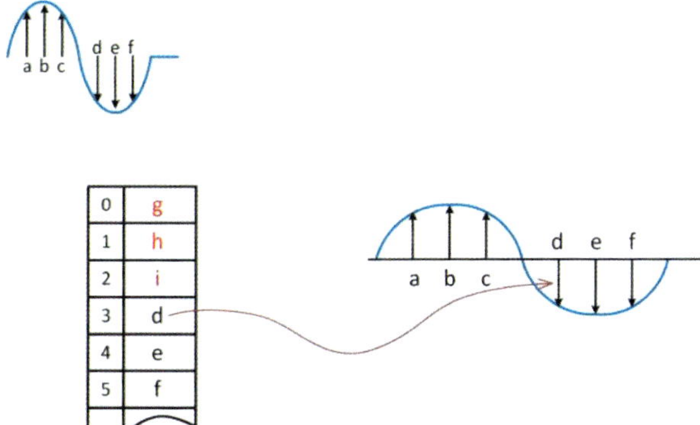

You can see that, by re-using memory locations, once we reach the end of our memory, and assuming that we're done with the beginning ones, we can start over at the beginning.

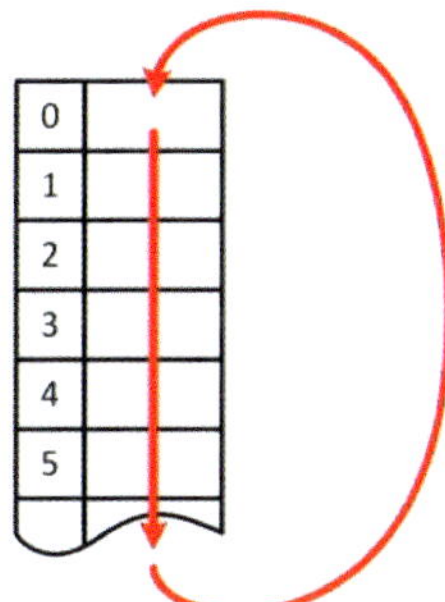

Thus, we can store—temporarily—a lot of samples with a smaller amount of memory. The key is that this is just temporary storage, that we're done with the earlier ones as we proceed around in a circle. In fact, we call this sort of memory a circular buffer.

Let's go back to our original packet sample storage. Note that the first sample that was stored ("a") was the first one read out.

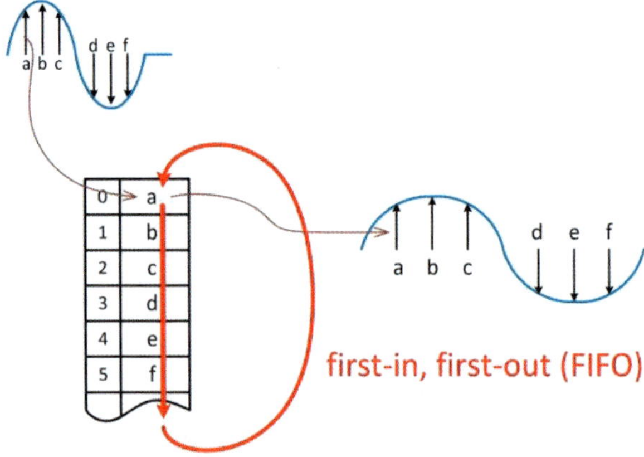

This type of circular buffer is called a FIFO—first-in, first-out.

FIFO Operation

We can see that a FIFO is a special application of a dual-port memory, where one side, the input, performs only writes to the internal memory, and the other side, the output, performs only reads.

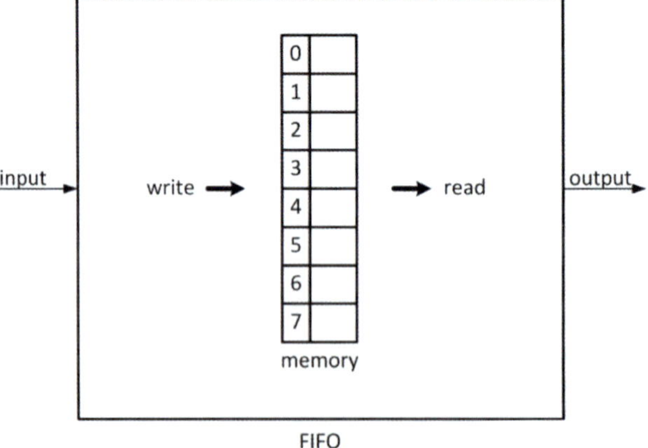

In this simple example, the FIFO is 8 entries deep, meaning that the internal memory has 8 locations.

The addressing is done internally, and the write addresses start at zero, and increment with each FIFO write until we reach the end of the internal memory.

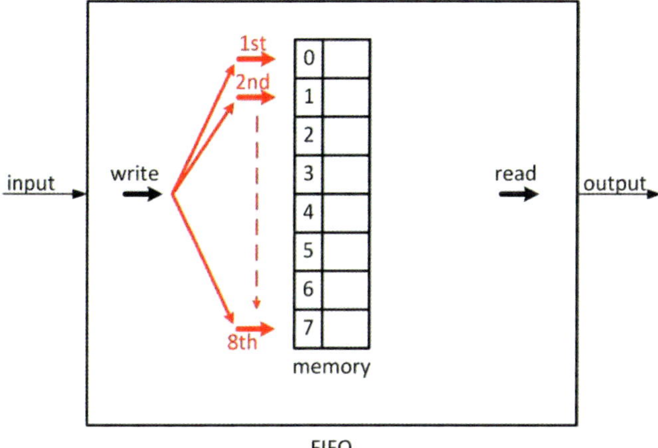

Once the end of the memory is reached, the write addresses start over again at zero.

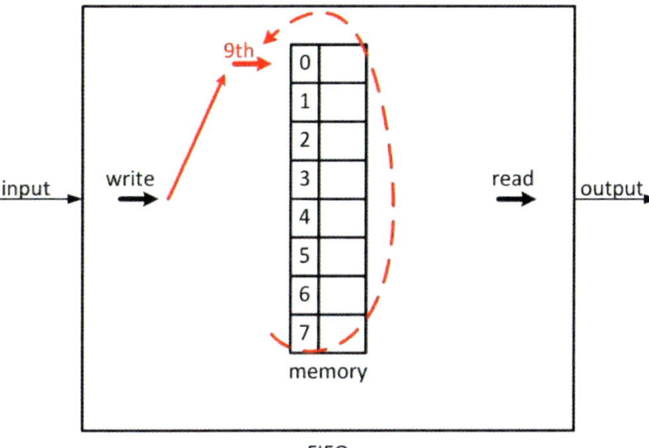

Meanwhile, and independent of the write side, the read addressing works the same way.

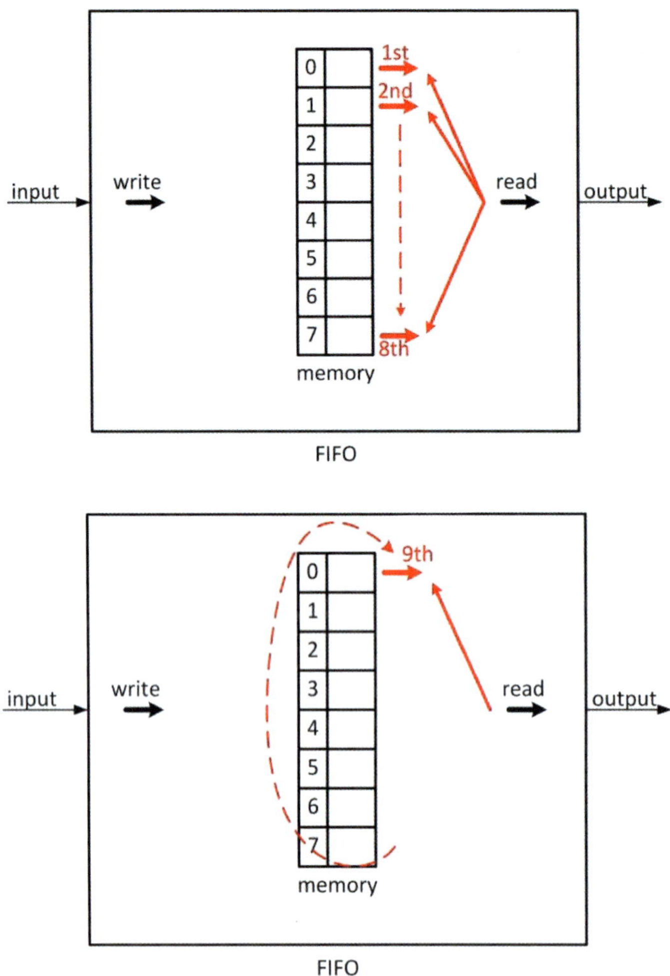

Imagine that we've written eight values, "a" through "h", into the memory.

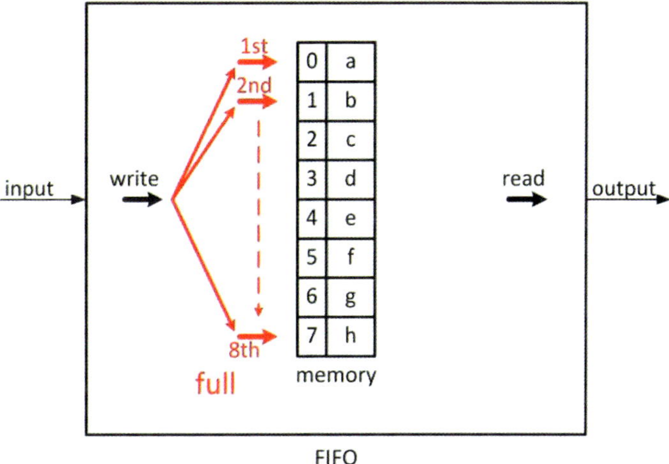

We say that the FIFO is now full, since the next input would write over the "a" value in address 0.

When using a FIFO, however, we are presumably also reading from it as well. So, the first read frees up address zero, and the next read frees up address one. The FIFO is no longer full—it contains only six unread values.

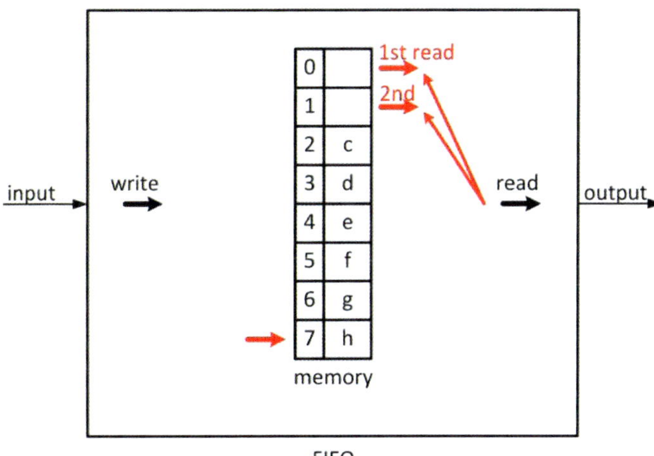

This allows the next two inputs to enter the FIFO without overwriting values that have not yet been read (and renders the FIFO again full). If we can keep the FIFO from getting full, we won't lose data.

On the other hand, if we read all the values that had been written, then we say that the FIFO is empty.

FIFO Implementation

Here's a black box representation of a simple FIFO. This one is an "8x8", meaning that it is one byte wide, and 8 entries deep. For each clock that the "we" input is active, the data on the "d_in" input is written into the FIFO, and for each clock that the "rd" input is active, the next entry in the FIFO is clocked out on "d_out".

Here's the VHDL code for this simple FIFO.

```
entity fifo is
port (
            clock         : in   std_logic;
            --
            we            : in   std_logic;
            d_in          : in   std_logic_vector(7 downto 0);
            --
            rd            : in   std_logic;
            d_out         : out  std_logic_vector(7 downto 0)
        );
end entity;

architecture behavioral of fifo is

   type array_8x8 is array (7 downto 0) of std_logic_vector(7 downto 0);
   signal mem_dat   : array_8x8;
   signal addr_write     : unsigned(2 downto 0) := "000";
   signal addr_read      : unsigned(2 downto 0) := "000";

begin

   fifo_memory : process(clock)
   begin
      if rising_edge(clock) then
         -- writes ---
         if ( we = '1' ) then
            mem_dat(to_integer(unsigned(addr_write))) <= d_in;
            addr_write <= addr_write + 1;
         end if;
         -- reads ---
         if (rd = '1') then
            d_out      <= mem_dat(to_integer(unsigned(addr_read)));
            addr_read <= addr_read + 1;
         end if;
      end if;
   end process;

end architecture behavioral;
```

We see that when the "we" input signal is high, the "d_in" input is written to the memory, and the write address, "addr_write", is incremented.

```
begin

   fifo_memory : process(clock)
   begin
      if rising_edge(clock) then
         -- writes ---
         if ( we = '1' ) then
            mem_dat(to_integer(unsigned(addr_write))) <= d_in;
            addr_write <= addr_write + 1;
         end if;
         -- reads ---
         if (rd = '1') then
            d_out      <= mem_dat(to_integer(unsigned(addr_read)));
            addr_read <= addr_read + 1;
         end if;
      end if;
   end process;

end architecture behavioral;
```

And when the "rd" input signal is high, the addressed memory location is clocked out, and the read address, "addr_read" is incremented.

```
begin

   fifo_memory : process(clock)
   begin
      if rising_edge(clock) then
         -- writes ---
         if ( we = '1' ) then
            mem_dat(to_integer(unsigned(addr_write))) <= d_in;
            addr_write <= addr_write + 1;
         end if;
         -- reads ---
         if (rd = '1') then
            d_out      <= mem_dat(to_integer(unsigned(addr_read)));
            addr_read <= addr_read + 1;
         end if;
      end if;
   end process;

end architecture behavioral;
```

FIFO Testbench

Here's a testbench I created to show the FIFO operation. We'll zoom in on the different parts, so don't strain your eyes yet.

```
architecture Behavioral of sim_4_4_tb is

   component fifo is
   port (
            clock          : in   std_logic;
            --
            we             : in   std_logic;
            d_in           : in   std_logic_vector(7 downto 0);
            --
            rd             : in   std_logic;
            d_out          : out  std_logic_vector(7 downto 0)
         );
   end component;

   signal clk              : std_logic := '0';
   signal we               : std_logic := '0';
   signal rd               : std_logic := '0';
   signal wr_dat           : unsigned(7 downto 0) := X"00";
   signal wr_d             : std_logic_vector(7 downto 0);
   signal rd_d             : std_logic_vector(7 downto 0);

begin

   clk <= '0' after 50 ns when clk = '1' else
          '1' after 50 ns;

   stimulus : process
      variable j  : integer;
   begin
      we      <= '0';
      --
      wait until rising_edge(clk);
      we <= '1';
      for j in 0 to 4 loop
         wait until rising_edge(clk);
         wr_dat <= wr_dat + 1;
      end loop;
      wait until rising_edge(clk);
      we <= '0';
      --
      wait until rising_edge(clk);
      rd <= '1';
      for j in 0 to 5 loop
         wait until rising_edge(clk);
      end loop;
      rd <= '0';
      --
      wait until rising_edge(clk);
      for j in 0 to 2 loop
         wait until rising_edge(clk);
         we <= '1';
         wr_dat <= wr_dat + 1;
      end loop;
      wait until rising_edge(clk);
      we <= '0';
      --
      wait until rising_edge(clk);
      rd <= '1';
      for j in 0 to 2 loop
         wait until rising_edge(clk);
      end loop;
      rd <= '0';
      --
      for j in 0 to 50 loop
         wait until rising_edge(clk);
      end loop;
   end process;

   wr_d    <= std_logic_vector(wr_dat);

   fifo_1 : fifo
   port map
      (
       clock   => clk,
       --
       we      => we,
       d_in    => wr_d,
       --
       rd      => rd,
       d_out   => rd_d
      );
end architecture Behavioral;
```

This is the FIFO module declaration, the same IO we just saw in the FIFO VHDL file:

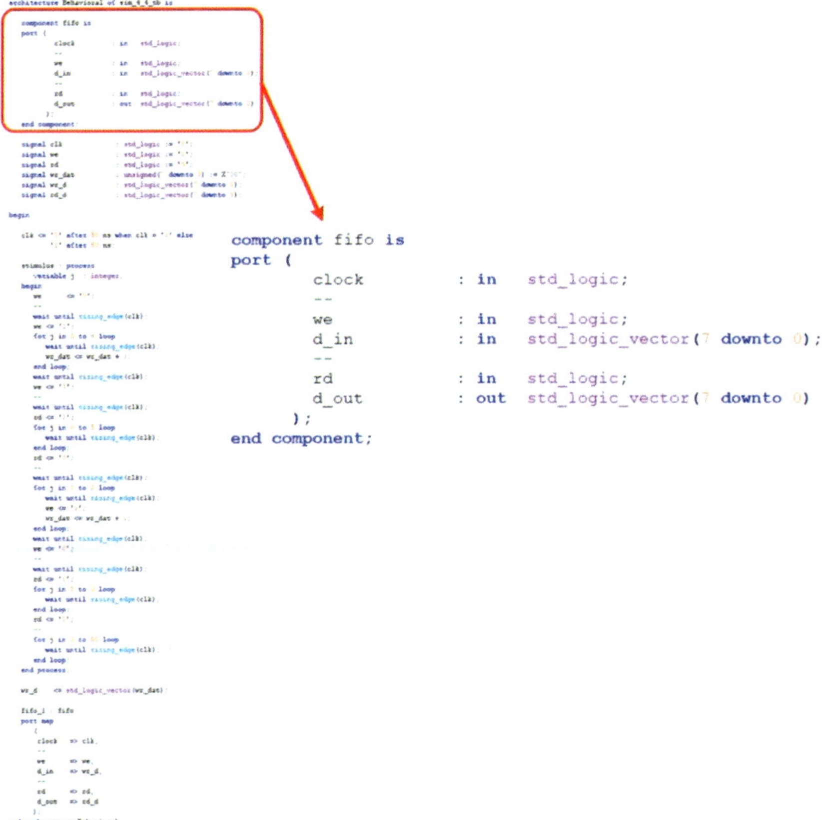

Here's the signal declarations:

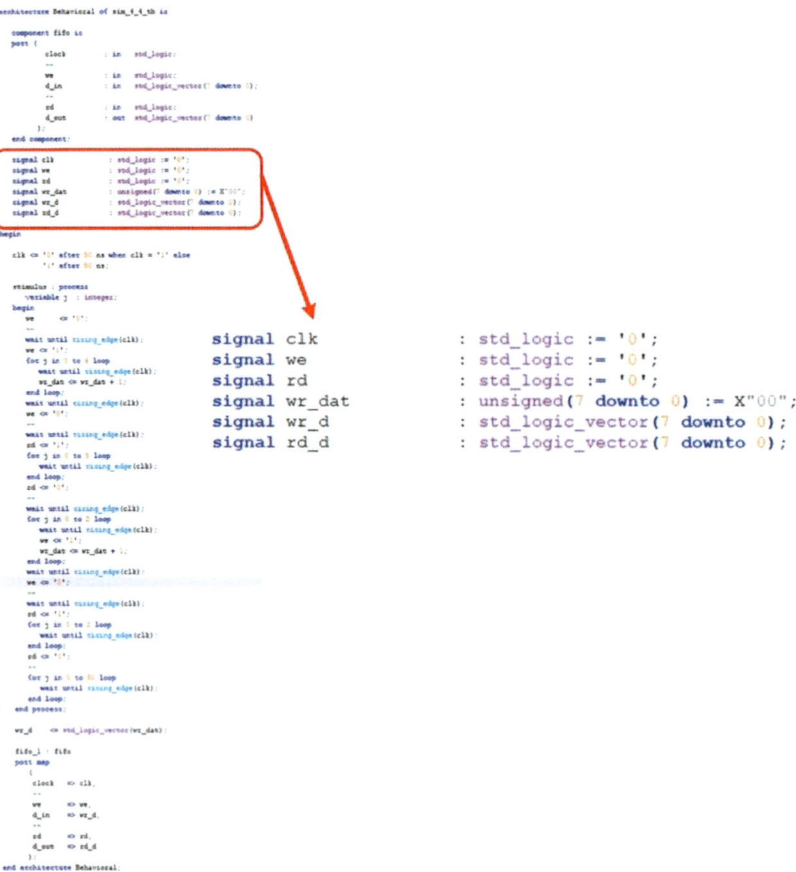

Notice that the "wr_dat" signal is unsigned—as we'll see, this is because we'll be incrementing it.

We jump down to look at the FIFO module instantiation, and see the connected signals that we just looked at.

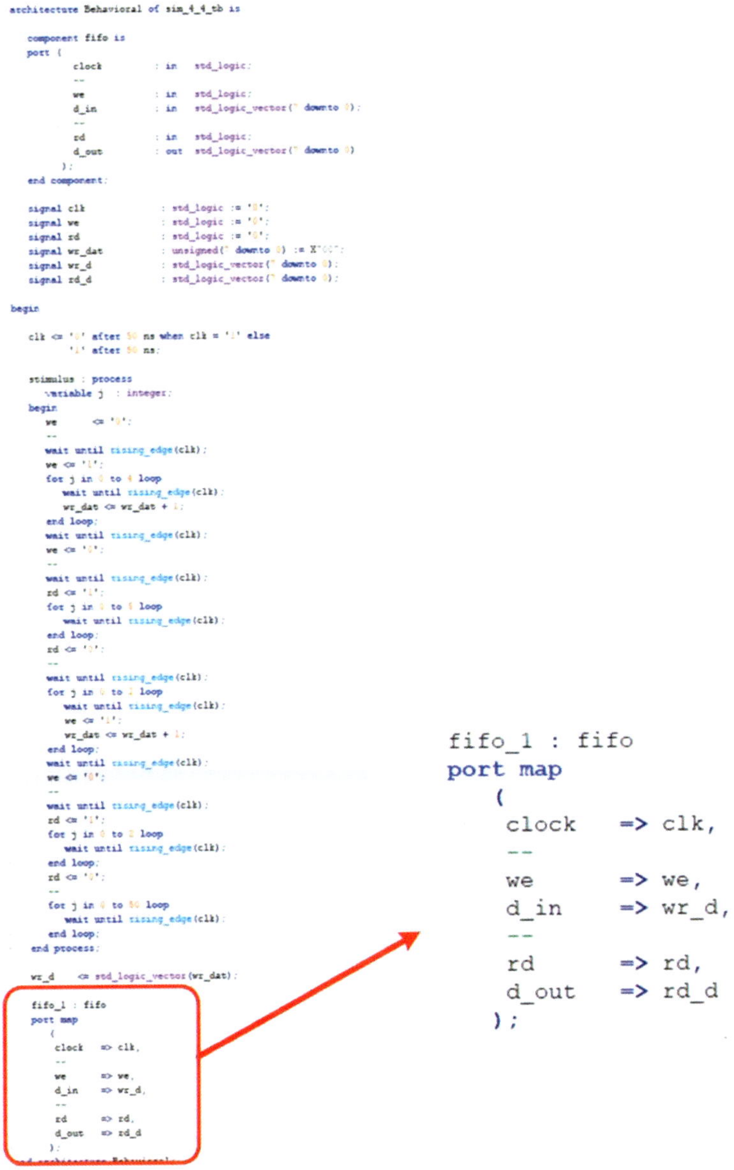

```
architecture Behavioral of sim_4_4_tb is

   component fifo is
   port (
         clock         : in   std_logic;
         --
         we            : in   std_logic;
         d_in          : in   std_logic_vector(7 downto 0);
         --
         rd            : in   std_logic;
         d_out         : out  std_logic_vector(7 downto 0)
      );
   end component;

   signal clk            : std_logic := '0';
   signal we             : std_logic := '0';
   signal rd             : std_logic := '0';
   signal wr_dat         : unsigned(7 downto 0) := X"00";
   signal wr_d           : std_logic_vector(7 downto 0);
   signal rd_d           : std_logic_vector(7 downto 0);

begin

   clk <= '0' after 50 ns when clk = '1' else
          '1' after 50 ns;

   stimulus : process
     variable j  : integer;
   begin
     we      <= '0';
     --
     wait until rising_edge(clk);
     we <= '1';
     for j in 0 to 4 loop
        wait until rising_edge(clk);
        wr_dat <= wr_dat + 1;
     end loop;
     wait until rising_edge(clk);
     we <= '0';
     --
     wait until rising_edge(clk);
     rd <= '1';
     for j in 0 to 5 loop
        wait until rising_edge(clk);
     end loop;
     rd <= '0';
     --
     wait until rising_edge(clk);
     for j in 0 to 1 loop
        wait until rising_edge(clk);
        we <= '1';
        wr_dat <= wr_dat + 1;
     end loop;
     wait until rising_edge(clk);
     we <= '0';
     --
     wait until rising_edge(clk);
     rd <= '1';
     for j in 0 to 1 loop
        wait until rising_edge(clk);
     end loop;
     rd <= '0';
     --
     for j in 0 to 50 loop
        wait until rising_edge(clk);
     end loop;
   end process;

   wr_d     <= std_logic_vector(wr_dat);

   fifo_1 : fifo
   port map
     (
      clock    => clk,
      --
      we       => we,
      d_in     => wr_d,
      --
      rd       => rd,
      d_out    => rd_d
     );
end architecture Behavioral;
```

```
fifo_1 : fifo
port map
    (
     clock    => clk,
     --
     we       => we,
     d_in     => wr_d,
     --
     rd       => rd,
     d_out    => rd_d
    );
```

We generate the clock the same as always:

```
architecture Behavioral of sim_4_4_tb is

   component fifo is
   port (
         clock        : in   std_logic;
         --
         we           : in   std_logic;
         d_in         : in   std_logic_vector(7 downto 0);
         --
         rd           : in   std_logic;
         d_out        : out  std_logic_vector(7 downto 0)
      );
   end component;

   signal clk              : std_logic := '0';
   signal we               : std_logic := '0';
   signal rd               : std_logic := '0';
   signal wr_dat           : unsigned(7 downto 0) := X"00";
   signal wr_d             : std_logic_vector(7 downto 0);
   signal rd_d             : std_logic_vector(7 downto 0);

begin

   clk <= '0' after 50 ns when clk = '1' else
          '1' after 50 ns;

   stimulus : process
      variable j  : integer;
   begin
      we       <= '0';
      --
      wait until rising_edge(clk);
      we <= '1';
      for j in 0 to 4 loop
         wait until rising_edge(clk);
         wr_dat <= wr_dat + 1;
      end loop;
      wait until rising_edge(clk);
      we <= '0';
      --
      wait until rising_edge(clk);
      rd <= '1';
      for j in 0 to 5 loop
         wait until rising_edge(clk);
      end loop;
      rd <= '0';
      --
      wait until rising_edge(clk);
      for j in 0 to 2 loop
         wait until rising_edge(clk);
         we <= '1';
         wr_dat <= wr_dat + 1;
      end loop;
      wait until rising_edge(clk);
      we <= '0';
      --
      wait until rising_edge(clk);
      rd <= '1';
      for j in 0 to 2 loop
         wait until rising_edge(clk);
      end loop;
      rd <= '0';
      --
      for j in 0 to 50 loop
         wait until rising_edge(clk);
      end loop;
   end process;

   wr_d     <= std_logic_vector(wr_dat);

   fifo_1 : fifo
   port map
      (
       clock    => clk,
       --
       we       => we,
       d_in     => wr_d,
       --
       rd       => rd,
       d_out    => rd_d
      );
end architecture Behavioral;
```

```
clk <= '0' after 50 ns when clk = '1' else
       '1' after 50 ns;
```

And we finally get to the meat—the process statement that generates the stimulus signals:

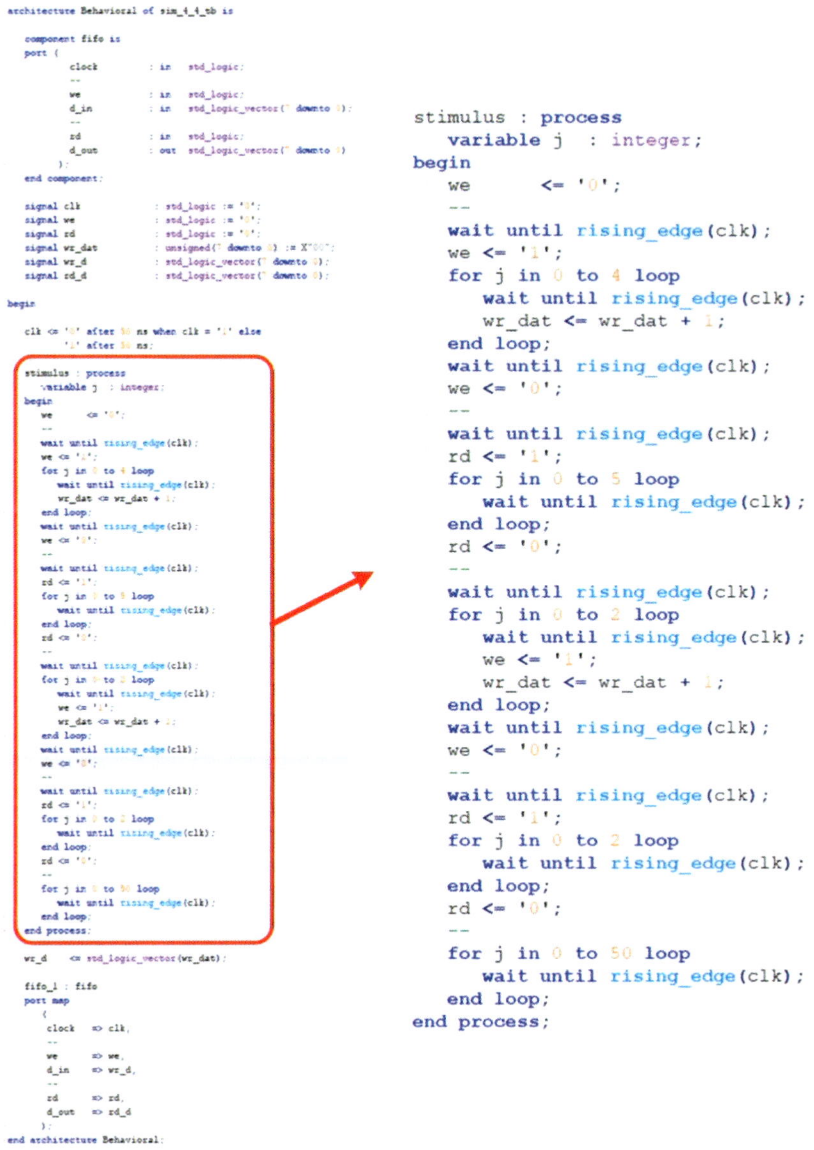

```
architecture Behavioral of sim_4_4_tb is

   component fifo is
   port (
          clock        : in   std_logic;
          --
          we           : in   std_logic;
          d_in         : in   std_logic_vector(7 downto 0);
          --
          rd           : in   std_logic;
          d_out        : out  std_logic_vector(7 downto 0)
        );
   end component;

   signal clk          : std_logic := '0';
   signal we           : std_logic := '0';
   signal rd           : std_logic := '0';
   signal wr_dat       : unsigned(7 downto 0) := X"00";
   signal wr_d         : std_logic_vector(7 downto 0);
   signal rd_d         : std_logic_vector(7 downto 0);

begin

   clk <= '0' after 50 ns when clk = '1' else
          '1' after 50 ns;

   stimulus : process
      variable j   : integer;
   begin
      we       <= '0';
      --
      wait until rising_edge(clk);
      we <= '1';
      for j in 0 to 4 loop
         wait until rising_edge(clk);
         wr_dat <= wr_dat + 1;
      end loop;
      wait until rising_edge(clk);
      we <= '0';
      --
      wait until rising_edge(clk);
      rd <= '1';
      for j in 0 to 5 loop
         wait until rising_edge(clk);
      end loop;
      rd <= '0';
      --
      wait until rising_edge(clk);
      for j in 0 to 2 loop
         wait until rising_edge(clk);
         we <= '1';
         wr_dat <= wr_dat + 1;
      end loop;
      wait until rising_edge(clk);
      we <= '0';
      --
      wait until rising_edge(clk);
      rd <= '1';
      for j in 0 to 2 loop
         wait until rising_edge(clk);
      end loop;
      rd <= '0';
      --
      for j in 0 to 50 loop
         wait until rising_edge(clk);
      end loop;
   end process;

   wr_d    <= std_logic_vector(wr_dat);

   fifo_1 : fifo
   port map
      (
       clock   => clk,
       --
       we      => we,
       d_in    => wr_d,
       --
       rd      => rd,
       d_out   => rd_d
      );
end architecture Behavioral;
```


```
stimulus : process
   variable j   : integer;
begin
   we       <= '0';
   --
   wait until rising_edge(clk);
   we <= '1';
   for j in 0 to 4 loop
      wait until rising_edge(clk);
      wr_dat <= wr_dat + 1;
   end loop;
   wait until rising_edge(clk);
   we <= '0';
   --
   wait until rising_edge(clk);
   rd <= '1';
   for j in 0 to 5 loop
      wait until rising_edge(clk);
   end loop;
   rd <= '0';
   --
   wait until rising_edge(clk);
   for j in 0 to 2 loop
      wait until rising_edge(clk);
      we <= '1';
      wr_dat <= wr_dat + 1;
   end loop;
   wait until rising_edge(clk);
   we <= '0';
   --
   wait until rising_edge(clk);
   rd <= '1';
   for j in 0 to 2 loop
      wait until rising_edge(clk);
   end loop;
   rd <= '0';
   --
   for j in 0 to 50 loop
      wait until rising_edge(clk);
   end loop;
end process;
```

Here, we activate "we", and feed six incrementing data values into the FIFO. Notice that although the loop only cycles five times, the "we" signal stays active one clock after the loop ends.

```
stimulus : process
   variable j  : integer;
begin
   we      <= '0';
   --
   wait until rising_edge(clk);
   we <= '1';
   for j in 0 to 4 loop
      wait until rising_edge(clk);
      wr_dat <= wr_dat + 1;
   end loop;
   wait until rising_edge(clk);
   we <= '0';
   --
   wait until rising_edge(clk);
   rd <= '1';
   for j in 0 to 5 loop
      wait until rising_edge(clk);
   end loop;
   rd <= '0';
   --
   wait until rising_edge(clk);
   for j in 0 to 2 loop
      wait until rising_edge(clk);
      we <= '1';
      wr_dat <= wr_dat + 1;
   end loop;
   wait until rising_edge(clk);
   we <= '0';
   --
   wait until rising_edge(clk);
   rd <= '1';
   for j in 0 to 2 loop
      wait until rising_edge(clk);
   end loop;
   rd <= '0';
   --
   for j in 0 to 50 loop
      wait until rising_edge(clk);
   end loop;
end process;
```

Note that down here, we convert the incrementing "wr_dat" signal to standard logic that we then connect to the FIFO input.

```
architecture Behavioral of sim_4_4_tb is

   component fifo is
   port (
            clock         : in   std_logic;
            --
            we            : in   std_logic;
            d_in          : in   std_logic_vector(7 downto 0);
            --
            rd            : in   std_logic;
            d_out         : out  std_logic_vector(7 downto 0)
         );
   end component;

   signal clk            : std_logic := '0';
   signal we             : std_logic := '0';
   signal rd             : std_logic := '0';
   signal wr_dat         : unsigned(7 downto 0) := X"00";
   signal wr_d           : std_logic_vector(7 downto 0);
   signal rd_d           : std_logic_vector(7 downto 0);

begin

   clk <= '0' after 10 ns when clk = '1' else
          '1' after 10 ns;

   stimulus : process
      variable j  : integer;
   begin
      we      <= '0';
      --
      wait until rising_edge(clk);
      we <= '1';
      for j in 0 to 4 loop
         wait until rising_edge(clk);
         wr_dat <= wr_dat + 1;
      end loop;
      wait until rising_edge(clk);
      we <= '0';
      --
      wait until rising_edge(clk);
      rd <= '1';
      for j in 0 to 5 loop
         wait until rising_edge(clk);
      end loop;
      rd <= '0';
      --
      wait until rising_edge(clk);
      for j in 0 to 2 loop
         wait until rising_edge(clk);
         we <= '1';
         wr_dat <= wr_dat + 1;
      end loop;
      wait until rising_edge(clk);
      we <= '0';
      --
      wait until rising_edge(clk);
      rd <= '1';
      for j in 0 to 1 loop
         wait until rising_edge(clk);
      end loop;
      rd <= '0';
      --
      for j in 0 to 50 loop
         wait until rising_edge(clk);
      end loop;
   end process;

   wr_d     <= std_logic_vector(wr_dat);

   fifo_1 : fifo
   port map
      (
       clock   => clk,
       --
       we      => we,
       d_in    => wr_d,
       --
       rd      => rd,
       d_out   => rd_d
      );
end architecture Behavioral;
```

wr_d   <= std_logic_vector(wr_dat);

After writing six bytes into the FIFO, we now read them from the output.

```
stimulus : process
   variable j   : integer;
begin
   we       <= '0';
   --
   wait until rising_edge(clk);
   we <= '1';
   for j in 0 to 4 loop
      wait until rising_edge(clk);
      wr_dat <= wr_dat + 1;
   end loop;
   wait until rising_edge(clk);
   we <= '0';
   --
   wait until rising_edge(clk);
   rd <= '1';
   for j in 0 to 5 loop
      wait until rising_edge(clk);
   end loop;
   rd <= '0';
   --
   wait until rising_edge(clk);
   for j in 0 to 2 loop
      wait until rising_edge(clk);
      we <= '1';
      wr_dat <= wr_dat + 1;
   end loop;
   wait until rising_edge(clk);
   we <= '0';
   --
   wait until rising_edge(clk);
   rd <= '1';
   for j in 0 to 2 loop
      wait until rising_edge(clk);
   end loop;
   rd <= '0';
   --
   for j in 0 to 50 loop
      wait until rising_edge(clk);
   end loop;
end process;
```

We then write three more bytes:

```
stimulus : process
   variable j   : integer;
begin
   we       <= '0';
   --
   wait until rising_edge(clk);
   we <= '1';
   for j in 0 to 4 loop
      wait until rising_edge(clk);
      wr_dat <= wr_dat + 1;
   end loop;
   wait until rising_edge(clk);
   we <= '0';
   --
   wait until rising_edge(clk);
   rd <= '1';
   for j in 0 to 5 loop
      wait until rising_edge(clk);
   end loop;
   rd <= '0';
   --
   wait until rising_edge(clk);
   for j in 0 to 2 loop
      wait until rising_edge(clk);
      we <= '1';
      wr_dat <= wr_dat + 1;
   end loop;
   wait until rising_edge(clk);
   we <= '0';

   wait until rising_edge(clk);
   rd <= '1';
   for j in 0 to 2 loop
      wait until rising_edge(clk);
   end loop;
   rd <= '0';
   --
   for j in 0 to 50 loop
      wait until rising_edge(clk);
   end loop;
end process;
```

... and finally read them out:

```
stimulus : process
   variable j   : integer;
begin
   we      <= '0';
   --
   wait until rising_edge(clk);
   we <= '1';
   for j in 0 to 4 loop
      wait until rising_edge(clk);
      wr_dat <= wr_dat + 1;
   end loop;
   wait until rising_edge(clk);
   we <= '0';
   --
   wait until rising_edge(clk);
   rd <= '1';
   for j in 0 to 5 loop
      wait until rising_edge(clk);
   end loop;
   rd <= '0';
   --
   wait until rising_edge(clk);
   for j in 0 to 2 loop
      wait until rising_edge(clk);
      we <= '1';
      wr_dat <= wr_dat + 1;
   end loop;
   wait until rising_edge(clk);
   we <= '0';

   wait until rising_edge(clk);
   rd <= '1';
   for j in 0 to 2 loop
      wait until rising_edge(clk);
   end loop;
   rd <= '0';

   for j in 0 to 50 loop
      wait until rising_edge(clk);
   end loop;
end process;
```

This is just a filler at the end (so the process statement doesn't start over again too soon):

```
stimulus : process
   variable j  : integer;
begin
   we      <= '0';
   --
   wait until rising_edge(clk);
   we <= '1';
   for j in 0 to 4 loop
      wait until rising_edge(clk);
      wr_dat <= wr_dat + 1;
   end loop;
   wait until rising_edge(clk);
   we <= '0';
   --
   wait until rising_edge(clk);
   rd <= '1';
   for j in 0 to 5 loop
      wait until rising_edge(clk);
   end loop;
   rd <= '0';
   --
   wait until rising_edge(clk);
   for j in 0 to 2 loop
      wait until rising_edge(clk);
      we <= '1';
      wr_dat <= wr_dat + 1;
   end loop;
   wait until rising_edge(clk);
   we <= '0';
   --
   wait until rising_edge(clk);
   rd <= '1';
   for j in 0 to 2 loop
      wait until rising_edge(clk);
   end loop;
   rd <= '0';

   for j in 0 to 50 loop
      wait until rising_edge(clk);
   end loop;
end process;
```

FIFO Simulation

This is the resulting waveform in Modelsim. The signals at the top are from the testbench, and the two at the bottom are the internal addresses inside the FIFO. I've changed the data buses to be hex, and the FIFO addresses to be integers.

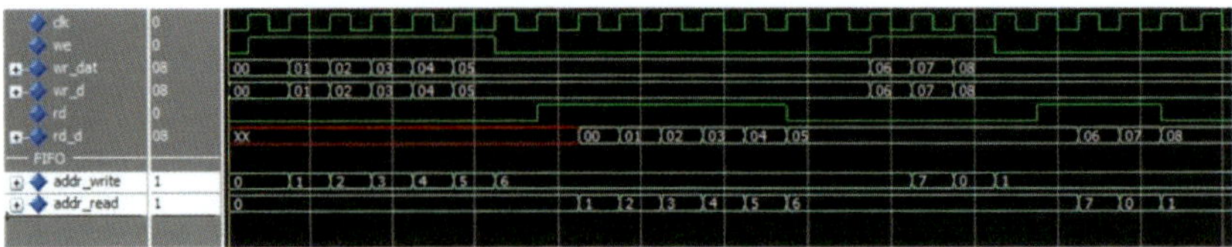

Here's the first writes to the FIFO, the values 0 through 5:

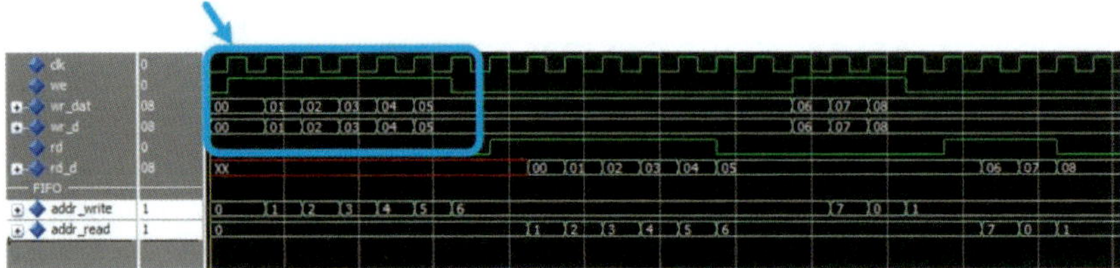

And here we read them out:

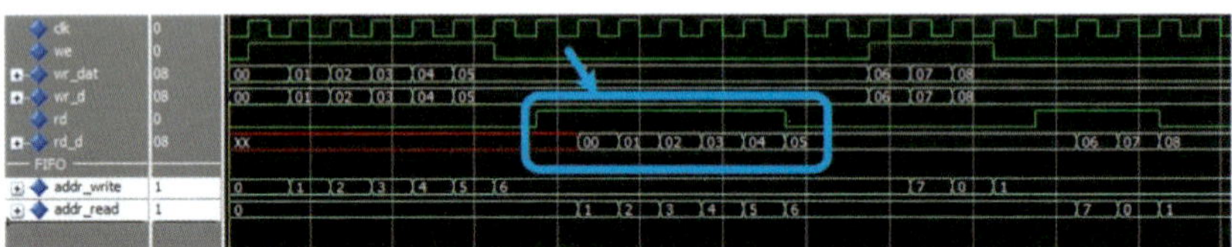

The X's inside the red borders do not indicate a high-impedance tri-stated value, but rather just that Modelsim doesn't know the value of this signal until we do the first read of the FIFO. This is a simulation artifact only—it has no relevance to a design compiled into an FPGA.

We write three more bytes into the FIFO.

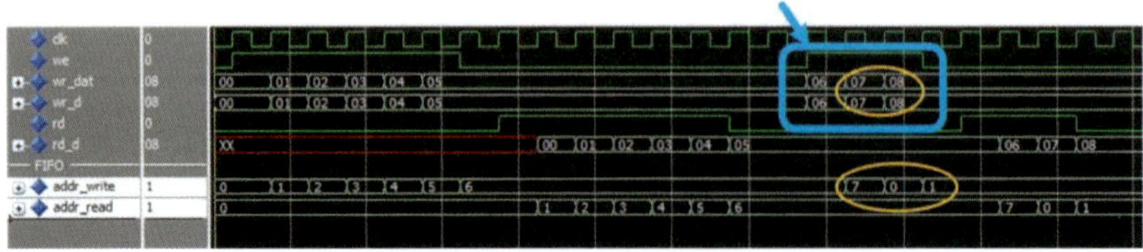

Note that the data value continues incrementing from 7 to 8, but the FIFO write address rolls over from 7 back to 0. This means that we'll overwrite the value in address 0, but we've already read that out—we're done with it.

Finally, we read those three bytes out of the FIFO:

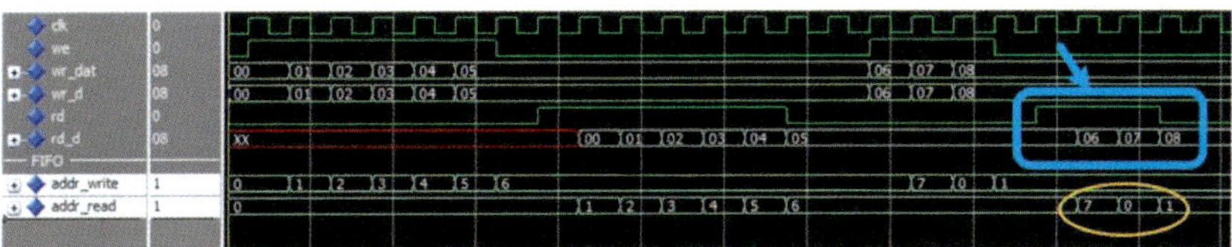

... and now with the read addresses rolling over (07 to 08).

Notice that here, we wrote a byte value of 2 into the FIFO address 2 (just a coincidence that they're the same):

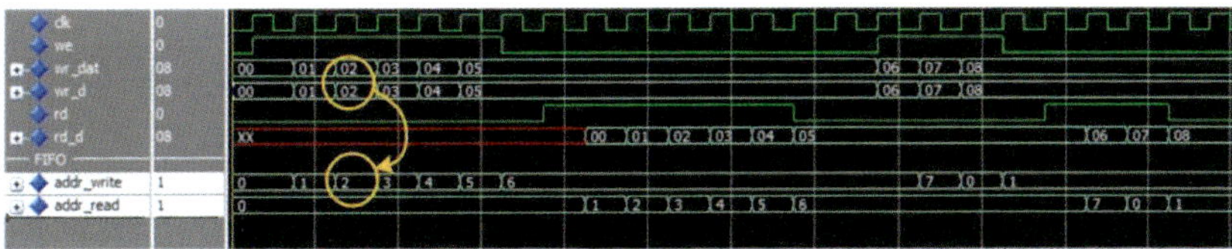

... and here we read that byte value from address 2—there's a one-clock delay.

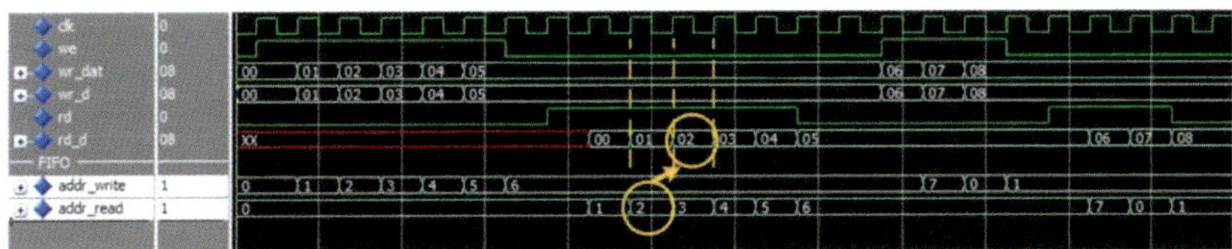

At the beginning of this read stream, we activated the read enable here:

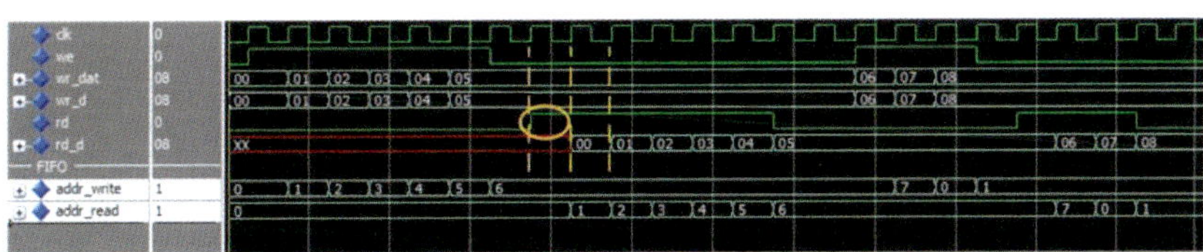

... and saw the resulting byte appear here:

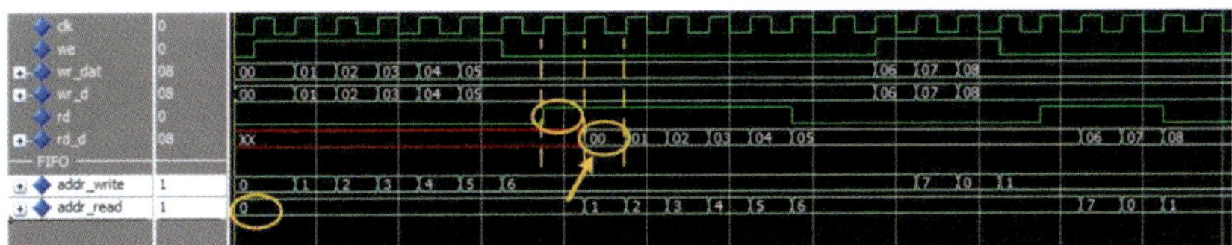

Exercises, Chapter 3

Exercise 3-1:
Add "empty" and "full" flags to this FIFO from the chapter.

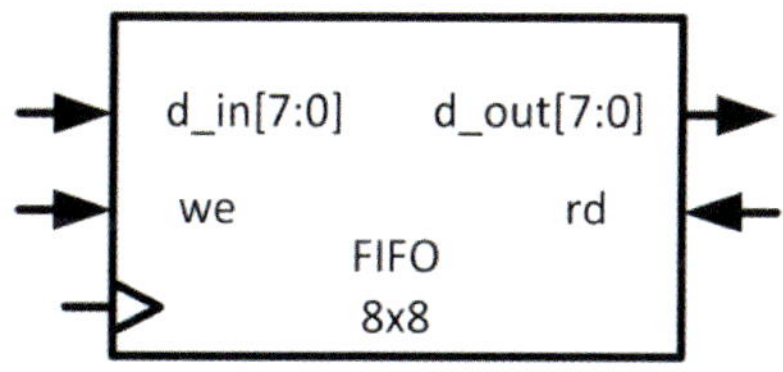


```
architecture behavioral of fifo is

   type array_8x8 is array (7 downto 0) of std_logic_vector(7 downto 0);
   signal mem_dat  : array_8x8;
   signal addr_write    : unsigned(2 downto 0) := "000";
   signal addr_read     : unsigned(2 downto 0) := "000";

begin

   fifo_memory : process(clock)
   begin
      if rising_edge(clock) then
         -- writes ---
         if ( we = '1' ) then
            mem_dat(to_integer(unsigned(addr_write))) <= d_in;
            addr_write <= addr_write + 1;
         end if;
         -- reads ---
         if (rd = '1') then
            d_out     <= mem_dat(to_integer(unsigned(addr_read)));
            addr_read <= addr_read + 1;
         end if;
      end if;
   end process;

end architecture behavioral;
```

Exercise 3-2:

Create a testbench to exercise the FIFO from the previous exercise using while loops to generate the write and read control signals. The testbench should write eight times and see the full flag go active, and then read eight times to see the empty flag go active.

This is an example result:

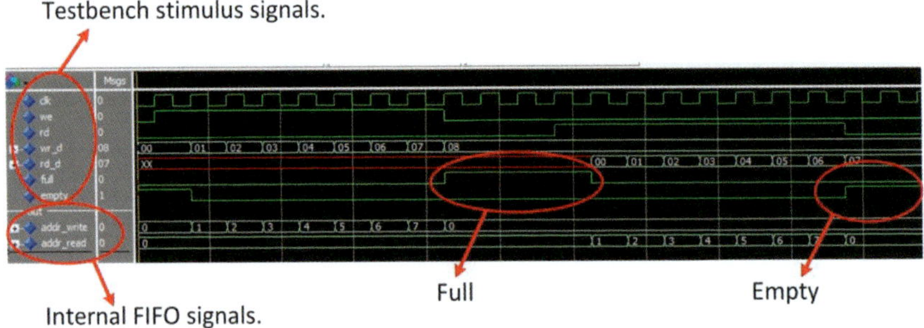

Chapter 4

Memory-mapped Buses

Mapped to Memory

Memories are an important component of almost any design of moderate complexity, and standards for interfacing with them have existed and evolved over the decades. The original DOS-based IBM PC started life with a proprietary 8-bit parallel memory bus that soon upgraded to the ISA bus.

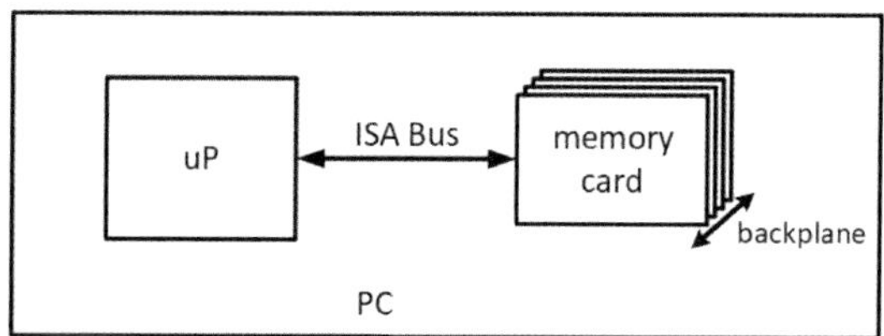

We refer to these as parallel buses, not because multiple memory cards reside on the same backplane, but because each bit position of the bus—the data and addresses sub-buses—have a dedicated signal path, in this case, a printed circuit board trace and connector pin. If the data bus is16-bits wide, there were 16 traces and connector pins. All the signals of the buss operate together in parallel. ISA stands for Industrial Standard Architecture, and, as the name indicates, was the first non-proprietary memory bus to achieve wide acceptance, and was instrumental in evolving the IBM PC into the ubiquitous multi-vendor Windows-based laptops and desktop PCs of today.

In addition to memory cards, the bus also allows the micro-processor to communicate with peripheral cards, such as Ethernet.

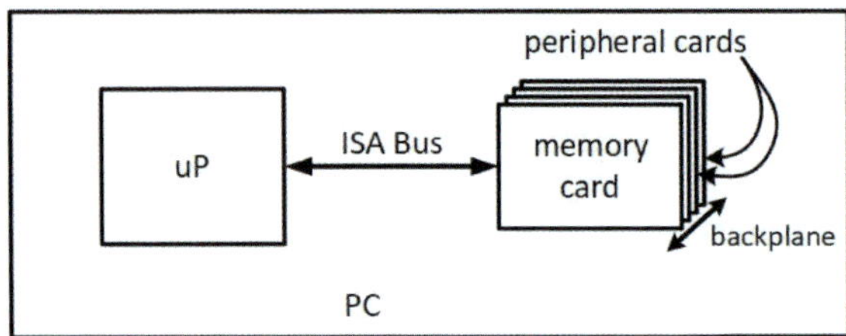

By the end of the millennia, the 32-bit PCI bus was fast replacing 16-bit ISA buses in PCs.

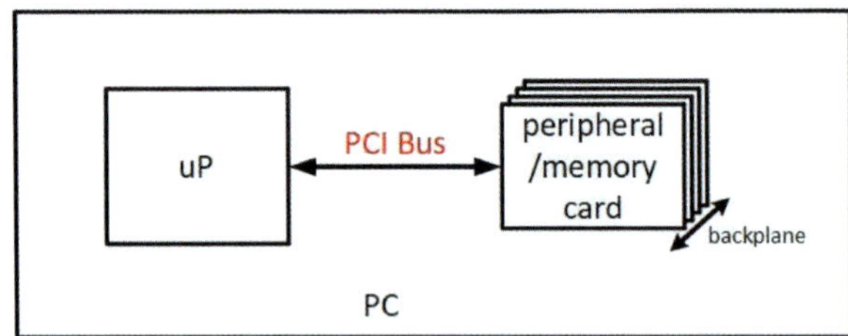

A major improvement afforded by the PCI bus over ISA is data bursting, which we'll get to in a bit.

Of interest to us are memory bus applications inside FPGAs, and a major use is connecting an internal FPGA-based micro-processor to internal memory.

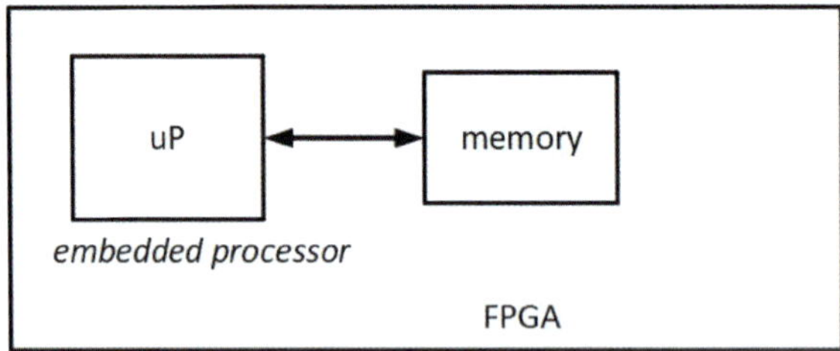

We refer to these FPGA-based micro-processors as embedded processors, and the programming of them as embedded software.

In addition to embedded micro-processors, also referred to as micro-controllers, our VHDL-generated RTL logic can interface with memories as well, either stand-alone, or via dual-port configurations, where both the micro-processor and the RTL both have access.

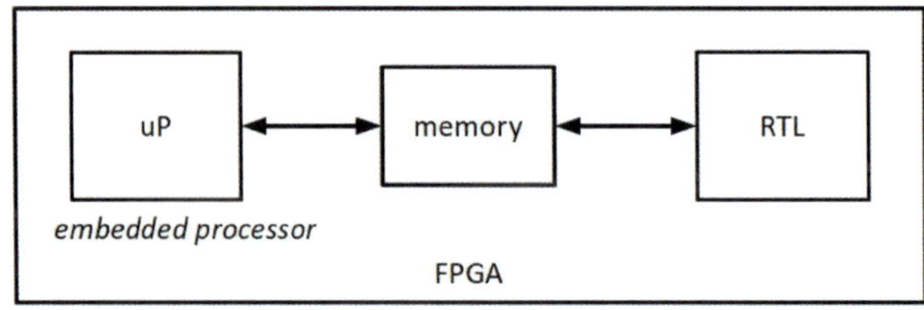

In order to facilitate a variety of FPGA memory applications, vendors have created standardized bus interfaces.

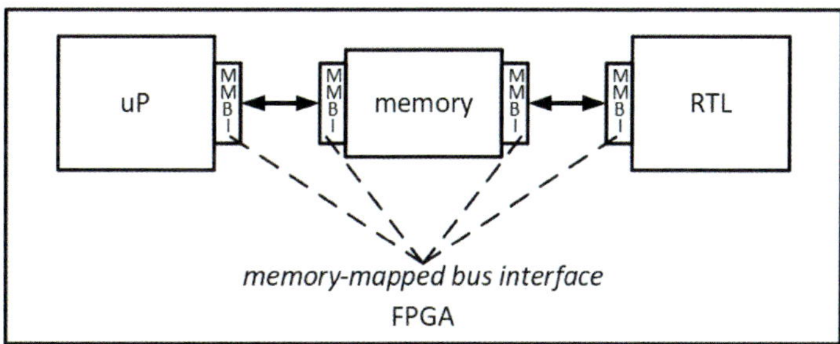

We call these Memory-Mapped Bus Interfaces, since they transfer data to and from locations as identified by unique addresses, as encountered in memories.

Memories are not the only functions used by standardized bus interfaces. Any function that handles location-specific data can benefit from their use.

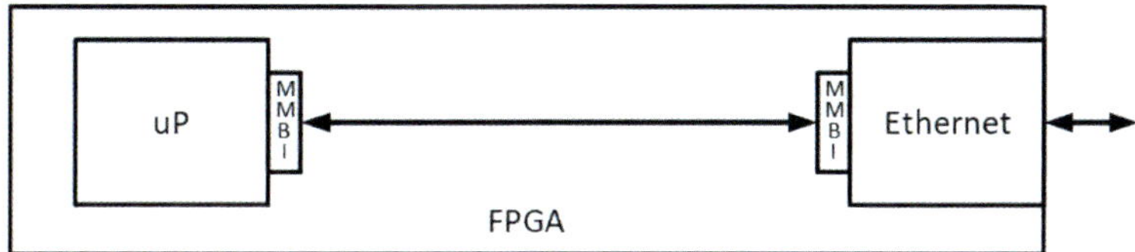

The example shown here is an external Ethernet interface, where the location-specific data might be, in addition to the stream of communicated link data values themselves, the variety of configuration registers associated with the Ethernet operation.

In some cases, our VHDL RTL logic may interface directly with these functions, and could also connect via the same bus.

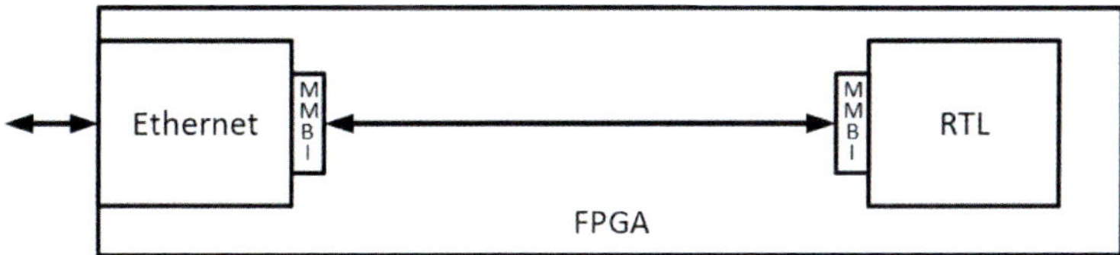

Memory-Mapped Buses include a master side, and a slave side.

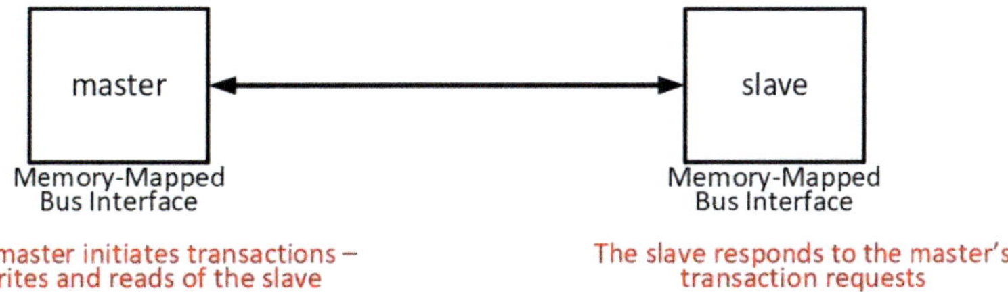

The Master side initiates all data transfers, often called transactions, and the slave responds to the write and read requests. Unlike a simple stand-alone memory, slaves often have the option to pause the transaction, to tell the master to hold on a bit, or even to reject the request.

We'll look at one example of FPGA-based memory-mapped bus interfaces, the Altera Avalon Bus. And, since Altera has been acquired by Intel, now known as the Intel Avalon Bus.

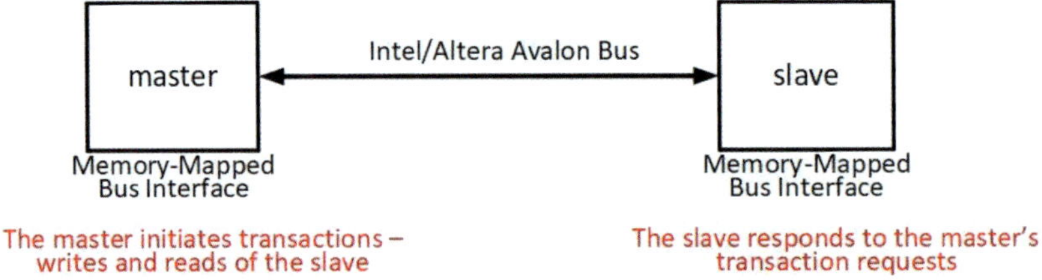

The Avalon Bus

Before looking at the various signals making up the Avalon Bus, we need to review some semantics. We previously talked about 8-bit write and read address and data buses.

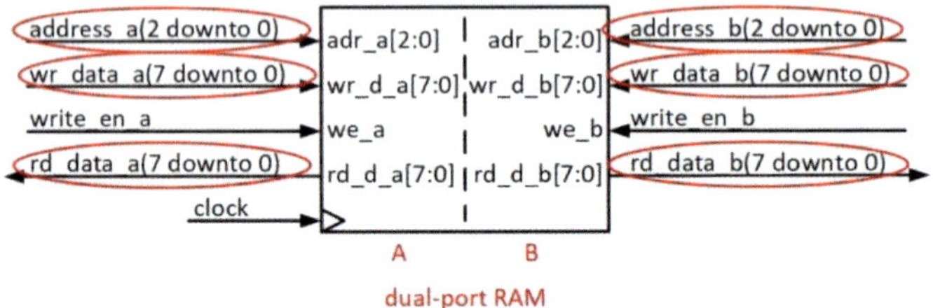

Any group of signals that comprise a single digital value, e.g., a VHDL vector, we call a bus. But now we're talking about a single entity that we're calling a memory-mapped "bus," which is actually a group or collection of individual buses and control signals that operate together.

It's as though we draw a circle around the signals we used to read and write the memories of chapter 2 and call them the "chapter 2 bus."

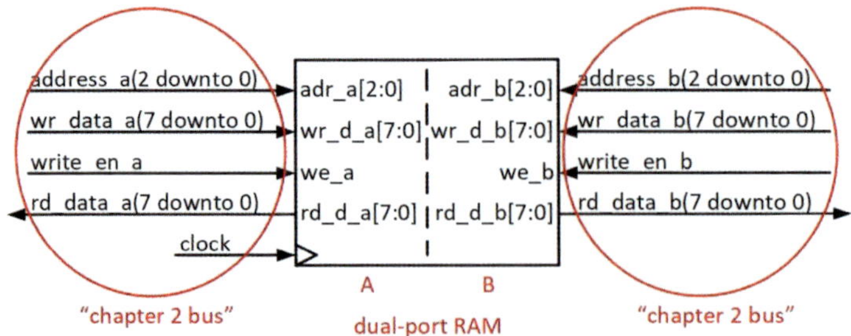

It can be confusing.

Returning to the constituent parts of the Avalon bus:

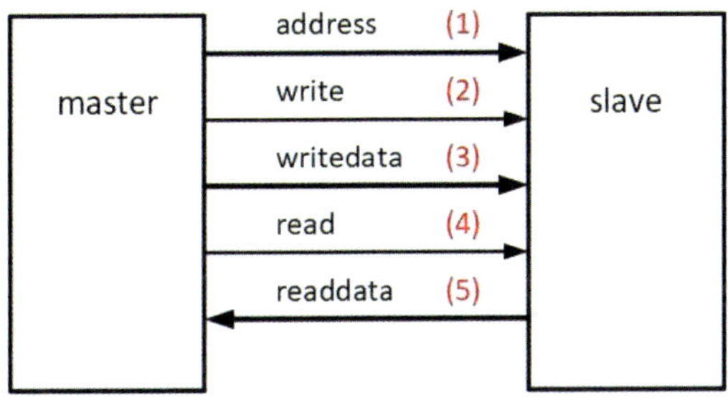

Intel/Altera Avalon Bus

(1) there's an address, of course, otherwise it couldn't be memory-mapped. The width of the address value is configurable, depending on the memory-mapped size of this slave. Many non-memory slave functions may have just a handful of mapped registers, and the address may be just a few bits wide;

(2) a write control signal with the same meaning as what we've already seen;

(3) an associated write data value. The width of this is also configurable, but, whereas the address value could be any width, the write data value must have a width that's a binary power number of bytes, meaning that the width can be 1, 2, 4, 8, 16, 32, 64, or 128 bytes. I.e., it could be 8 bits, 16 bits, 32 bits, etc., up to 1024 bits wide;

(4) a read control signal, and,

(5) a read data bus, whose width is also constrained to binary powers of bytes.

When the write control signal is active, the value on the writedata bus is written into the slave location as defined by the address,

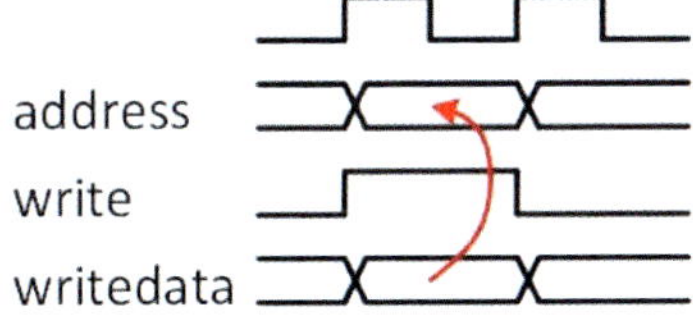

and when the read control signal is active, the data located at the slave's address location is presented on the readdata bus.

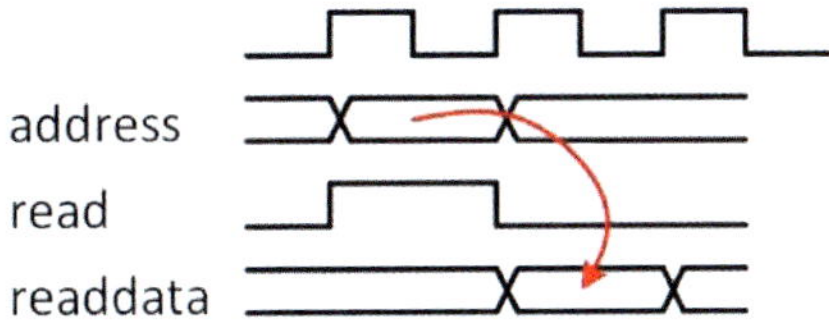

This operation is essentially the same as we saw when writing and reading directly from memory.

We'll note that it is possible to configure the Avalon bus to have the slave return the read data in the same clock period,

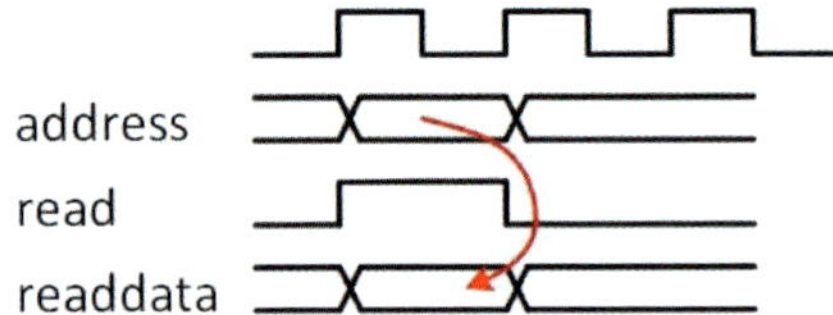

but this is rarely done, since it limits the maximum clock frequency.

We'll now go beyond the operation we saw when interfacing directly with memory as we introduce the Avalon bus waitrequest signal, which allows the slave to hold off the master.

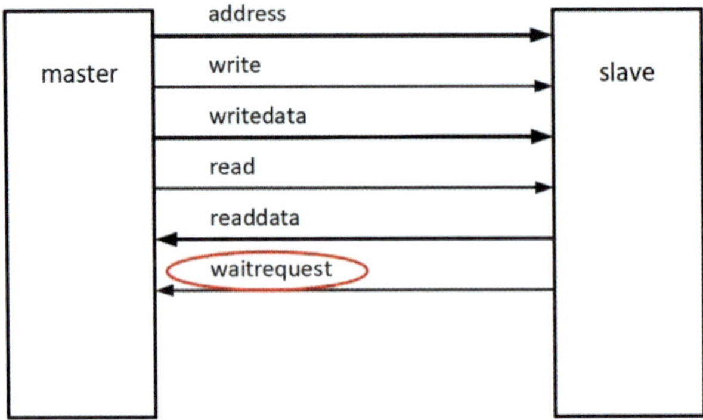

Intel/Altera Avalon Bus

Here, we see the same write operation, but now the slave uses the waitrequest to hold off the actual write for two clocks.

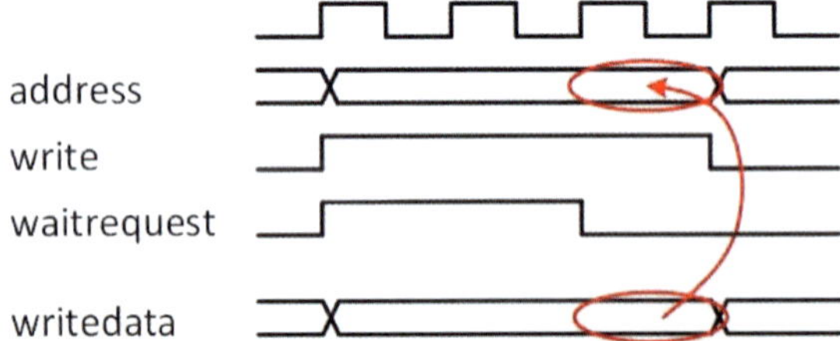

The slave may not be ready to accept the data right away. According to Avalon bus specifications, the slave can hold off the master for an indefinite amount of time.

Here's the waitrequest as it holds off a Master's read request.

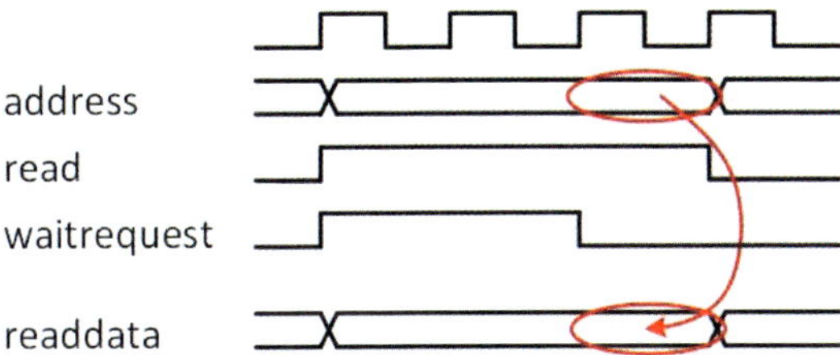

We next expand the Avalon operation further with the byte enable signal.

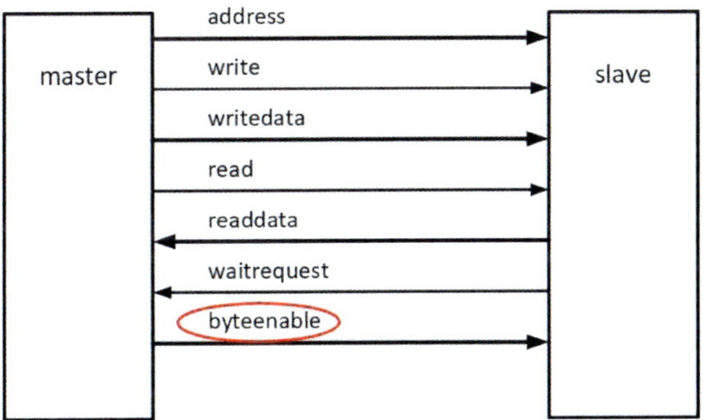

Intel/Altera Avalon Bus

The byteenable signal is only relevant when the write and read data buses are more than one byte wide. It allows the master to direct the write or read transaction to specific bytes in the multi-byte wide write or read data bus. Byte enables are rarely used for reads (the slave generally just returns the entire width of the read data bus) so we'll only look at its operation with writes.

The byteenable is a vector signal, and each bit location is associated with a byte location of the write data bus. In this example, the writedata bus is 32 bits wide, or four bytes, which means that the byte enable signal is four bits wide:

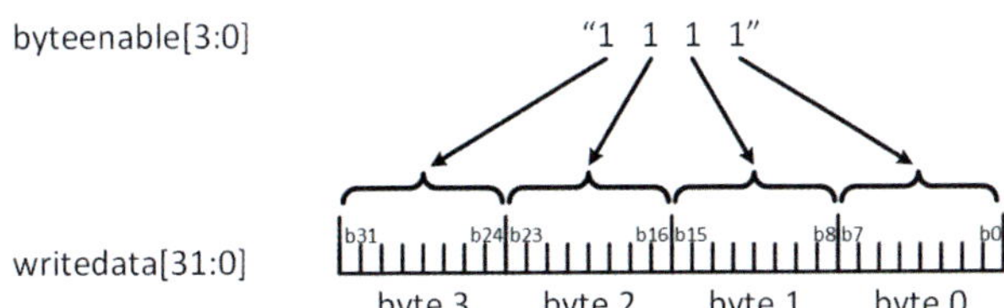

Here, we see how each bit of the byteenable corresponds to one byte of the write data. If just bit 0 of the byte enable is asserted, then only the LS byte of write data is written:

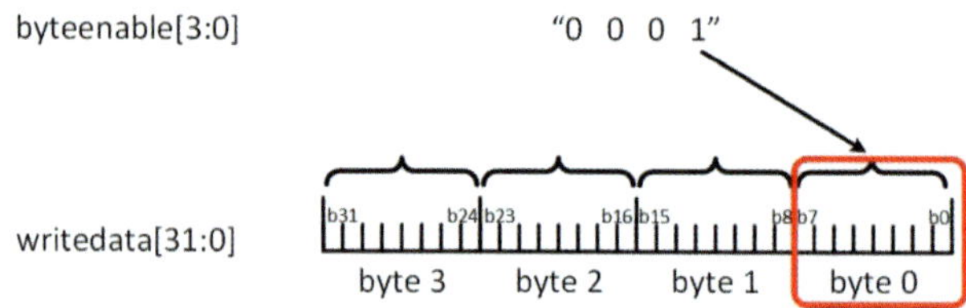

If bits 1 and 2 of byte enable are asserted, then the second and third bytes of write data are written:

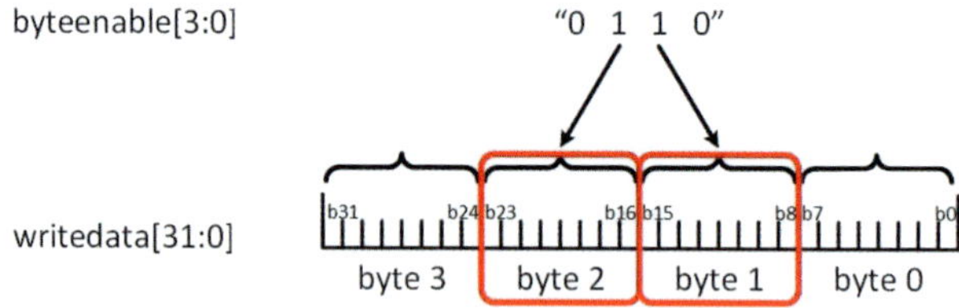

Let's see how this would work if the slave were an 8 by 32-bit memory:

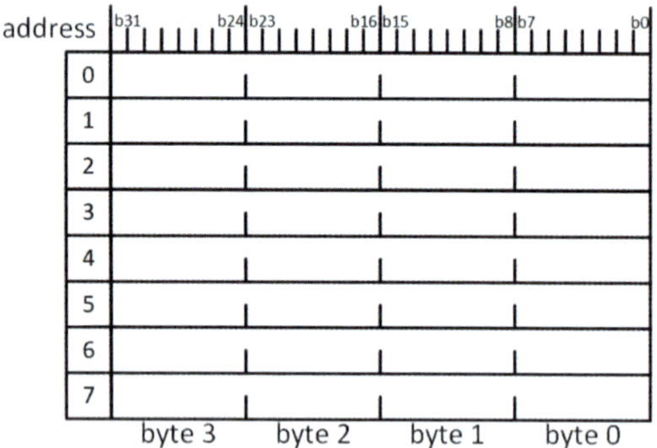

Suppose we do a write to address two, with a byte enable value of "0110".

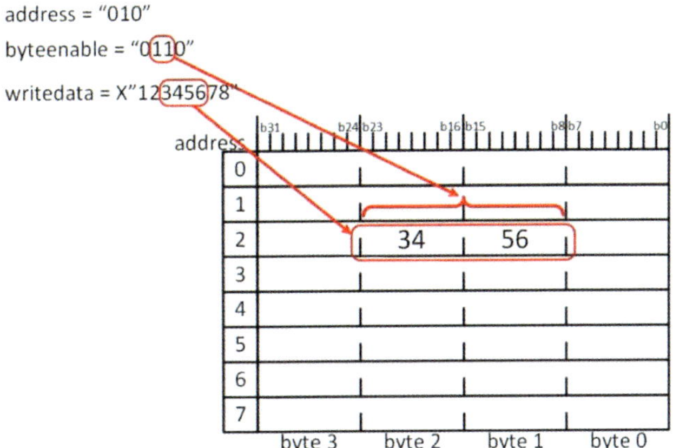

This means that the middle two bytes of the write data are written into the address two location. Note that the first and fourth bytes of that address are unaffected.

So far, other than the byte enables, the Avalon bus operates very much like the memory interface we saw in chapter 2. Now, though, with the introduction of the burst count, we encounter a whole new level of functionality.

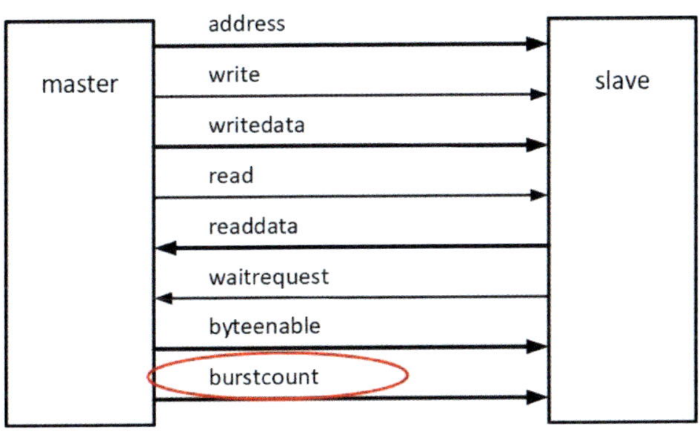

Intel/Altera Avalon Bus

Bursting allows a sequence of writes or reads to proceed with just one starting address.

Here, we begin a write burst with an address of hex A0, some data value (labeled "data0"), and a burstcount of 4:

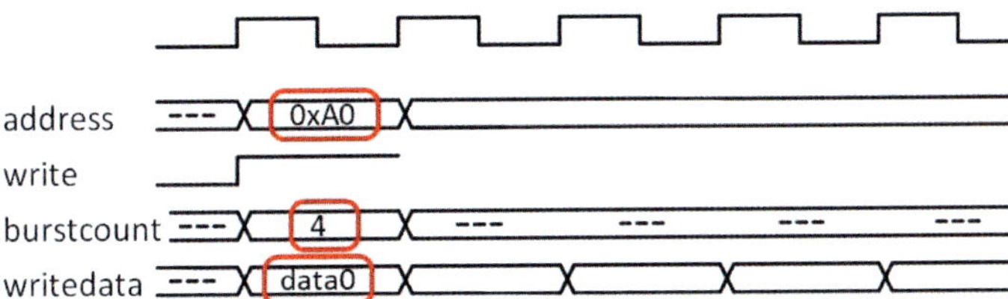

After the first data value ("data0"), the master places the next three data values ("data1" through "data3") in sequence each clock.

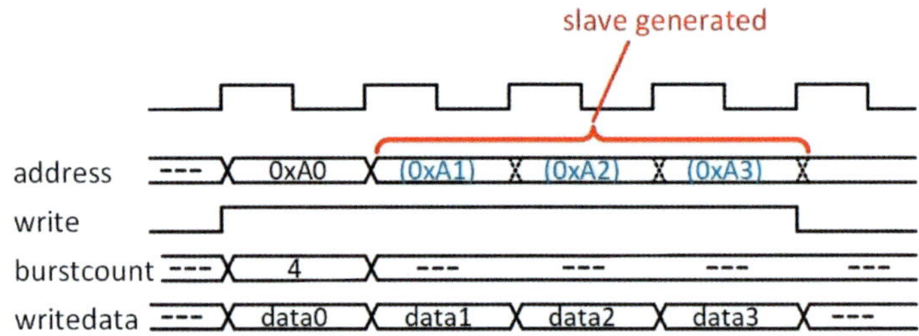

The slave must increment the address it uses each clock cycle, so that the sequence of data values is written into an associated sequence in the slave's memory, or set of registers.

Note that the burstcount value is only valid during the first clock cycle.

The waitrrequest signal is valid for bursts:

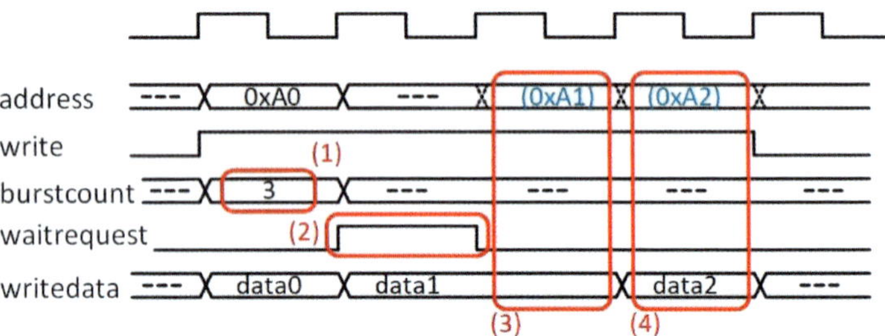

(1) here, we define a burst of 3 cycles;

(2) the slave asserts waitrequest;

(3) so that the write of the second data value occurs here,

(4) and the third value, here.

Note that the burstcount signal is optional on the bus, and when it's not included, it is assumed to be one (as in previous writes).

The AXI Bus

Let's compare the Altera/Intel Avalon bus with another common memory-mapped one, the AXI bus, which is somewhat more sophisticated, and therefore more complicated. The AXI bus was originated for use with the ARM processor family, and since this has been a staple embedded processor for the Xilinx FPGA vendor, the AXI bus became the de-facto interconnect bus for Xilinx. The AXI has become somewhat universal in high-end designs.

Like the Avalon bus, the AXI assumes a master and a slave. And, like the Avalon bus, writes and reads comprise the two basic types of transactions, and data bursts are the mainstay.

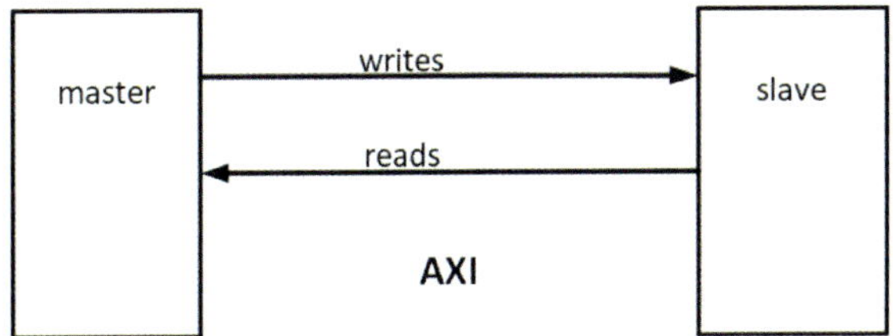

Whereas the Avalon bus has one address shared between both writes and reads, however, the AXI bus has two address paths, one dedicated for writes, and the other dedicated for reads.

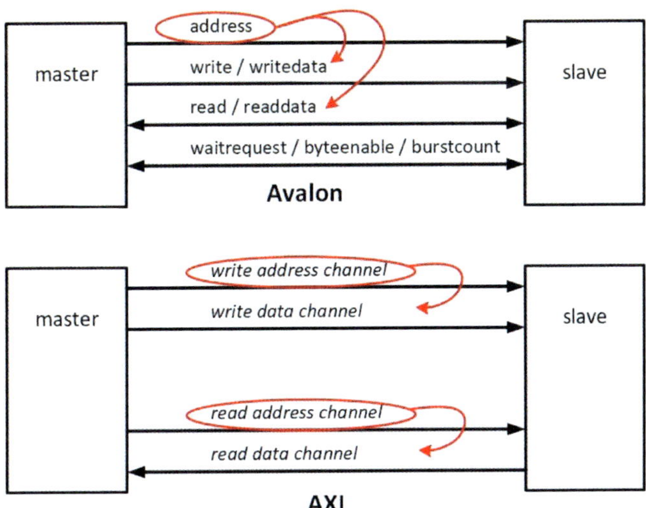

Also, whereas the Avalon bus has one vector signal each for address, write data, and read data, the AXI bus includes multiple vector signals for each type (write address, write data, read address, and read data), and groups them together in what it calls channels. Each channel has its own set of control signals (when data is valid, when the slave is read—the analog to the Avalon waitrequest—and other signals associated with bursts).

One obvious advantage of having separate addresses for writes and reads is that we can have both operating at the same time, and, further, by including bus control signals with the addresses as well as the read and write data paths, and by including ID information associated with each burst, multiple data bursts can be requested without waiting for a previous one to be finished. We call this type of request "posted."

Here, we see a series of write bursts proceeding, while acknowledgements of earlier bursts are still arriving:

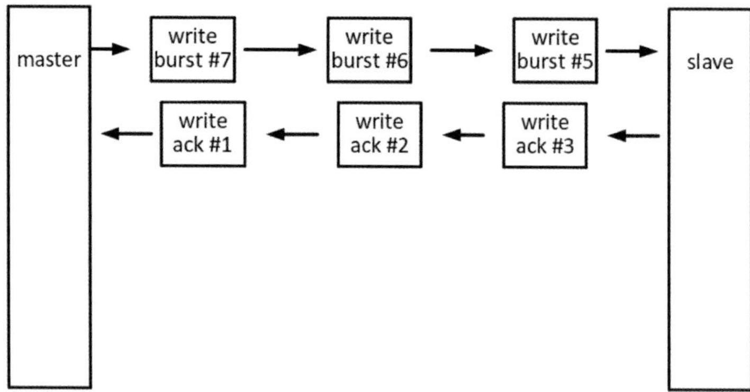

Simultaneously to these writes, the master could issue multiple read requests, while returned read data bursts from the slave associated with earlier requests are still arriving.

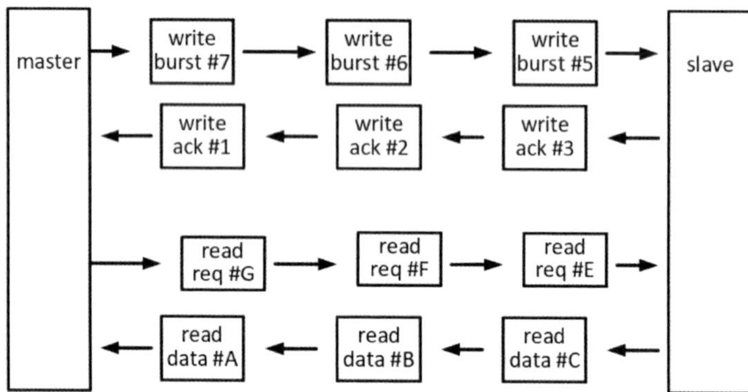

This begins to look like the packet data transfers, which is exactly what the ARM engineers were after:

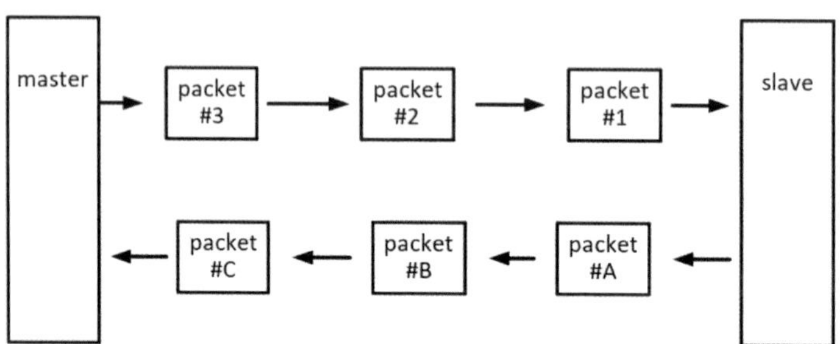

As is probably obvious, the AXI bus doesn't connect directly to a memory. Rather, significant amounts of logic at each end process the data, which may then indeed eventually be written to or read from memory, but often is handed off to higher-level functions, perhaps a gig-ethernet port, or a block of DSP (digital signal processing).

Before despairing too much, however, you may be relieved to know that a reduced function version of the AXI bus called "LITE" is available that strips the operation down to something a bit more like a straight memory interface.

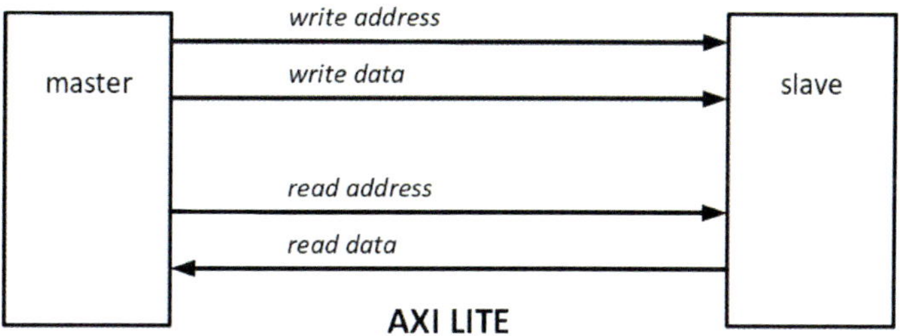

DMA

Finally, we'll take a quick look at an important memory management operation that's found in many high-end designs. DMA (direct memory access) is a means to transfer data into or from a memory directly, i.e., by means other than that of the normal user of the memory, usually a micro-processor.

Imagine a PC microprocessor that has regular access to its local memory, and also to a peripheral board—say gig-ethernet—via a high-speed PCIe link. The peripheral board in this case has an FPGA in which the RTL (VHDL code) has access to its own local memory.

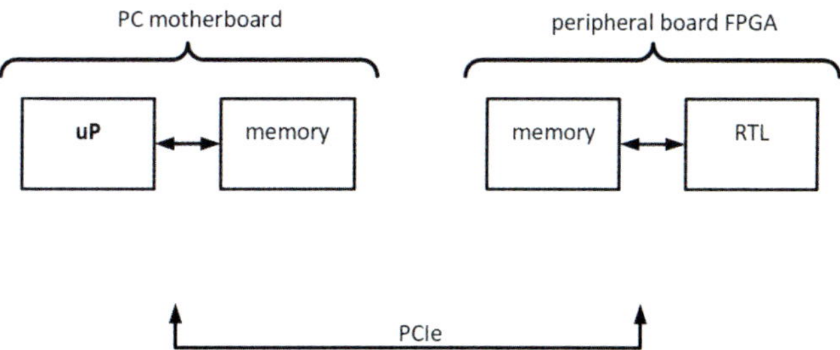

A DMA controller could read blocks of received gig-ethernet data and transfer them directly into the uP memory across the PCIe link.

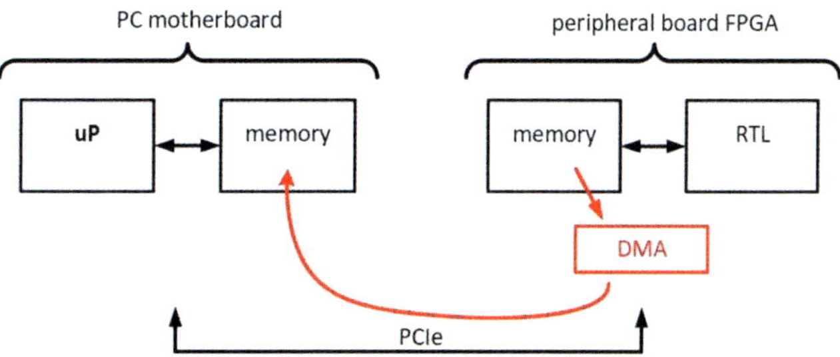

You can see how dual-port memories would be advantageous here.
Alternatively, the DMA controller could read data from the microprocessor memory and copy it to the RTL memory in preparation for gig-E transmission.

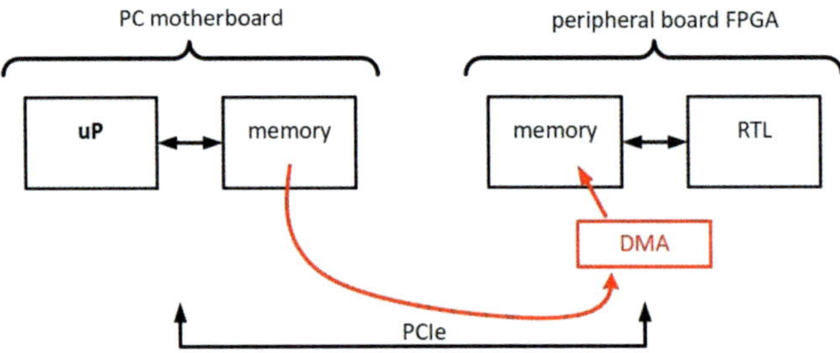

The DMA controller needs information related to the transfer—for example, where to get the data from, and where to write it to. Some of the DMA controller configuration is provided directly by the microprocessor ahead of time:

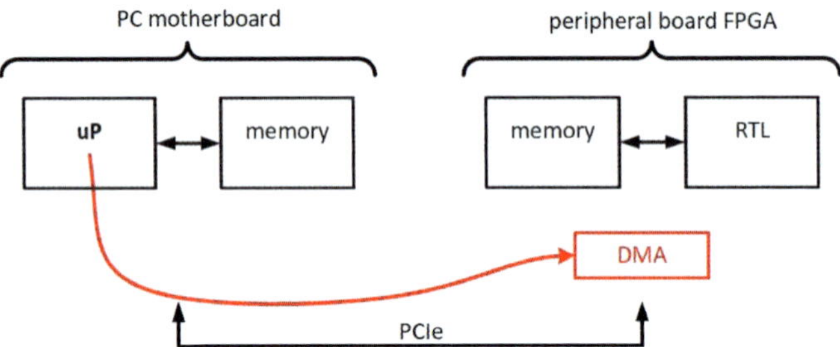

… but often the DMA controller reads the "real-time" transfer information from dedicated areas in the microprocessor memory called descriptor tables (the microprocessor loads the information into the descriptor tables ahead of time).

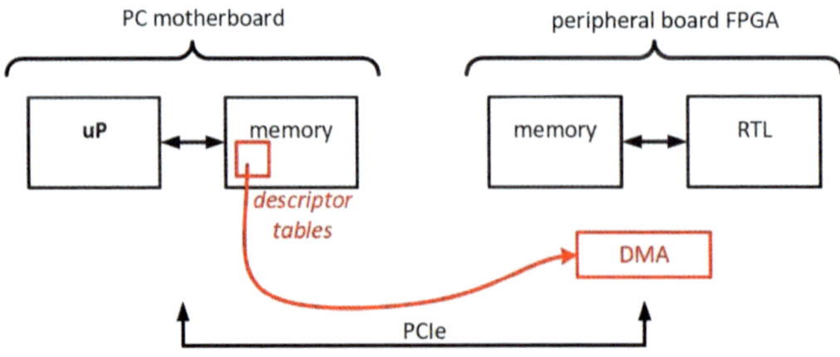

Additionally, it's possible that the DMA controller interfaces directly with the RTL.

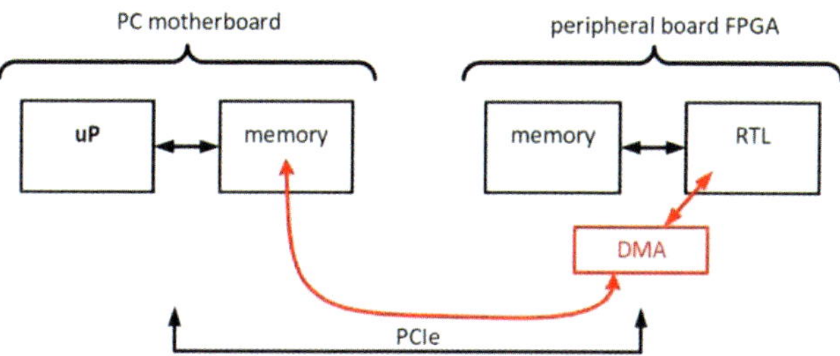

The DMA transfer may be from contiguous blocks of memory:

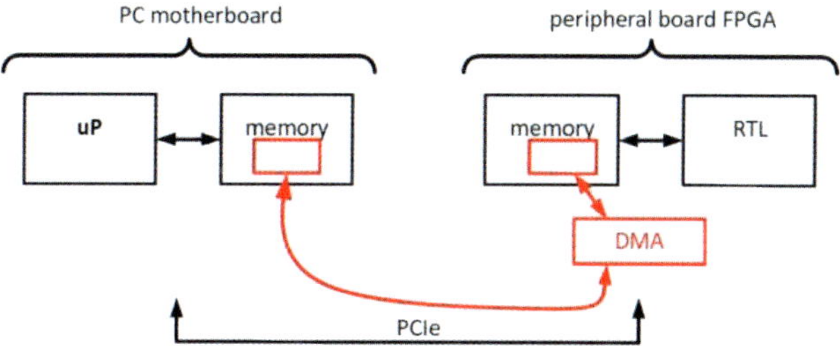

... or it may collect or write to non-sequential places in one or both sides.

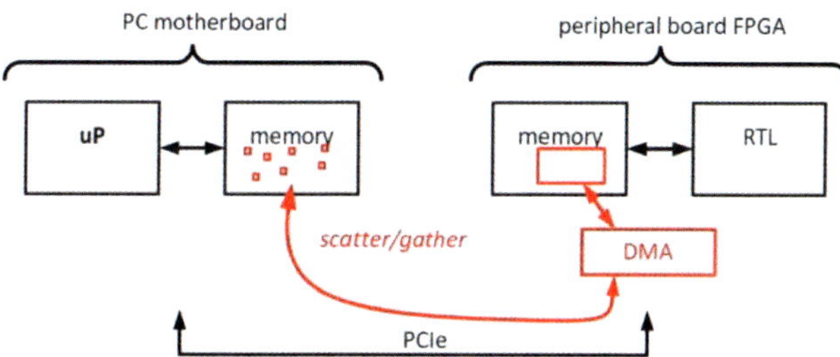

This is referred to as "scatter/gather" operation.

You won't be expected to design a DMA controller, unless perhaps you work for Xilinx or Altera, as these are provided as IP by the vendors, and in some cases, also third-party vendors.

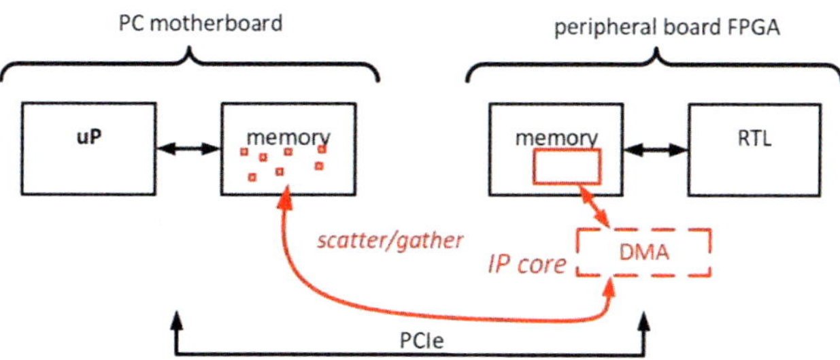

Exercises, Chapter 4

Exercise 4-1:

Fill in the locations in the memory with the data written below.

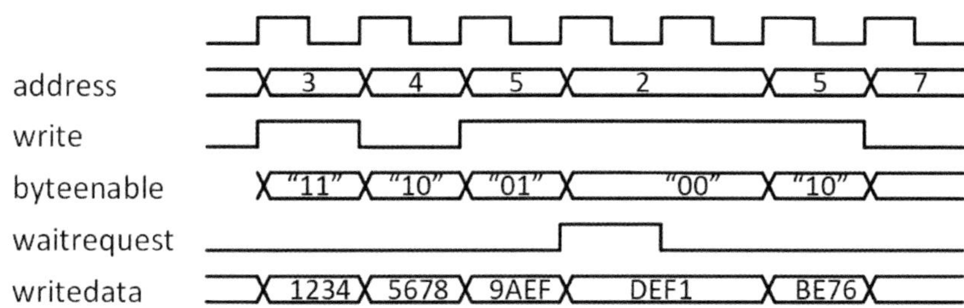

(values are in hex)

address	byte 1	byte 0
0	00	00
1	00	00
2	00	00
3	00	00
4	00	00
5	00	00
6	00	00
7	00	00

Exercise 4-2:

Fill in the read data from the memory below.

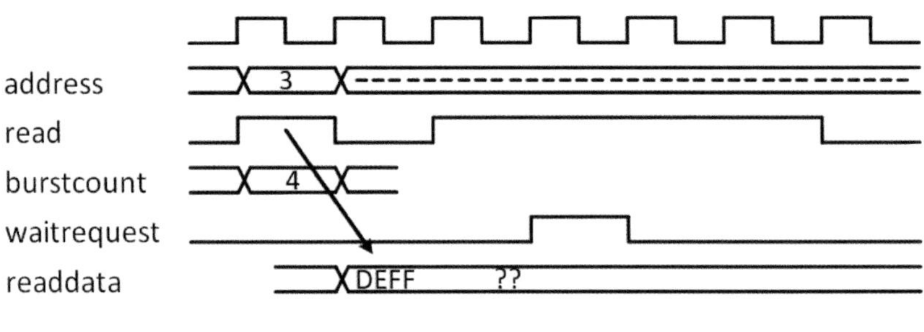

(values are in hex)

address	byte 1	byte 0
0	12	34
1	56	78
2	9A	BC
3	DE	FF
4	ED	CB
5	A9	87
6	65	43
7	21	00

Chapter 5

DSP

DSP is an abbreviation of Digital Signal Processing. This is where the traditional analog signal processing of early electronics is implemented in the digital arena, and then expanded to perform mathematical feats unattainable in the analog realm.

When we refer to signal processing, we're talking about signals that are different from those we've been using over the span of the previous chapters. The "signals" of signal processing are the sampled real-world signals we introduced in the very first chapter of volume 1 (time-based representations of real-world events).

Examples might be the PC microphone audio we saw in the first chapter of volume 1, or, perhaps, the electromagnetic energy of a software-controlled radio:

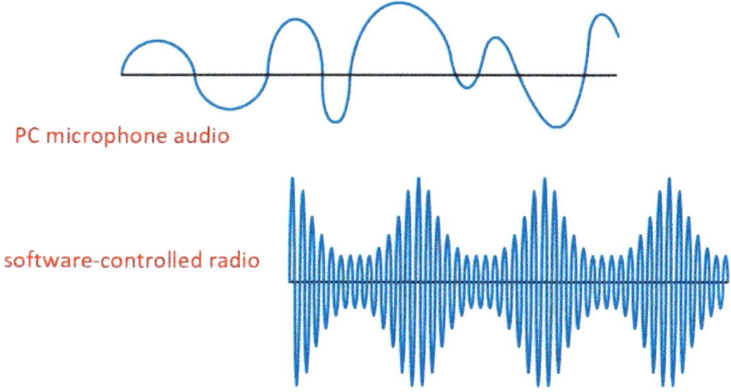

Pretty much anything that can be sampled is a candidate for digital signal processing.

We might imagine an analog signal processing chain where two input signals are modulated together, filtered, and passed on. Perhaps this is one component in a proprietary communication system.

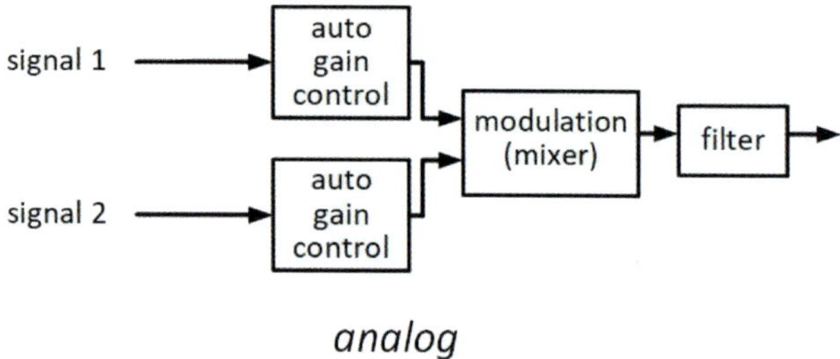

analog

Here's what the processing would look like in the digital domain.

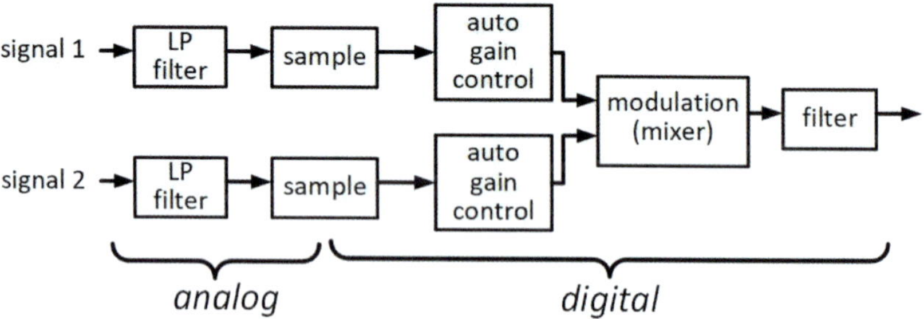

Front-end low-pass filtering is standard before sampling the signal with our processing clock:

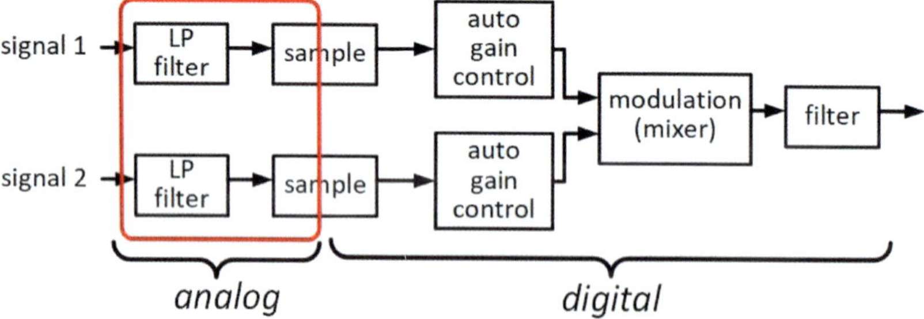

From then on, all the processing is done digitally, perhaps in an FPGA:

DSP in FPGAs

FPGAs are prime platforms for DSP. At low frequencies—low-quality audio, for example—a micro-processor can often handle the math operations, but most applications include far too many math steps for a linear-stepping micro-processor to manage. An FPGA, on the other hand, can perform many parallel math operations at the same time.

You might be adept at higher math, and be able to both define the signal processing algorithms as well as develop the VHDL implementation:

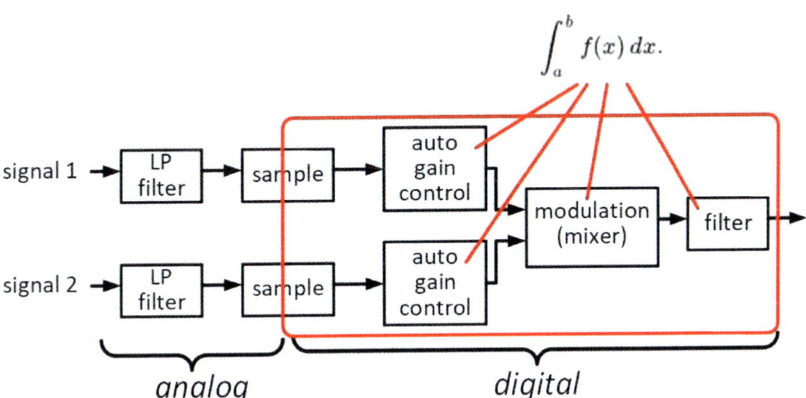

… but more likely, a systems engineer would provide detailed descriptions of the basic math steps needed, and you would then code them up, with perhaps only a vague understanding of the math underpinnings:

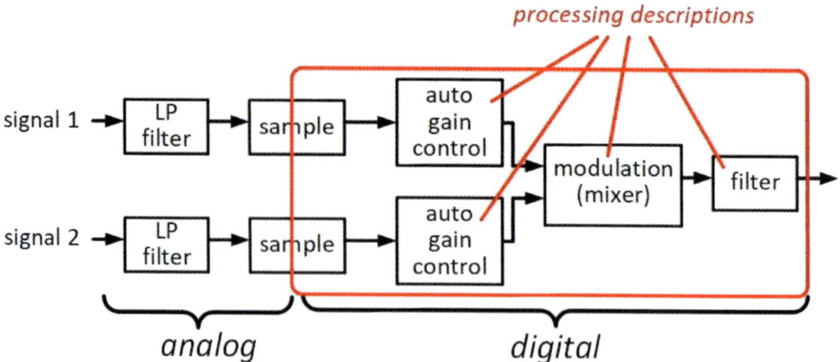

You might be given something like this:

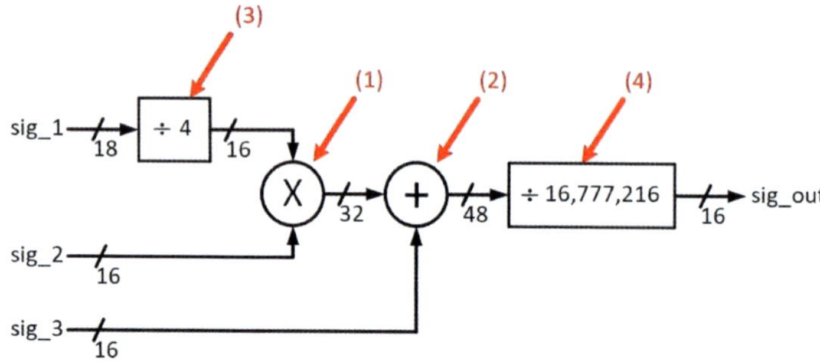

where:

(1) this is multiplication,

(2) this is addition, and

(3,4) are division. Notice, though, that both of these divisions are binary powers, where (3) is 2-to-the-2, (4) is 2-to-the-24.

Dividing by binary powers is simply shifting right that many places. For example, 40 divided by 8 is 5. The binary power of 8 is 3 ($2^3 = 8$). In binary, 40 is "101000". Shifting right three places yields "000101", which, is 5. This is very common in DSP processing.

Division by non-binary powers is a whole other story that we won't go into here, but let's look at the multiplication. Like division, multiplication by binary powers would be shifting left, and we do this sometimes. But in this case, we're multiplying two numbers that can be any value, so we would need to do a true arithmetic multiplication. In VHDL, this is easily coded by using the universal multiplication symbol, the asterisk. Thus, for this operation:

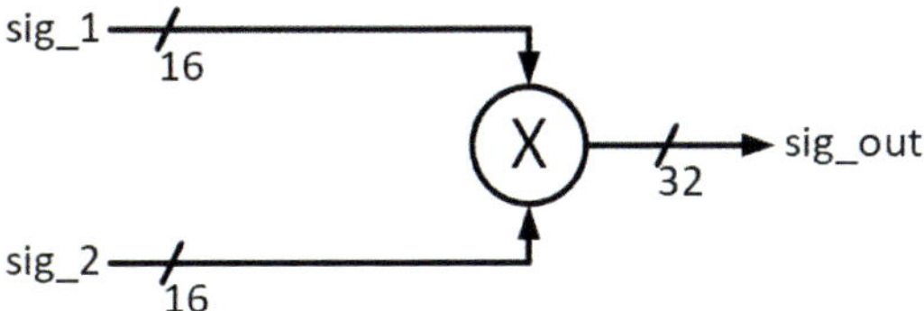

the VHDL code would be:

```
sig_out <= sig_1 * sig_2;
```

The vendor's compiler will dutifully implement this, but multiplication, and, to a lesser extent, addition, consume a fair amount of FPGA logic. Fortunately, almost all modern FPGAs include hard-coded DSP blocks that perform addition, subtraction, and multiplication (no division, though). Here's a diagram of the most basic operation of a typical DSP block.

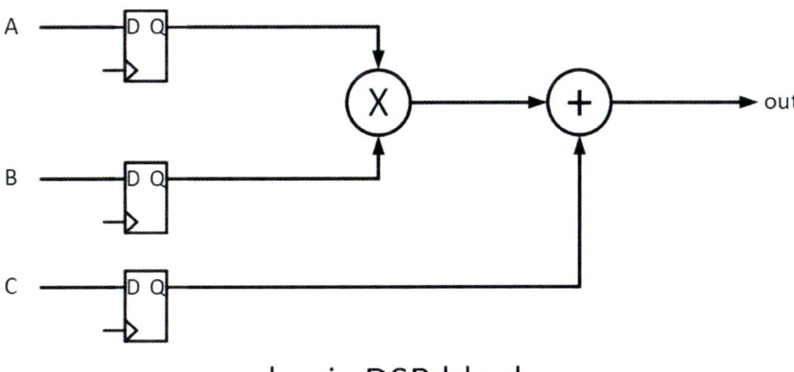

basic DSP block

And here, we substitute the multiplier and adder in the original figure above with this hard-coded block:

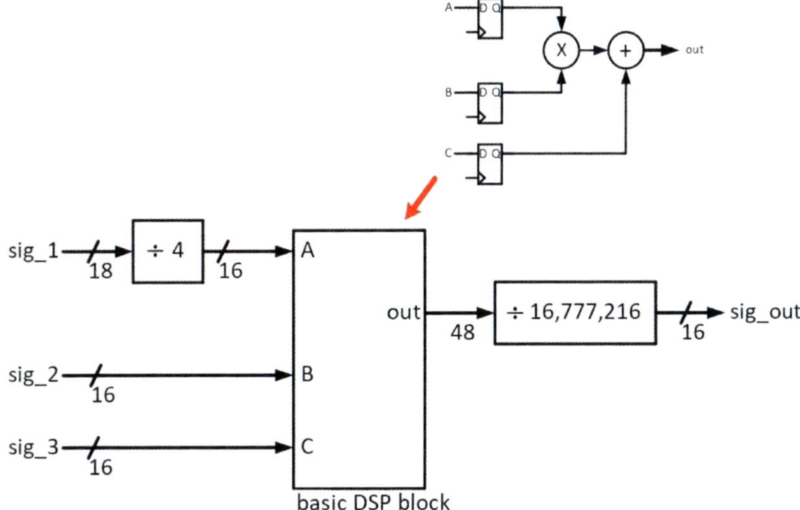

Simple enough, but what about these registers?

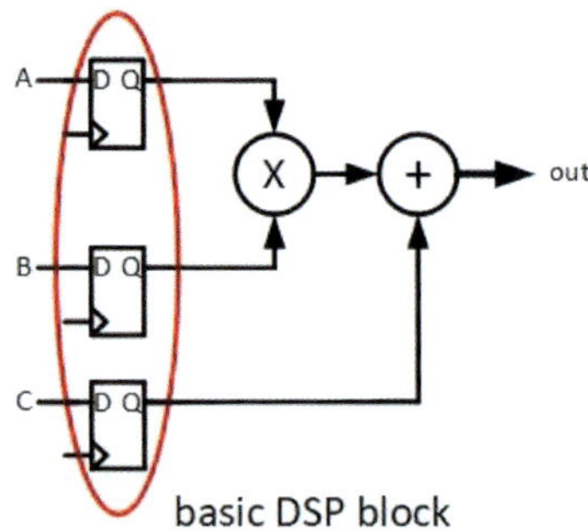

Well, they're inserted as an integral part of the IP block. We have no choice, and our algorithm will have to adapt to accommodate them. The reality is that at any reasonable clock rate, they would probably be necessary anyway in order to meet timing.

Typical DSP blocks have other features as well. They almost always have a selectable feedback path, which allows an accumulation of multiplications and additions, a common function in DSP.

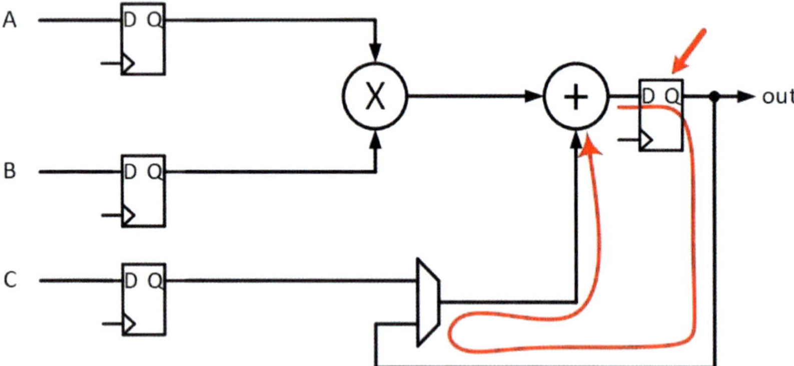

Of course, in order to do a proper accumulation, we need a register in the path, otherwise our adder will shoot off like a squealing guitar in a full-up amp, adding away as fast as the timing of the path will allow.

And, while we're adding registers in the paths, we can include one at the multiplier output as well.

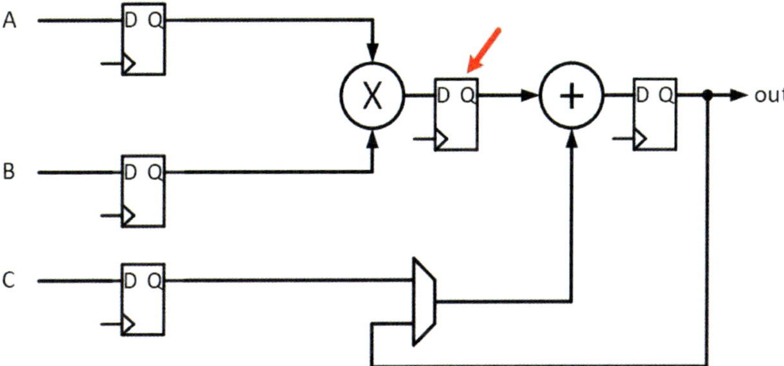

Note that many of these registers, and the various paths as well, are selectable, meaning that we use the vender's GUI tool to set up the configuration.

Some DSP blocks have two built-in multipliers.

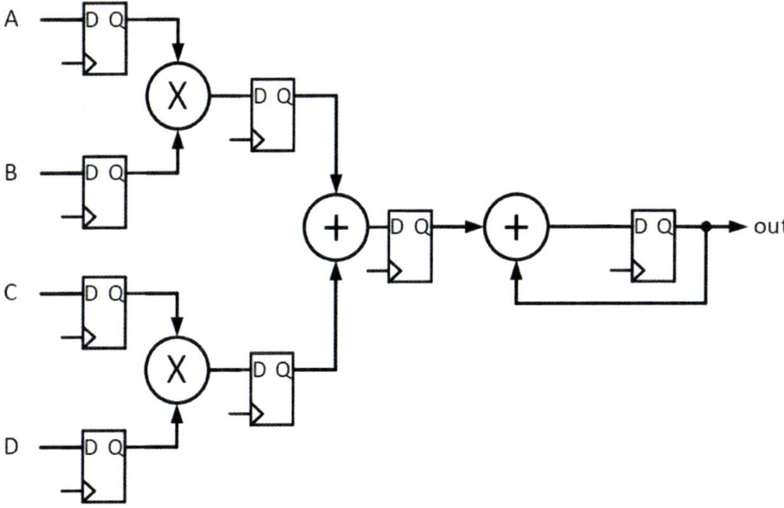

Each vendor has dedicated manuals describing how to use their DSP IP.

Signed Arithmetic

A word about signed numbers is in order as we work with DSP operation. We've been using unsigned numbers in our counters, and we know that they look the same as standard logic vectors. When we declare a signal as unsigned, we're really just telling the compiler that it can use it with arithmetic operations.

VHDL allows signed numbers as well. But, whereas for an 8-bit signal, standard logic vectors and unsigned numbers have 8 bits of amplitude, a signed version has just 7.

std_logic_vector	b7 b6 b5 b4 b3 b2 b1 b0
unsigned	b7 b6 b5 b4 b3 b2 b1 b0
signed	S b6 b5 b4 b3 b2 b1 b0

sign bit

The MS bit is a sign bit, indicating whether the other 7 bits are positive (when the sign bit is a 0), or negative (when the sign bit is 1).

Negative signed numbers are a little more complicated, however, than just having a sign bit tacked on to a positive amplitude.

Here's how an eight is represented as both a positive and negative binary number:

	S	b6	b5	b4	b3	b2	b1	b0
positive	0	0	0	0	1	0	0	0
negative	1	1	1	1	1	0	0	0

The negative eight looks nothing like any binary 8 that we're familiar with. That's because negative signed numbers are represented as two's-compliments of their positive brothers.

A two's compliment number is created by first inverting the positive number (including the sign bit), and then adding a binary one.

	S	b6	b5	b4	b3	b2	b1	b0
positive	0	0	0	0	1	0	0	0
↓ invert								
	1	1	1	1	0	1	1	1
↓ add 1								
	1	1	1	1	0+1	1+1	1+1	1
+	0	0	0	0	0	0	0	1
negative	1	1	1	1	1	0	0	0

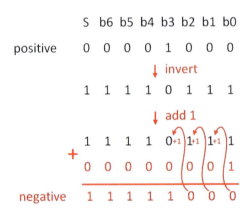

Just as we carry a one when the addition exceeds 9 when adding decimal numbers, in binary numbers we carry a one when the addition exceeds one.

Two's compliment numbers may seem unreasonably complicated, but they have the great advantage that you never have to actually subtract numbers, it's all addition.

Adding a positive binary number to a negative two's compliment effectively subtracts the two's compliment number. Also, the sign bit works out on its own. If your addition of a negative and a positive number ends up negative, the sign bit naturally ends up a one.

Exercises, Chapter 5

Exercise 5-1:
Perform the following binary arithmetic.

i. add these two unsigned numbers

```
  0 1 1 1 1 1 1 1
+ 0 0 0 0 0 0 0 1
  ---------------
  ? ? ? ? ? ? ? ?
```

ii. add these two signed numbers

```
  S
 [0] 0 1 1 1 0 1 1
+[0] 0 0 1 0 0 0 1
  -----------------
  ? ? ? ? ? ? ? ?
```

iii. add these two signed numbers

```
  S
 [0] 1 0 0 0 1 0 0
+[1] 1 1 1 1 0 0 0
  -----------------
  ? ? ? ? ? ? ? ?
```

iv. subtract these two signed numbers

```
  S
 [0] 1 0 0 0 1 0 0
-[0] 0 0 0 1 0 0 0
  -----------------
  ? ? ? ? ? ? ? ?
```

Exercise 5-2:
Complete the timing diagram.

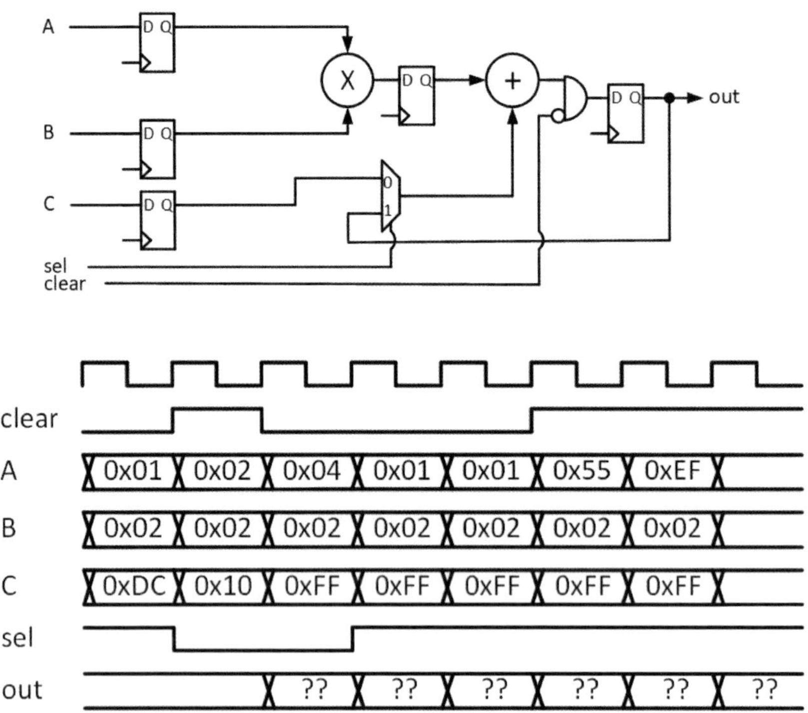

Chapter 6

Serial Interfaces, the UART

A PC motherboard processes a massive amount of data, to a large part thanks to the parallel nature of the interconnections. The peripheral ICs and memory are interconnected by hundreds of traces comprising data buses many bytes wide.

When communicating outside the PC enclosure, however, a parallel bus is not practical, at least not since the days of boa constrictor-sized parallel printer cables.

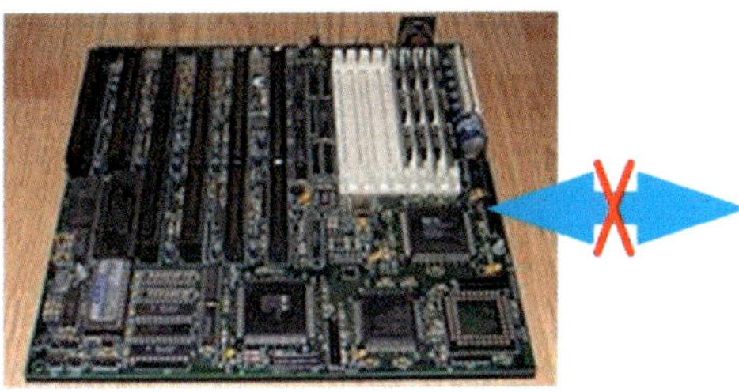

The bandwidth needs of peripheral interfaces, the bits per second, is often not great—think of what a keyboard needs to communicate—and so instead of a parallel bus, we often use just a single wire:

… sometimes one in each direction:

These are called serial interfaces since each direction is just one signal. Serial interfaces can be extremely sophisticated, such as gigabit Ethernet, and single-lane PCIe (both of which may have more than one signal path, but are still considered serial interfaces). At the low-bandwith end, we find the old workhorse, RS232.

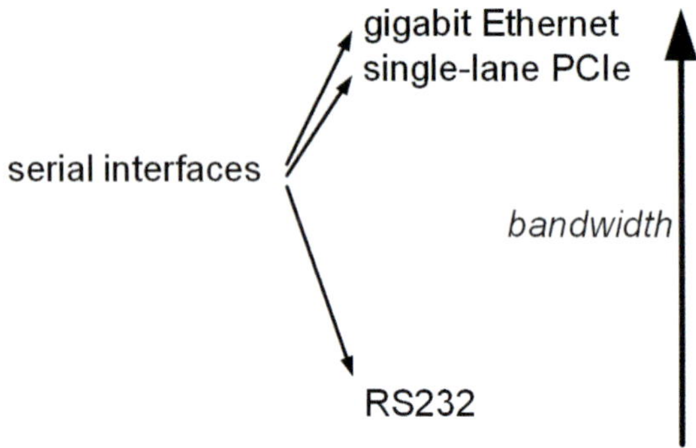

RS232

Although the RS232 serial interface has been replaced in many applications (such as keyboards) with USB and wireless connections (e.g., Bluetooth), it's still used in many non-commercial applications, and is a good place to begin to understand the basic principles of serial interfaces.

The RS232 interface originated in 1960, before we sent the first man into space. It was created to connect the click-clacking electromechanical teletypewriters to mainframes via modems. The RS232 specification defines the electrical characteristics: between 3 and 15 volts, positive and negative. The specification also defines the meanings of a handful of signals, most of which were only relevant to applications using modems, and most of which eventually fell out of use, leaving just two, or sometimes, four.

It also defines the connector pinouts, originally a 25-pin DB25 connector, and later, the more common DB9 connector. When not connected to modems (that keyboard, for example), the original set of RS232 signals were reduced just these four—Received Data (RxD), Transmit Data (TxD), Ready to Send (RTS), and Clear to Send (CTS):

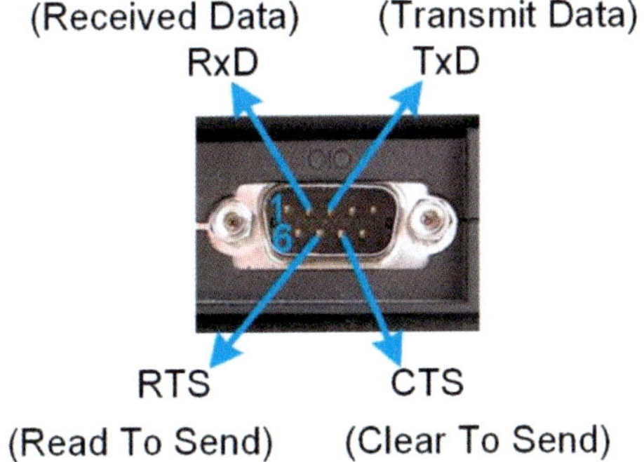

In the end, many applications (again that keyboard, for example), had no need for even the RTS and CTS control signals, and pared down to just two serial data connections, transmit and receive data.

There's often some confusion about what RS232 is, and what it is not. The RS232 specification defines electrically how bits get from source to destination. On its own, the specification is not sufficient to communicate data. We need a protocol that uses RS232 signals, i.e., a protocol that rides on top of RS232.

UART

Early on, just such a character-based protocol was established, and was originally implemented by a dedicated integrated circuit that was known as a UART, a "Universal Asynchronous Receiver/Transmitter." We can think of a character as a byte, since characters are defined by their 7-bit (or 8-bit) values. For some years now, the functions provided by these ICs have been integrated at a system level, within the PC's micro-processor chip set, or, of interest to us, within FPGAs. A UART communicates on a byte-by-byte (character-by-character) basis, sending and receiving individual characters.

Let's say we wanted to send the word "hello" across a serial link. The software would first convert the word into ASCII, where each letter, or character, is converted into its ASCII code, which consists of one byte for each letter.

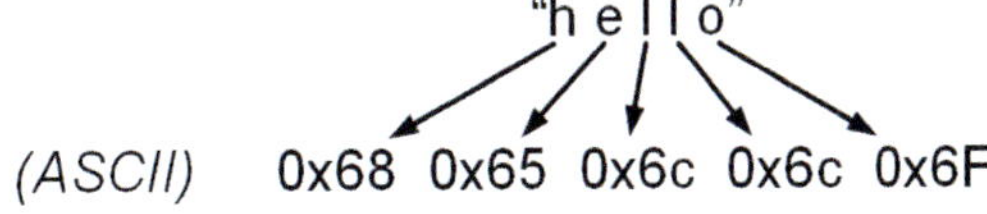

The UART function then transmits each letter, or character, serially, one byte at a time. First, the "h":

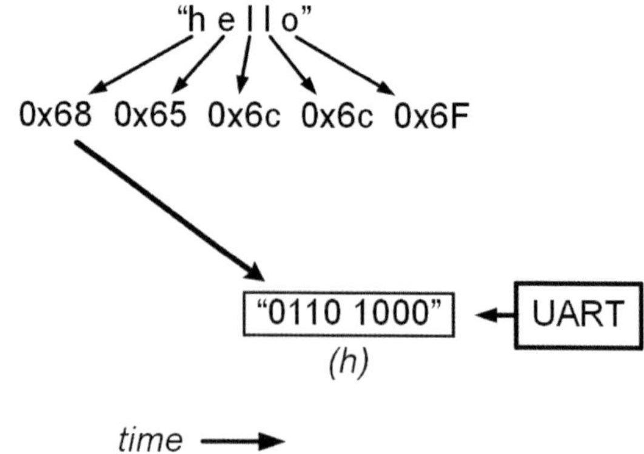

then the "e";

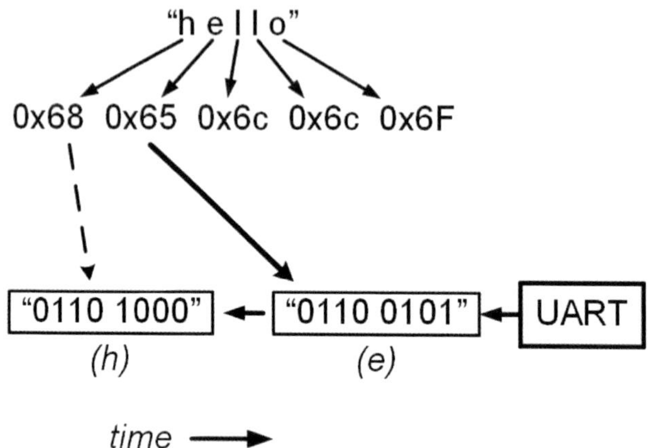

and so on:

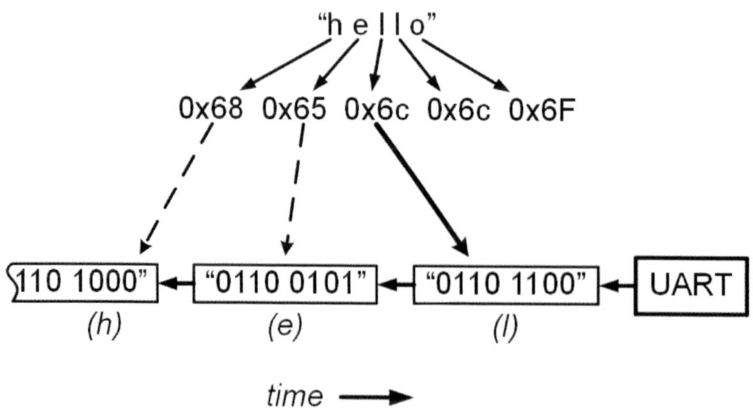

It's important to understand that the UART operates one character at a time. From the protocol's perspective, the characters in the serial transmission are self-contained—one character has no relation to another.

That said, we should note that many UARTs include FIFOs such that the software can queue up whole words or even sentences. But the fact remains that when these words and sentences are sent, they are transmitted one unassociated character at a time.

Let's see how that first character, "h," would be transmitted. First, though, we recognize that ASCII characters, at least the ones we generally use, are only seven bits wide (note that the MS bit of all the characters in "hello" are zero):

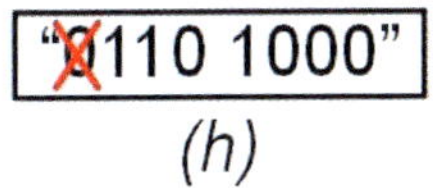

Here's the transmission as generated by the UART.

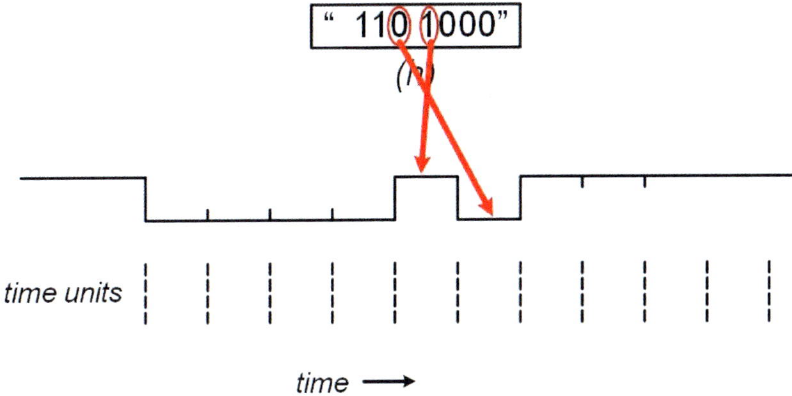

A couple of things to note. First, we're showing the bits as logical values: asserted (a logical one), and de-asserted (a logical zero). The sense of the signals of the RS232 transmission on the external RS232 wire (the RxD and TxD signals) is actually inverted, and the voltage levels are negative for ones and positive voltages for zeros.

Also, the time units—what we usually think of as the clock period—are relative, or perhaps virtual. UART communications do not include a clock, thus the "asynchronous" part of UART acronym. It's up to the receiving side to sample the signal at the middle of the bit times.

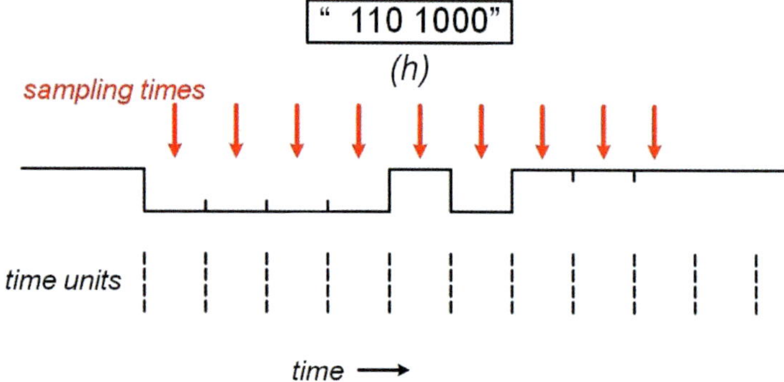

The idle time, i.e., when there no character transmission in progress, is a logic one.

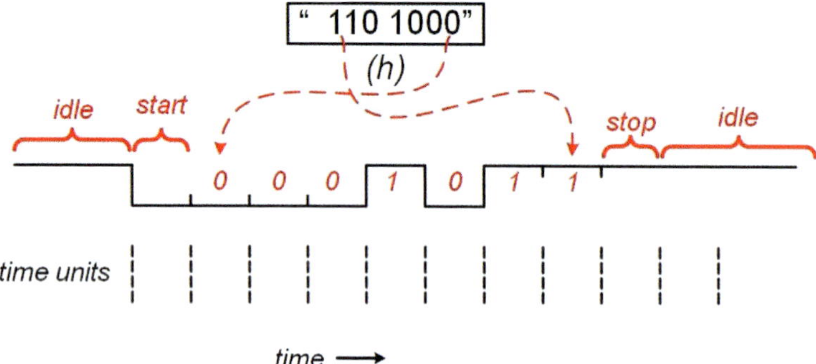

The first bit time is always a logic zero, and is called the start bit—it flags the UART receiver that a character transmission is starting. The start bit is followed by the seven bits of the character. Note that the bits are sent LSB-first. The last bit of the transmission is always a logic one, and is called the stop bit. In our case, since there's not an immediate next character, the stop bit is followed by the idle state.

Although the transmission is asynchronous (each bit is not associated with a clock edge), the receiver uses a clock to sample the bits of the serial stream. The receive sampling clock is a higher frequency than the virtual transmission time units, usually at least 8 times, and often 16 times or higher. This is called over-sampling.

Here's how the receiving side of the UART might work using a times-eight oversampling clock.

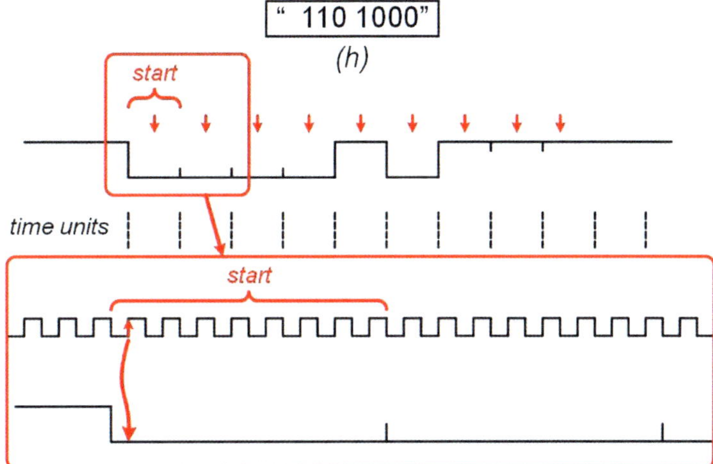

The receive logic is in an idle state, waiting for the line to go low, indicating a start bit. When it sees the line go low, this causes a counter to be cleared, which then counts until it reaches seven, the eighth count, and starts over.

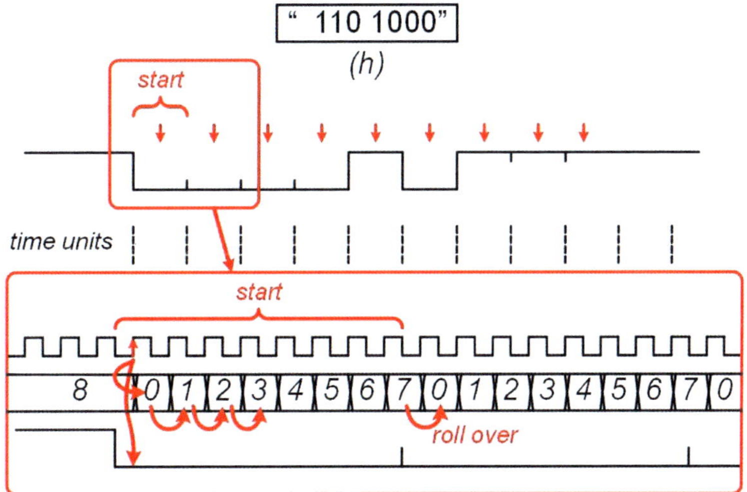

You can see that each cycle through the count spans approximately one bit time, and that the count of three marks the approximate middle, where the UART receive logic will sample the bit.

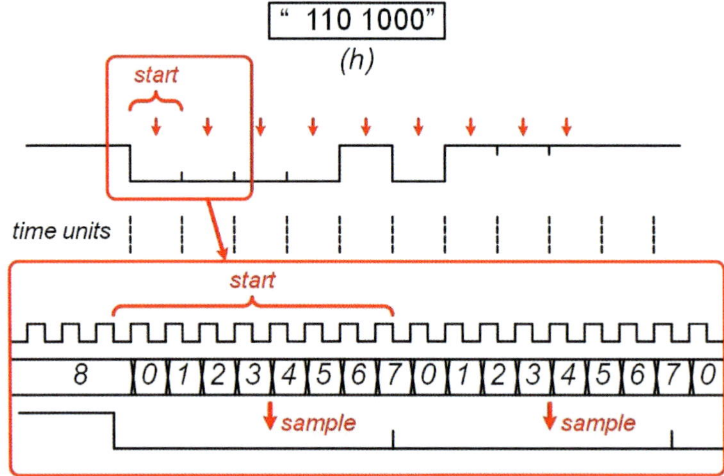

You can see that even if the start timing is a little off, say, for some reason the logic starts a little late, the receiver still catches the correct level.

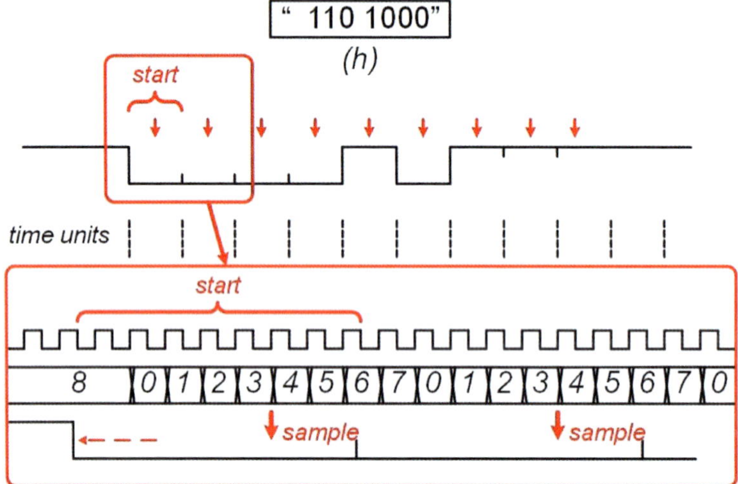

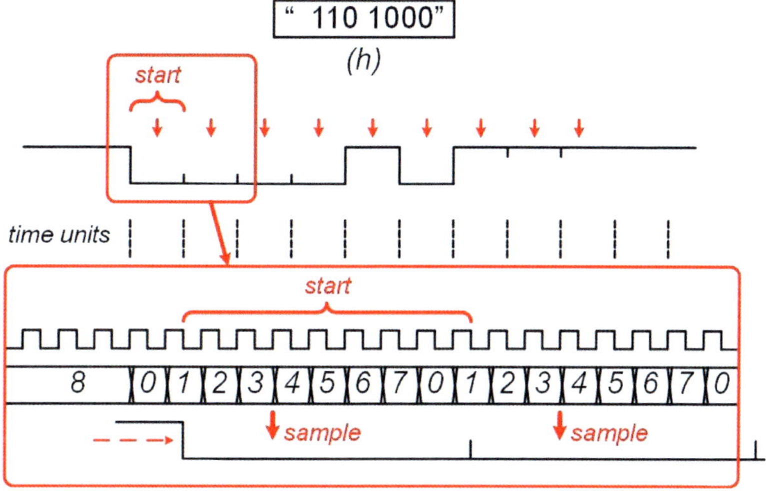

You can also see that the frequency of the receiver sampling times needs to be coordinated with the virtual bit times of the transmitter. This is done by having both sides configured to operate with the same virtual sampling clock.

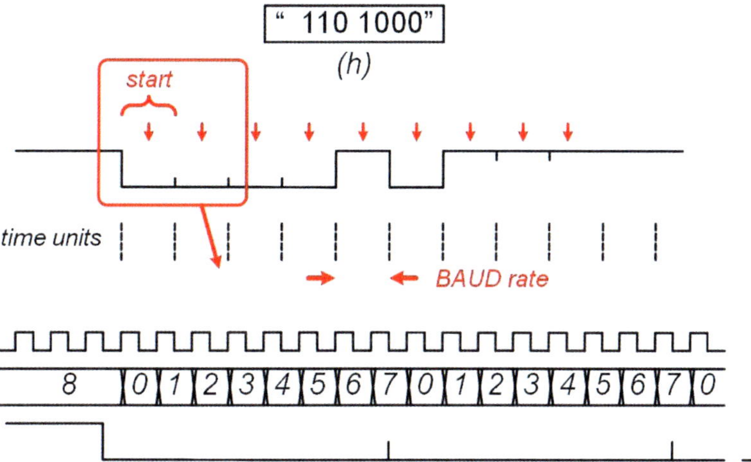

All possible virtual sampling clocks are chosen from a fixed set of frequencies, referred to as the BAUD rate, measured in cycles per second. There are dozens of established BAUD rates, but here's a few of the most common.

9,600
56,000
115,200

Note that, while an established BAUD rate is shared at both ends, the receive sampling clock is not prescribed, although an even binary multiple of the BAUD rate, e.g., times eight or sixteen, clearly yields the simplest logic solution.

UART Implementation

Here's a basic implementation of the UART receive side, noting that an actual UART would include complexities that we've been ignoring so far.

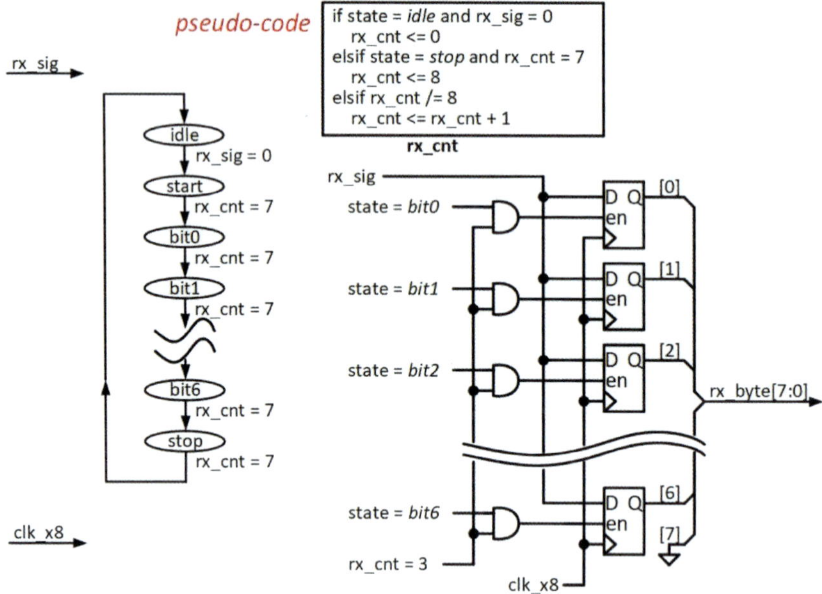

I've used pseudo-code—essentially VHDL, minus a lot of details not needed for human understanding—to describe the counter "rx_cnt."

The idle state machine kicks off as soon as the rx_sig signal goes low. This corresponds to this point:

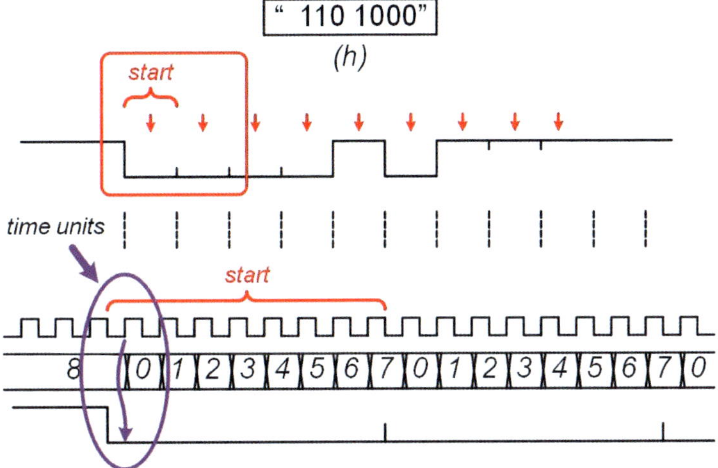

At the same time, we clear the counter:

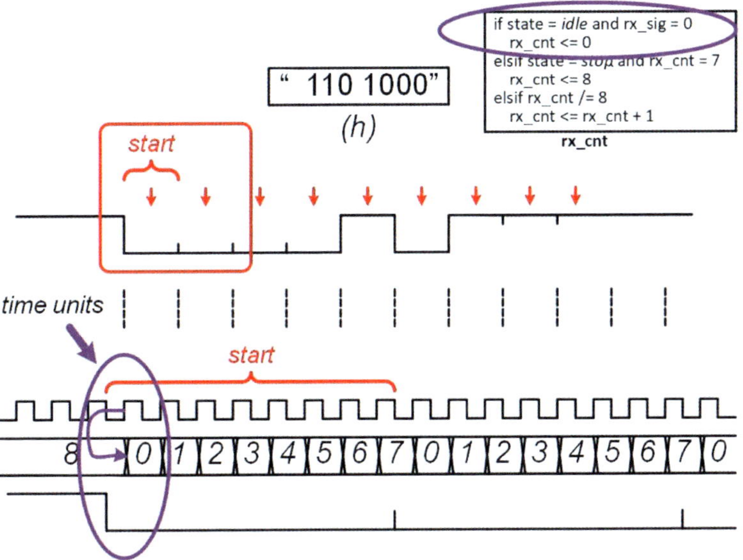

The state machine steps through the bits at the end of each count (0 to 7):

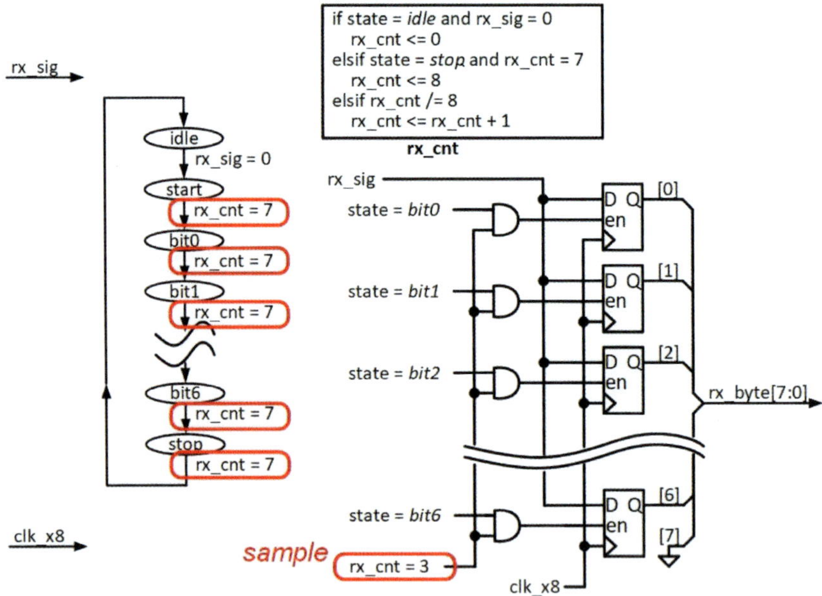

and samples the incoming data bits in the middle of their virtual clock cycles, at the count of three, as shown here:

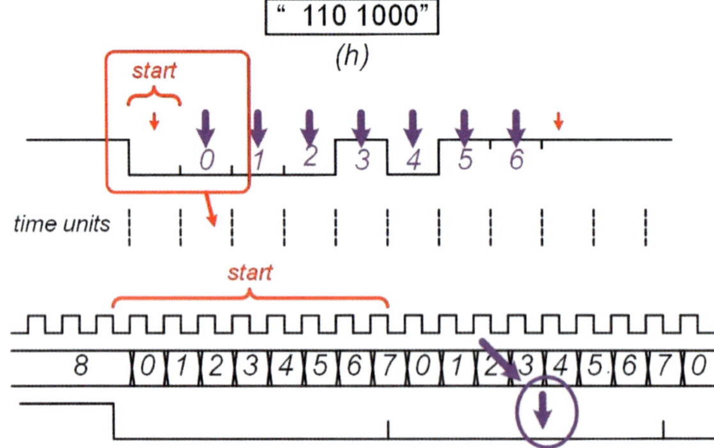

Note that I've tacked on a zero to the MSB of rx_byte (the triangle is a ground symbol, and used in VHDL diagrams to denote a logical zero), since the received character is only seven bits wide:

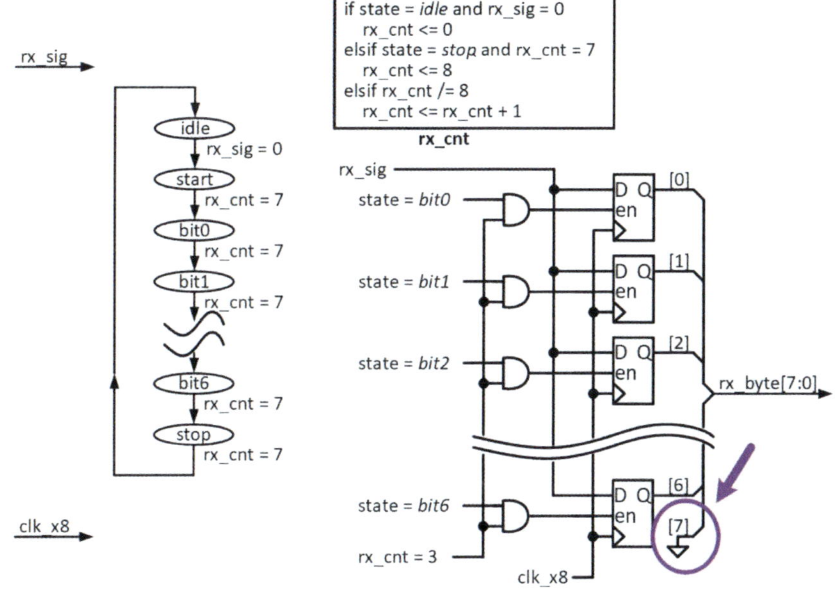

Here's a simulation, with the states transitioning at count 7 …

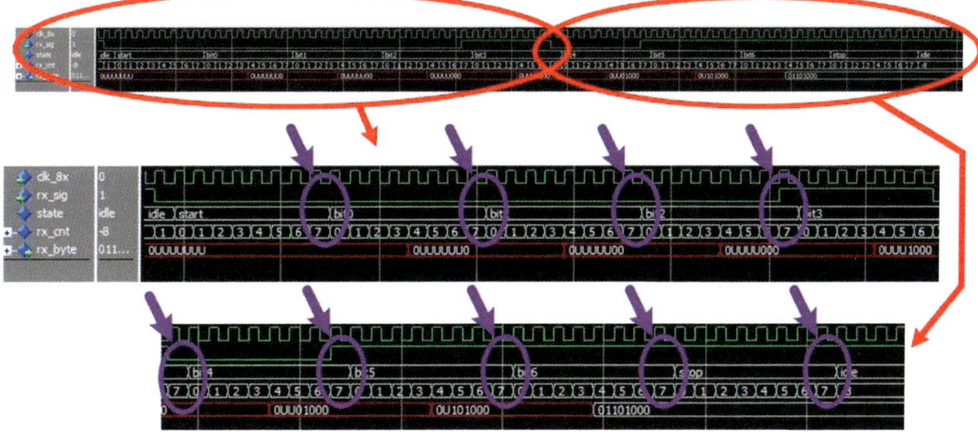

and the rx_sig being sampled at count 3:

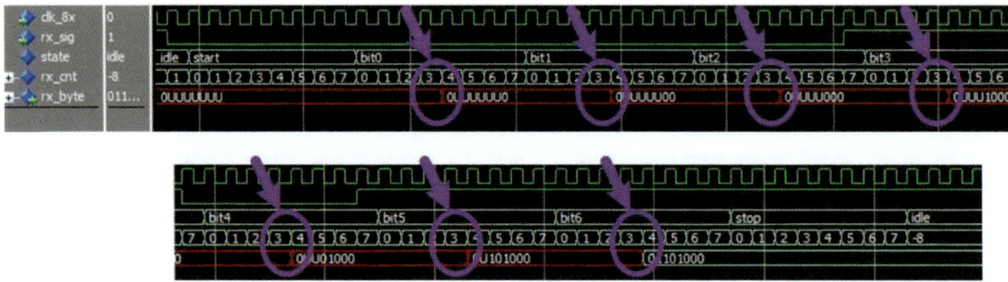

Note that the received vector value (rx_byte) is shown as indeterminate until all seven bits have been registered (remember that I've fixed the eighth bit of the byte as zero in the VHDL code).

Here's the VHDL code, with the entity and architecture declarations enlarged:

```
entity uart_rx is
port (
            clk_8x          : in   std_logic;
            --
            rx_sig          : in   std_logic;
            rx_byte         : out  std_logic_vector(7 downto 0)
        );
end entity;

architecture behavioral of uart_rx is

  type state_type is (idle, start, bit0, bit1, bit2, bit3, bit4,
                      bit5, bit6, bit7, stop
                      );
  signal state : state_type;
  signal rx_cnt     : unsigned(3 downto 0) := X"0";

begin

  uart_rx : process(clk_8x)
  begin
     if rising_edge(clk_8x) then
        --state machine
        case (state) is
           when idle  => if (rx_sig = '0')  then  state <= start; end if;
           when start => if (rx_cnt = X"7") then  state <= bit0;  end if;
           when bit0  => if (rx_cnt = X"7") then  state <= bit1;  end if;
           when bit1  => if (rx_cnt = X"7") then  state <= bit2;  end if;
           when bit2  => if (rx_cnt = X"7") then  state <= bit3;  end if;
           when bit3  => if (rx_cnt = X"7") then  state <= bit4;  end if;
           when bit4  => if (rx_cnt = X"7") then  state <= bit5;  end if;
           when bit5  => if (rx_cnt = X"7") then  state <= bit6;  end if;
           when bit6  => if (rx_cnt = X"7") then  state <= stop;  end if;
           when stop  => if (rx_cnt = X"7") then  state <= idle;  end if;
           when others => null;
        end case;
        --rx_cnt
        if (    state = idle
             and rx_sig = '1'
           ) then
           rx_cnt <= X"0";
        elsif (    state = stop
                and rx_cnt = X"7"
              ) then
           rx_cnt <= X"8";
        elsif (rx_cnt = X"7") then
           rx_cnt <= X"0";
        elsif (rx_cnt /= X"8") then
           rx_cnt <= rx_cnt + 1;
        end if;
        -- byte assembly
        if (rx_cnt = X"3") then
           case (state) is
              when bit0  => rx_byte(0) <= rx_sig;
              when bit1  => rx_byte(1) <= rx_sig;
              when bit2  => rx_byte(2) <= rx_sig;
              when bit3  => rx_byte(3) <= rx_sig;
              when bit4  => rx_byte(4) <= rx_sig;
              when bit5  => rx_byte(5) <= rx_sig;
              when bit6  => rx_byte(6) <= rx_sig;
              when others => null;
           end case;
        end if;
     end if;
  end process;

  rx_byte(7) <= '0';

end architecture behavioral;
```

```
entity uart_rx is
port (
            clk_8x          : in   std_logic;
            --
            rx_sig          : in   std_logic;
            rx_byte         : out  std_logic_vector(7 downto 0)
        );
end entity;
```

```
type state_type is (idle, start, bit0, bit1, bit2, bit3, bit4,
                    bit5, bit6, bit7, stop
                    );
signal state : state_type;
signal rx_cnt     : unsigned(3 downto 0) := X"0";
```

… and the architecture body:

```
entity uart_rx is
port (
        clk_8x      : in   std_logic;
        --
        rx_sig      : in   std_logic;
        rx_byte     : out  std_logic_vector(7 downto 0)
     );
end entity;

architecture behavioral of uart_rx is

   type state_type is (idle, start, bit0, bit1, bit2, bit3, bit4,
                       bit5, bit6, bit7, stop
                       );
   signal state : state_type;
   signal rx_cnt    : unsigned(3 downto 0) := X"0";

begin

uart_rx : process(clk_8x)
begin
   if rising_edge(clk_8x) then
      --state machine
      case (state) is
         when idle  => if (rx_sig = '0')  then  state <= start; end if;
         when start => if (rx_cnt = X"7") then  state <= bit0;  end if;
         when bit0  => if (rx_cnt = X"7") then  state <= bit1;  end if;
         when bit1  => if (rx_cnt = X"7") then  state <= bit2;  end if;
         when bit2  => if (rx_cnt = X"7") then  state <= bit3;  end if;
         when bit3  => if (rx_cnt = X"7") then  state <= bit4;  end if;
         when bit4  => if (rx_cnt = X"7") then  state <= bit5;  end if;
         when bit5  => if (rx_cnt = X"7") then  state <= bit6;  end if;
         when bit6  => if (rx_cnt = X"7") then  state <= stop;  end if;
         when stop  => if (rx_cnt = X"7") then  state <= idle;  end if;
         when others => null;
      end case;
      --rx_cnt
      if (     state = idle
           and rx_sig = '0'
         ) then
         rx_cnt <= X"0";
      elsif (     state = stop
              and rx_cnt = X"7"
            ) then
         rx_cnt <= X"8";
      elsif (rx_cnt = X"7") then
         rx_cnt <= X"0";
      elsif (rx_cnt /= X"8") then
         rx_cnt <= rx_cnt + 1;
      end if;
      -- byte assembly
      if (rx_cnt = X"3") then
         case (state) is
            when bit0  => rx_byte(0) <= rx_sig;
            when bit1  => rx_byte(1) <= rx_sig;
            when bit2  => rx_byte(2) <= rx_sig;
            when bit3  => rx_byte(3) <= rx_sig;
            when bit4  => rx_byte(4) <= rx_sig;
            when bit5  => rx_byte(5) <= rx_sig;
            when bit6  => rx_byte(6) <= rx_sig;
            when others => null;
         end case;
      end if;
   end if;
end process;

rx_byte(7) <= '0';

end architecture behavioral;
```

We'll finish our look at UARTs with an alternative implementation of the received byte assembly. Instead of bit-enabled registers, we use a shift register:

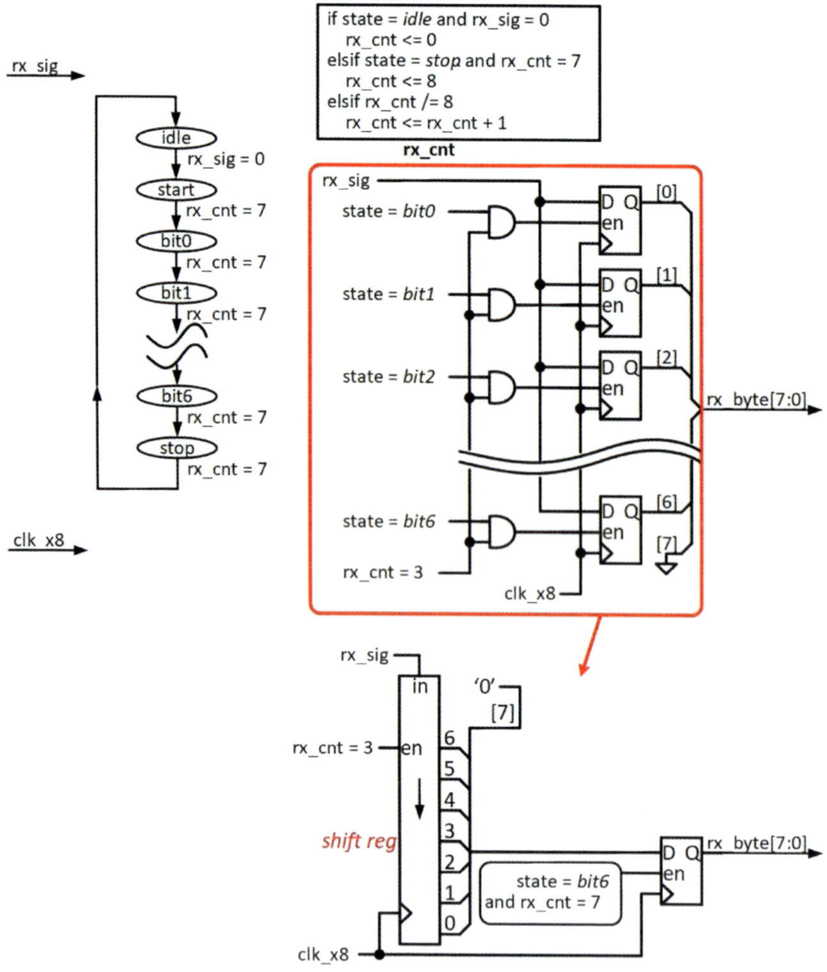

The shift register is enabled at the middle of each virtual clock cycle (when the counter is three), and the assembled byte is latched (when the output register is enabled) at the end of the seventh arriving bit (when the counter is seven at the seventh bit state, "bit6").

Alternatively, the latched output register could be eliminated by limiting the shift register's enables to just the data bit times:

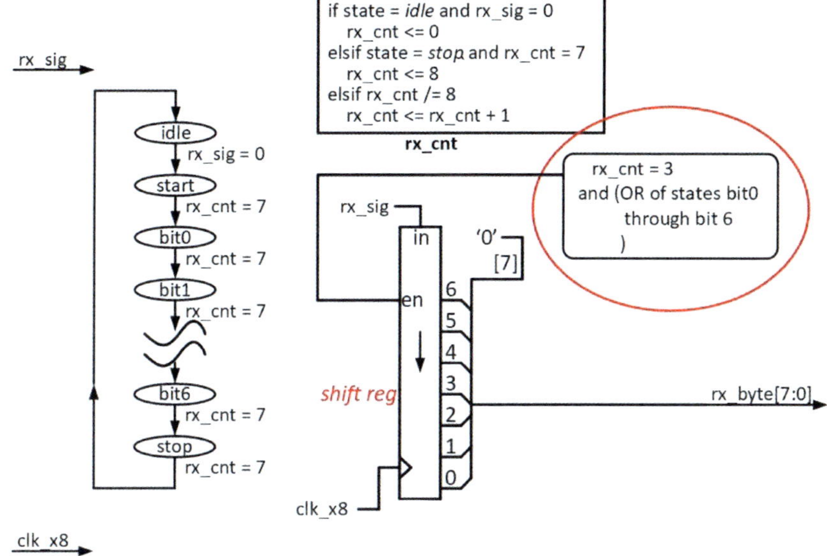

Before we leave the UART, we note that, in addition to the simple 7-bit transmission, UARTs allow the addition of a parity bit for reliability:

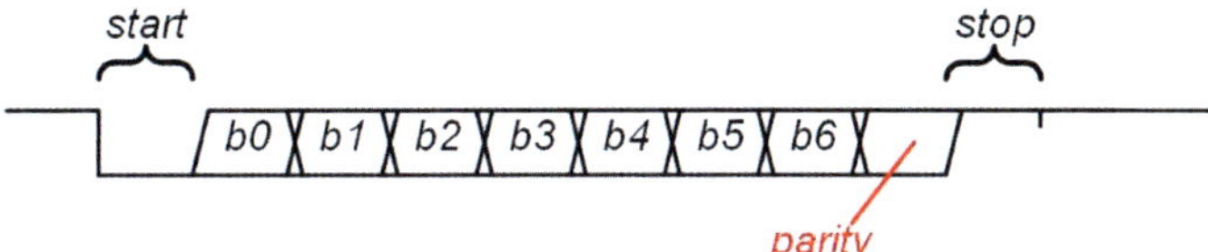

A parity bit is simply an indication of whether there are an even or an odd number of ones in the preceding bits. Thus, if one of the bits changes state, say by a noise glitch, the parity bit will be wrong. Parity bits can detect single-bit errors, but not double (the even/odd amount toggles back with two bit errors).

Additionally, the UART can generally be configured to carry 8 bits instead of 7:

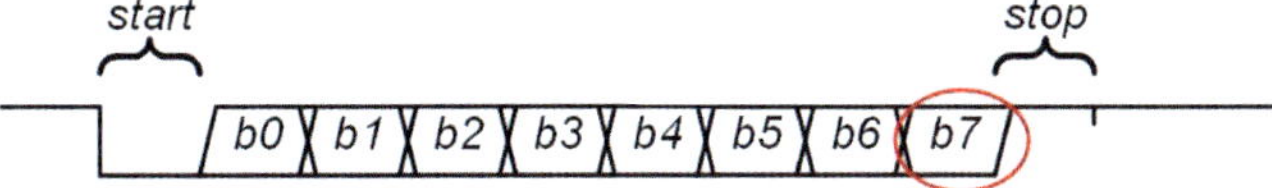

and can optionally carry a parity as well:

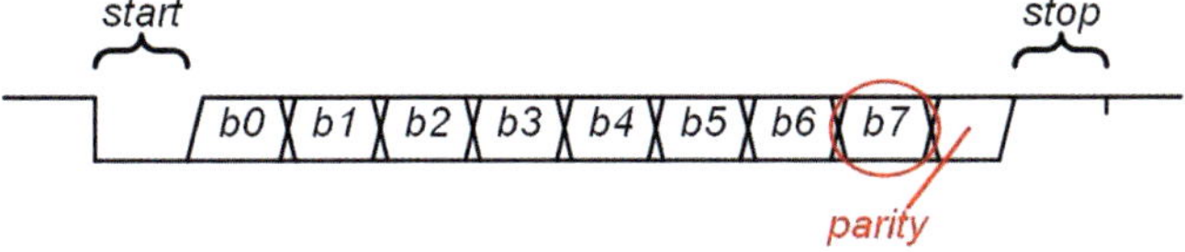

Finally, the UART can be configured to use two stop bits. This simply means that back-to-back transmissions must be at least two virtual clock cycles apart.

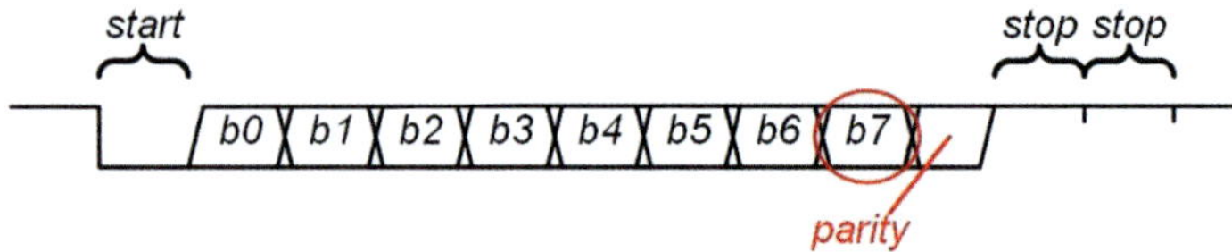

Exercises, Chapter 6

Exercise 6-1:
Add parity to the Rx half of the implemented UART. The "par_err" signal goes active when parity errors are detected. The parity is even when "par_even" is asserted, otherwise it is odd.

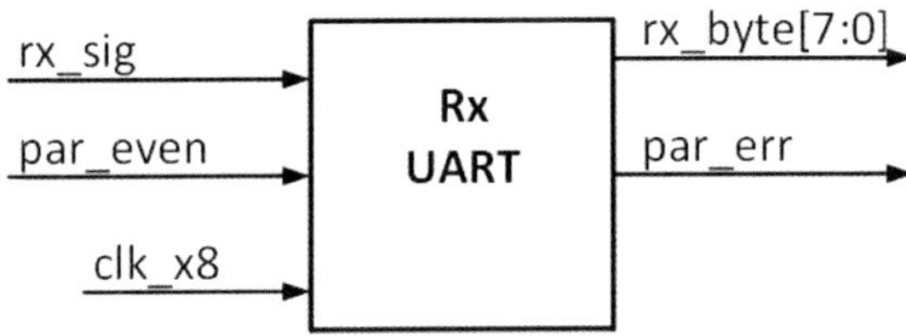

Chapter 7

Serial Interfaces, I²C and SPI

I²C

We'll next take an overview look at another popular serial interface, the I²C bus, also known as the I2C bus, also known as the IIC bus after its original name, the "Inter-Integrated Circuit" bus.

Whereas UARTs have serial lines in each direction, the I2C bus is a single bi-directional line.

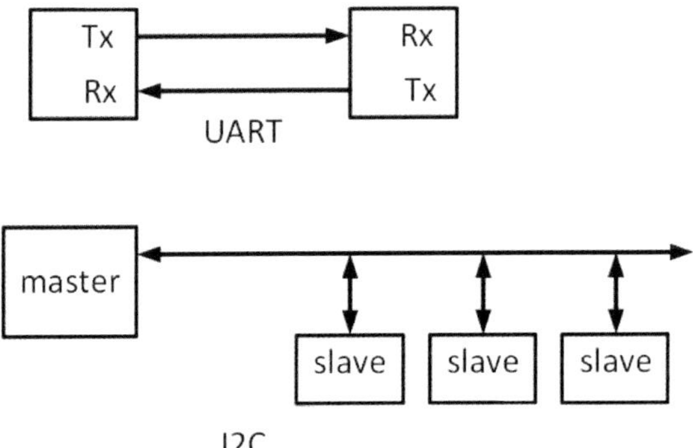

A master initiates all transactions, and one of a possible multiple slaves responds by either accepting write data, or returning requested read data.

Like a UART, a transaction begins with a start indication, followed by seven bits of slave device address, followed by a single bit that identifies this transaction as either a write (0) or a read (1).

The slave device address selects one of the slaves for the transaction.

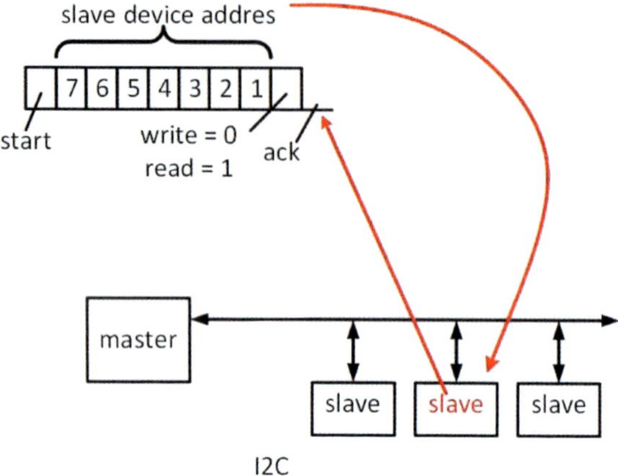

The selected slave then responds by driving the line low, which constitutes an acknowledgement, letting the master know that the slave is there and able to respond.

Upon receiving the acknowledge, the master transmits a data byte to the selected slave, which must again acknowledge.

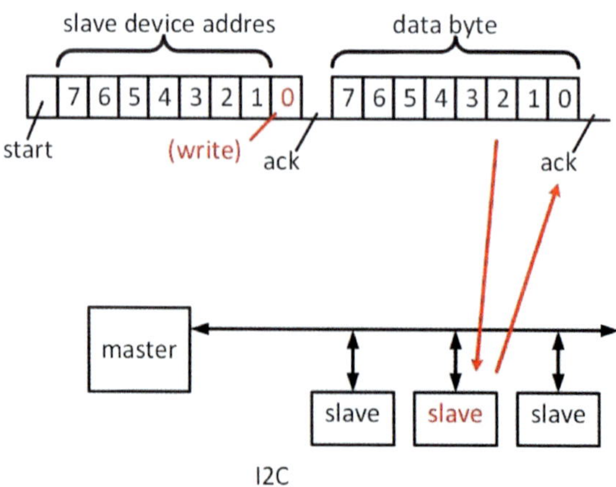

From here, the master can either end this current transmission by sending a stop indication:

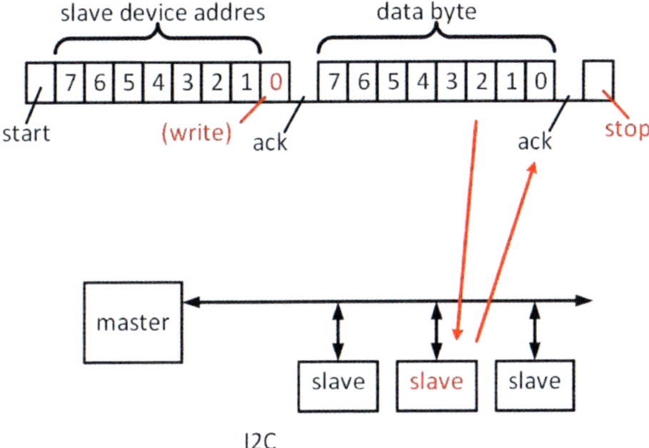

… or the master can continue sending more write bytes to that selected slave:

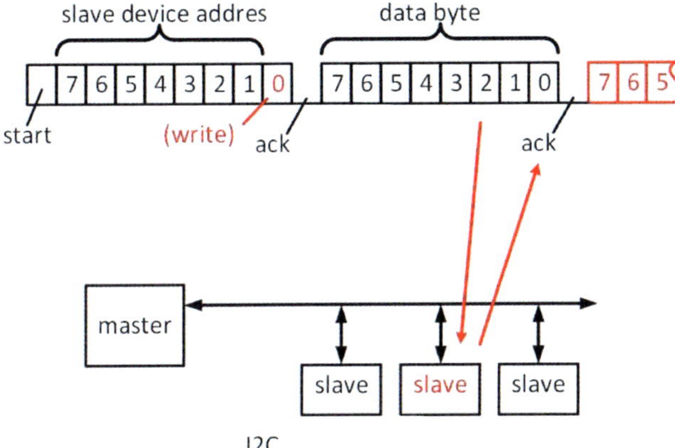

There is no limit to how many subsequent write bytes the master can send (although the last byte must be followed by a stop indication). After each byte, the slave must send an acknowledgment.

Reads operate the same way, except that the slave drives the bi-directional bus with the requested read data (after the master drives the line with the slave address and sets the write/read bit high), and it is the master who must acknowledge after each read byte.

Whereas both sides of a UART transmission use an agreed-upon timing—a virtual clock—the I²C bus carries a specific clock along on a separate signal.

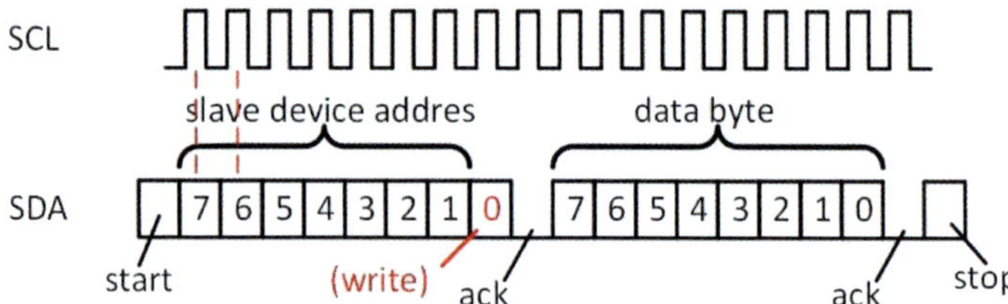

This clock is called "SCL" and the data line is called "SDA." The SCL clock line pulses high and back low while the SDA data line is held stable. I.e., the SDA line never changes while the SCL is high. This is important, because breaking this rule is what defines both the start and stop indications.

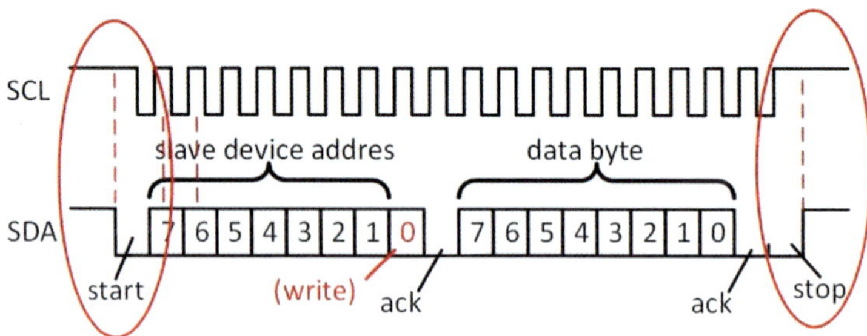

A start indication consists of the master driving the SDA line low while holding the SCL line high, and a stop indication consists of the master letting the SDA line go high while holding the SCL line high.

Notice that the master "drives" the SDA line low, while "letting" the SDA line go high. This is because I²C defines the lines as "open-drain" (also sometimes called "open-collector"), meaning that the transmitting source drives the line low for zeroes, but allows the line to be high-impedance for ones. A pullup resistor pulls the line high when nothing is driving it low. This easily allows multiple sources to drive the same line (i.e., the slave when driving acknowledgements, or returning read data) without fear of line contention (when one source is driving high, while the other driving low).

The I²C bus uses the open-drain operation for another feature—clock stretching. Here, a slave can pause a transaction if it needs more time to process the data by holding the SCL line low (possible, since the master is simply letting the line float otherwise). The master observes this, and must wait until the slave releases the SCL line, allowing it to go high.

The I²C bus also allows multiple masters on the same set of lines (although this is not usually the case). Normally, a master can see that anther is conducting a transaction, and waits for a stop indication before taking the line, but it is possible for two to coincidentally begin transactions at the same time. In this case, a master will eventually see the line low (when the other one is driving it), when it expects the line to be high. That master interprets that discrepancy as interference by another master, and terminates its transaction (the first master to detect the condition losses ownership for this transaction period).

Note that I2C does not define what the data bytes mean. Often times, the first data byte is a slave register address (after the slave device is selected via the slave device address), or perhaps a command. The meanings of the bytes are specific to the addressed slave device.

SPI

For a final overview of low-speed serial interfaces we look at the SPI bus, which stands for the Serial Peripheral Interface—descriptive enough.

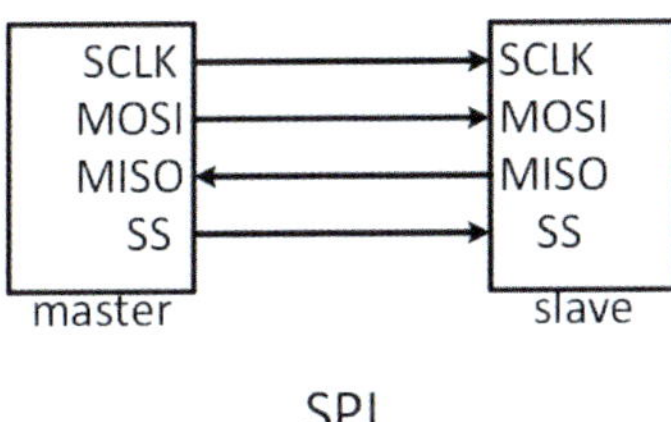

SPI

Like the I^2C bus, SPI defines a master and one or more slaves. Also like I^2C, SPI is synchronous, i.e., uses a clock, called the "Serial Clock," or SCLK. This clock can be free-running, or can transition only during data transactions.

Two unidirectional serial data lines, MOSI and MISO, carry serial data between the master and slave. The MOSI line, the "Master Out, Slave In," carries data from master to slave, and the MISO, the "Master In, Slave Out," carries data in the other direction.

The SS line is the "Slave Select," and the slave responds only when this line is active.

The SPI interface uses a unique method of transferring data between the master and slave. For every clock that the SS select is active, both the master and slave drive one bit of data onto their respective output lines.

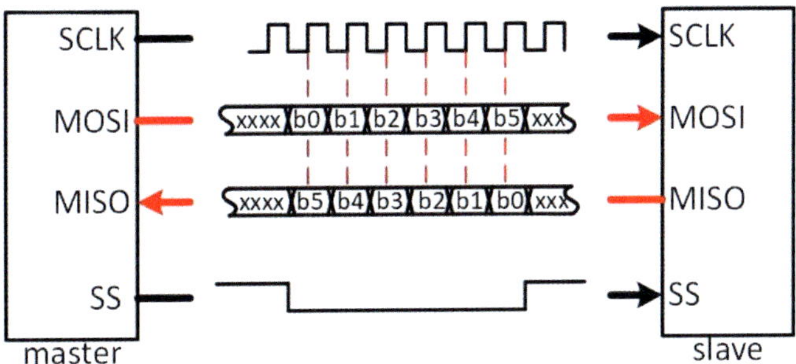

That's it. The SPI interface does not define how many bits each transaction contains, nor how registers in the slave are addressed. That's left for a higher level of protocol, defined by the slave.

This type of bi-directional data transfer allows slaves to be chained. This example chains three slaves together:

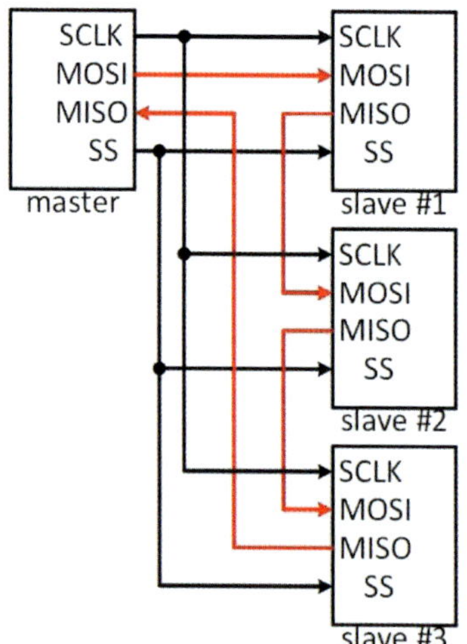

The master's clock and slave select drives all the chained slaves, and the MOSI and MISO lines are chained among the slaves. For every clock when the SS select is active, bits of data are transferred from the master to slave #1, from slave #1 to slave #2, from slave #2 to slave #3, and finally from slave #3 back to the master. In this configuration it is usually expected that each slave will transmit one byte of its data, and then loop its MOSI back to its MISO. In this way, after 24 clocks, the master will have received one byte from each of the three slaves.

Alternatively, the master can drive a separate SS select to each slave, and only the selected slave responds on the shared MOSI and MISO lines. The others (unselected) tri-state their MISO outputs, allowing just the selected slave to return data.

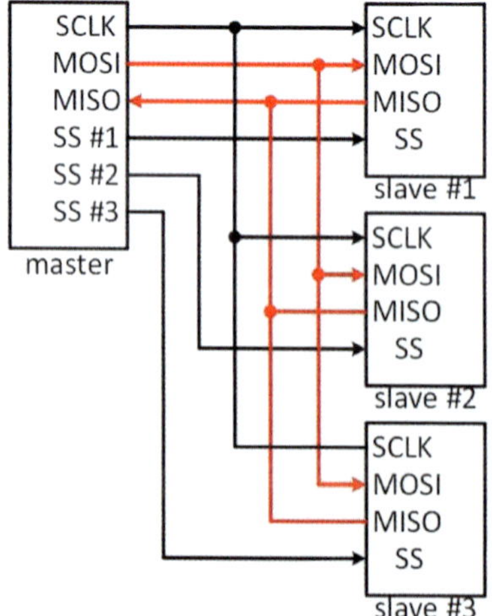

This is by far the more common multi-slave configuration.

Finally, an extension of the serial SPI has evolved that replaces the unidirectional MOSI and MISO signals with four bi-directional signals, rendering the interface a half-duplex 4-bit parallel bus.

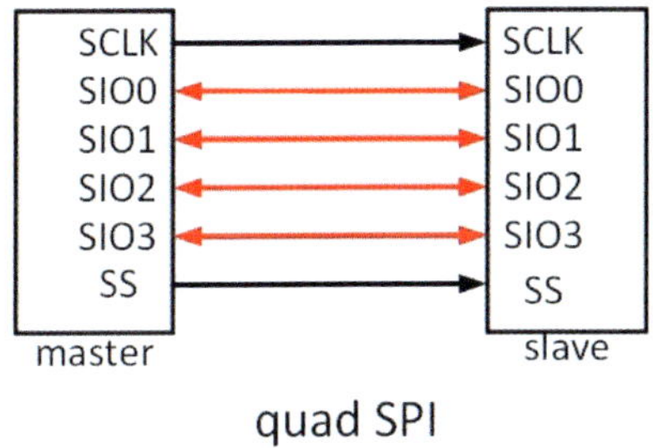

quad SPI

This type of interface is popular with external flash memory devices, allowing higher-bandwidth access, for example for loading the program file to FPGA chips after power-up.

Although functionally quite different from the original SPI interface once up and running, upon power-up, a quad-SPI capable master first uses the SIO0 and SIO1 lines as the original MOSI and MISO signals to query the slave to determine how it wants to operate, and if the slave is not a quad-SPI device, the master proceeds with original MOSI/MISO SPI operation.

Low-speed Serial Interface Comparison

Here's a comparison of the three low-speed serial interfaces.

	pins	half/full-duplex	max practical rate	connection
UART/RS232	two (plus GND)	full	<200 Kbps	intra-PCB, inter-system
I2C	two	half	400 Kbps	intra-PCB
SPI	four+	full (sort of)	multi-Mbps	intra-PCB

Low-speed Serial Interfaces

Note that the rate and connection columns are for typical practical applications. You can always find instances where these conditions are exceeded. For example, the I^2C interface is sometimes used for connecting prototype development boards together, such as Arduino. Intra-PCB means interfacing ICs on the same PCB, while inter-system would be cables connecting different enclosures. I say that SPI is sort of full duplex because the return MISO signal is only active when the MOSI line is as well.

High-speed Serial Interfaces

We end this chapter with an overview look at how the higher-level interfaces are implemented in FPGAs, and by higher-level, I mean both in protocol complexity and bandwidth. PCIe (PCI Express) and multi-gigabit Ethernet are two popular examples.

These gigabit data rates require very specialized interface circuitry, for both the external analog signaling, and the internal logic. The FPGA vendors that support these rates provide dedicated circuitry called high-speed transceivers, which your RTL instantiates and uses internally. These transceivers reside at the edge of the die, and interface to external pins on one side, and internally to your RTL on the other.

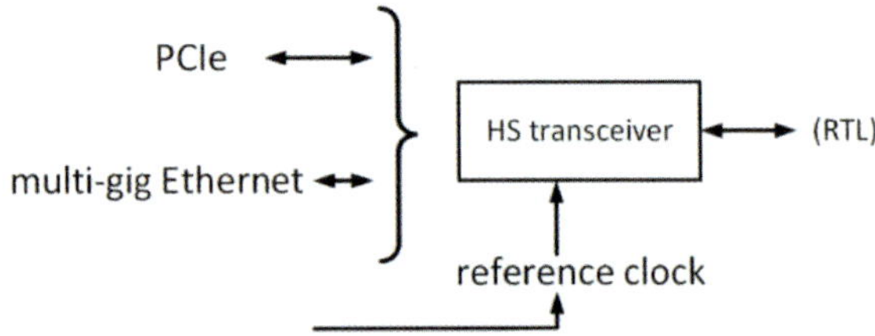

Processing an interface at gigabit rates (sub nanosecond) requires specialized, precise phase-locked loops, and these are included inside the vendor's transceivers. An accurate reference clock is often required, though, and must be provided from an external source.

Besides handling complex interface-specific protocol, such as 8B10B encoding, the transceivers provide the critical role of assembling the high-speed serial stream into parallel words at rates that the RTL can handle:

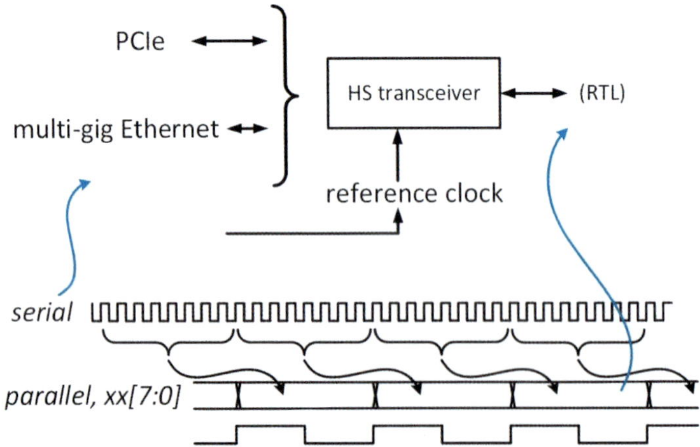

However, a serial-to-byte, 8-to-1 reduction of a 10 gigabit serial line yields an internal clock rate of 1.25 gHz, far above what is practical for RTL processing. Thus, at multi-gigabit rates, the transceiver usually hands off data that's 32 bits wide, or even higher.

Finally, we want to be careful to understand that these transceivers deliver raw bytes or words internally to the RTL. Both PCIe and gigabit Ethernet comprise complex levels of protocol (mind-numbing in the case of PCIe) that must be handled.

Fortunately, the FPGA vendors often provide IP for this as well (although due to the great complexities involved, using black-box protocol IP often constitutes a project onto itself).

Exercises, Chapter 7

Exercise 7-1:

Using the I2C protocol below, show the I2C transaction when the master reads a 0x5A data value from slave device 0x52 at register address 0x03.

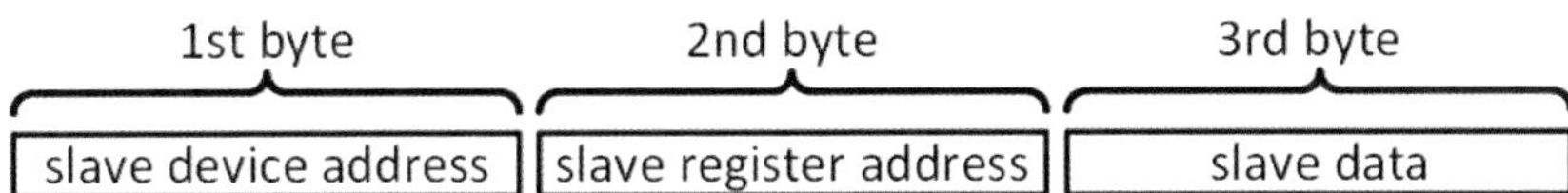

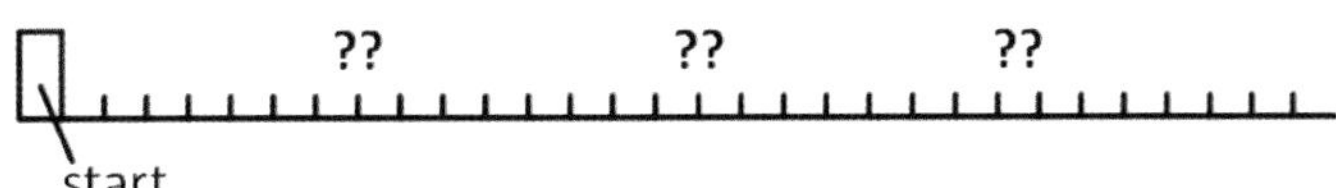

Chapter 8

Timing

Timing Constraints

In chapter two of volume 1, we saw how the clocked operation of logic in FPGAs relies on the fact that it takes time for signals to travel from one register to another—specifically, that the hold time of a register is met by guaranteeing some minimal amount of propagation time. We also saw, however, that the setup time of a register requires that the propagation time not be too large (otherwise, the changing signal won't make it to the next register before the next clock edge). Typical propagation times in modern FPGAs are often a couple of nanoseconds or less. At clock rates of 50MHz (20ns periods) or less, the accumulated propagations of layers of logic are generally not an issue. For clock rates above, say, 150MHz, however, it is essential that the FPGA compilation tool takes proper care (specifically the placement and routing stages, where register locations on the FPGA IC die are determined, and the signal paths established).

We support the FPGA vendor tool in achieving viable timing at higher clock frequencies within two categories:

1) by defining and constraining the timing for the vendor compiler tool, and

2) by then designing our logic to meet that timing.

Every FPGA design of any reasonable size uses clocks, sometimes two or more. When the vendor compiler tool lays out the logic parts, and then routes the connections between them, it needs to know how much time it has to get from each register source to all its clocked destinations. You, the designer, define the clock frequencies, and communicate this to the tool via a project file called a "constraints" file. The Intel/Altera Quartus FPGA tool uses constraint files with ".sdc" extensions ("Synopsis Design Constraints"), while the Xilinx tools use either ".ucf" files (older Xilinx tools), or ".xdc" files (the newer Vivado tool). The SDC and XDC files use the tcl scripting language to define the timing constraints, while the older UCF file definitions are specific to those Xilinx tools (i.e., IDE).

Each vendor provides documentation about using tcl to constrain the design's timing. Additionally, the vendors provide GUI methods of developing timing constraints.

The most important timing constraint is simply telling the tool the frequency of each clock. Additionally, however, we often include other timing constraint information, but in these cases much of it is the opposite of constraining. We can ease the burden of the tool's layout and router functions by telling them if some paths are don't-cares, i.e., "false paths." In these cases, the tool need not worry about the timing for the declared signals. One obvious example is static configuration information—logic levels that don't change during operation.

And, even if some logic is not static, but changes during operation, it may not require just one clock cycle to get from source to destination. We call these "multi-cycle paths," and an example might be an enable signal that goes active some number of clocks before a block of processing. In the case of multi-cycle paths, we tell the tool via the constraints file which signals are multi-cycle, and how many clocks it can assume to get from source to destination.

Metastability

There's a whole other category of signals that bridge two different clocked areas, what we call clock domains.

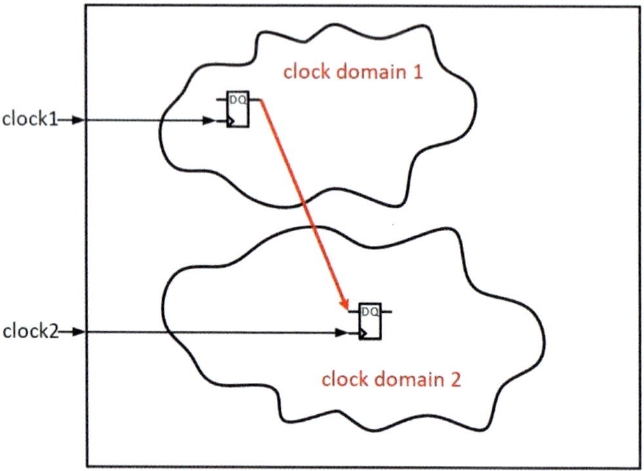

With some luck, and perhaps foresight, your design will not require timing constraints for cross-clock-domain signals, and these connections can become false-paths, part of the anti-constraints.

One way to guarantee that the cross-clock-domain signals are don't-care (timing-wise) is to design the crossing connection such that it's guaranteed to be free of potential metastability, which can occur when the input to a register is transitioning during the clock edge. This can produce uncertainty in the design operation (imagine carrying a byte across clock domains, where some bits are the new value, and some the old).

In the following example, imagine that we need to communicate to the clock2 domain that a one-clock pulse has occurred in the clock1 domain. Further, say that clock2 is significantly slower than clock1.

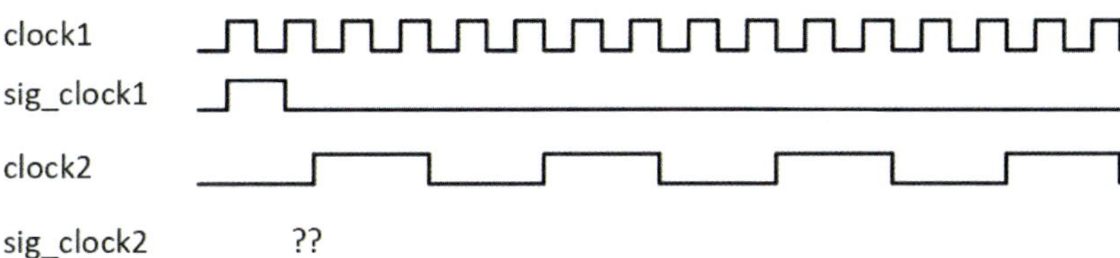

One method might be to simply stretch the clock1 signal, and catch it with clock2. The problem with this is that we still have a risk of metastability (when the later clocks happen to line up).

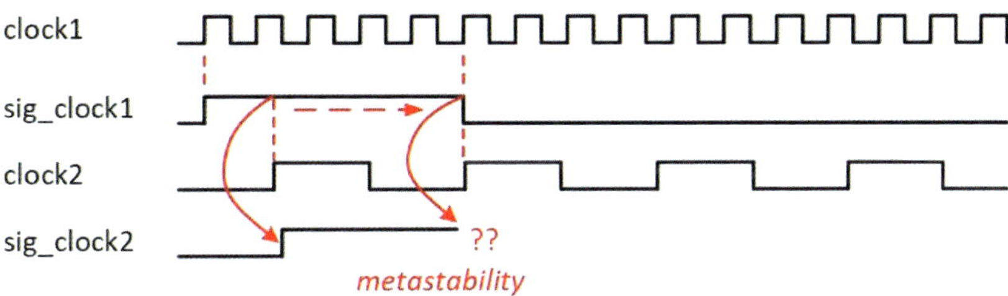

If we make the stretch too short, we may miss it with clock2:

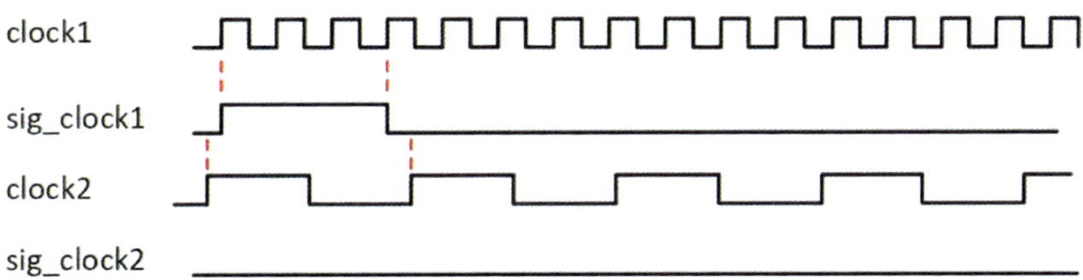

… but if we make it too long, then clock2 may get a pulse that's two clocks long (we wanted to communicate a one-clock pulse), and there's still the problem of potential metastability:

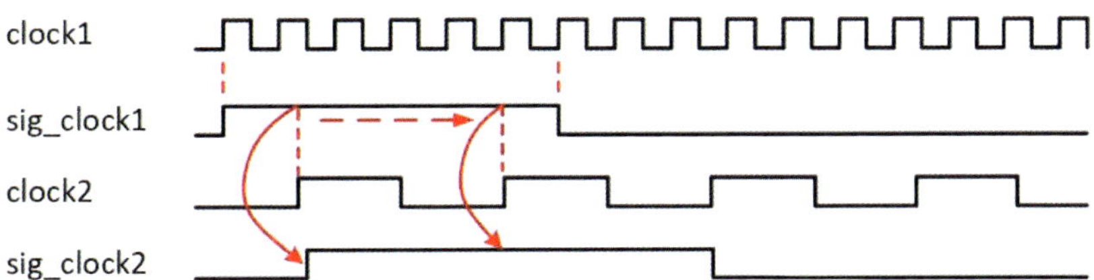

Finally, even ignoring problems with metastability, once the stretch is adjusted for the two clocks (no easy task, considering all possible alignments), you're locked into those two clock frequencies.

Here's a method that gets around these problems:

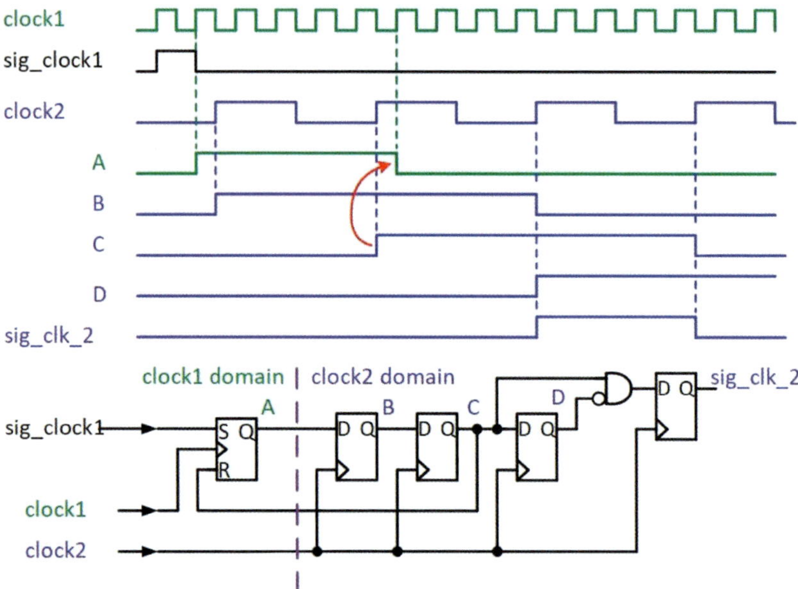

The clock1 input signal (sig_clock1) sets the set/reset latch A high upon a sig_clock1 occurrence. The clock2 clock catches this at point B. Note that this signal at d-flop B may be metastable, but this is isolated from the rest of the operation. D-flop C goes high (after the possible oscillations from the B metastability), and this resets the set/reset latch A. At this point, we've securely captured the clock1 pulse event in the clock2 domain.

The event is passed on to d-flop D, as B goes low, and the clock2 final pulse, sig_clk_2, goes active (since d-flop D was low just before this). Finally, sig_clk_2 goes inactive as d-flop D is now high. D-flop C goes low at this same clock, but even if it hadn't, sig_clk_2 is guaranteed to be a one-clock pulse due to the AND gate which only recognizes the one clock before a high on C is passed to D.

Note that this scheme only works when the clock1 input pulses are at least two clock2 periods apart (otherwise the second clock1 pulse tries to set the set/reset latch that is already in a set state). On the other hand, two clock1 pulses that occur in a shorter span than a clock2 period is invisible to clock2 in any case (if you looked outside your house only once a day, you would think that it's always either day or night).

Phase-Locked Loops

Instead of feeding two clock domains with two clocks from external sources, it may be possible to synthesize the second clock, using the first one as a reference.

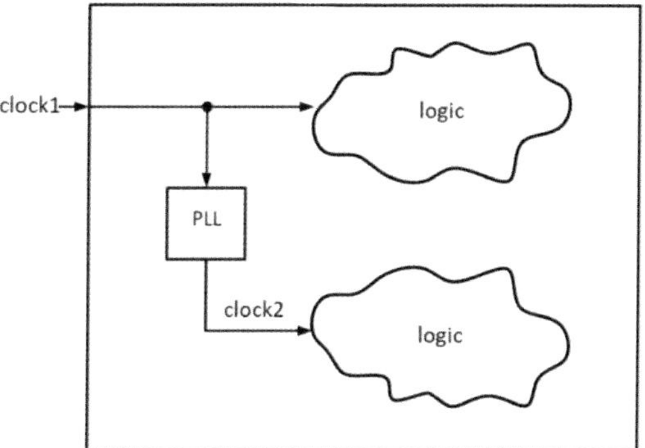

Most vendor's FPGAs include PLLs—phase-locked loops—that can create clocks whose frequencies are multiples of the source clock frequency. For example, times 2, or 10, or, many times, even by fractions, like 4/5ths. The second clock could be higher or lower in frequency, as long as it can be described as some fractional multiple of the first clock (within limits).

We could synthesize multiple clocks, each with its own fractional relationship.

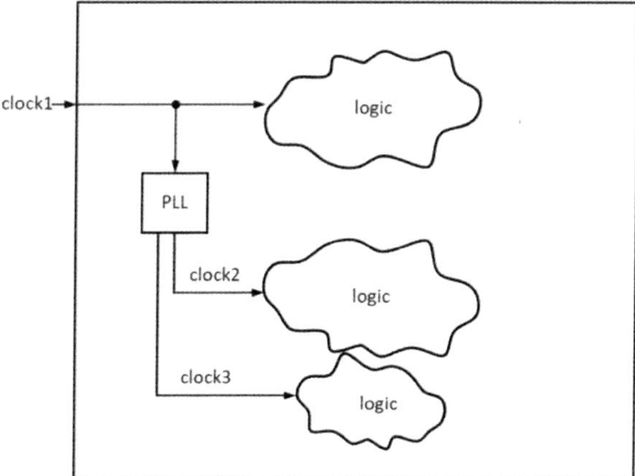

Or, we could synthesize them both with the same frequency, but different phases.

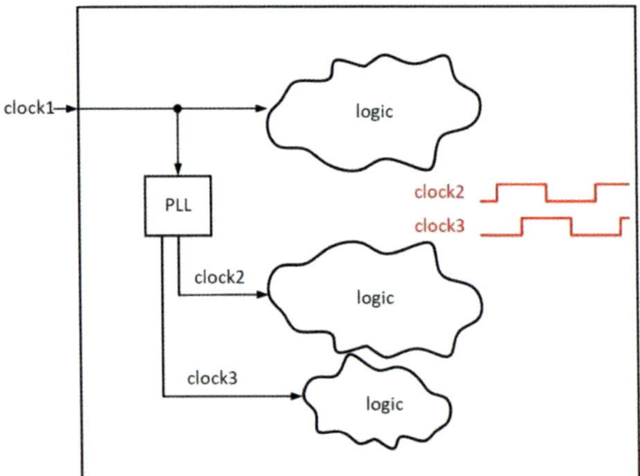

Clock synthesis can serve as a powerful tool in complex designs.

A note about timing constraints and PLLs: provided the feature is enabled, the vendor tool can often deduce the frequency of the PLL outputs, and apply that to its timing operations. This relieves you from defining the frequency of the PLL outputs in the constraints file, but you're still on the hook for any subsequent false-paths or multi-cycle declarations.

Pipelines & Filters

We often use PLLs to boost the frequency of a generic input clock—50MHz, for example—up to a higher frequency clock that we maximize for optimizing internal processing. Internal clocks may operate at 500MHz or even higher. At these frequencies, the placement and routing become critically important, and, even with the best efforts of the supremely competent tools, the timing cannot be met, meaning that the tool is not able to get the signal from one register to another in the 2ns available.

In these cases (and, really as a general rule), you, the designer, need to take steps in the design to relieve the effort. One important tool at your disposal is the pipeline.

We'll demonstrate the use of timing pipelines in a filter application. We sometimes find that we need to filter an input, i.e., we need to smooth out short-term variances. For example, say you're a manager at a retail store, and the owner calls toward the end of each shift to find out if she needs to assign more help for the floor. If you only report what you see when she calls, you are likely to give a false impression. Perhaps at that moment, the store, by pure coincidence, happens to be empty, or maybe the entire high school band happens to be sweeping through, looking for hats. Clearly, you would give your boss more of an average of the customer flow over the last hour. You would be mentally filtering the number of customers.

In the example that follows, we take a small portion of each input (1/8, in this case), and add it to a stored value that's reduced by that same small portion (7/8). The stored value is the result of the previous calculation, the output. In this way, a change in the input signal contributes a small amount to the output, but if the change is long-term, then the output will eventually (slowly) change to this new value.

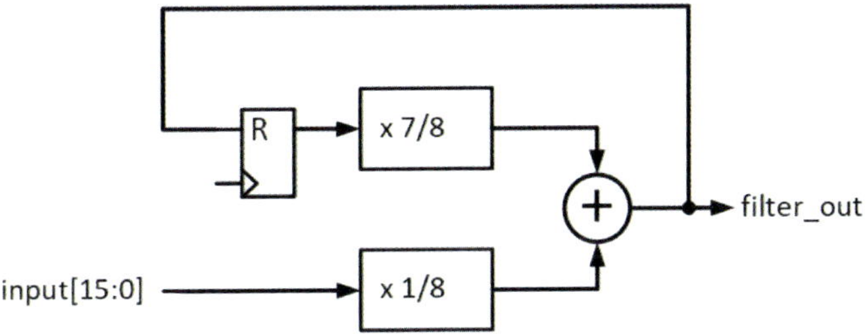

As we've recognized, division in FPGAs is difficult (imagine writing code to divide two 16-bit numbers). At least, division by arbitrary amounts is difficult. As we've seen, however, division by binary powers is easy as pie. That's why we chose fractions of eighths. Reducing a value by binary powers in logic is simply shifting to the right.

Say we have an eight-bit value of hex 0x84, or binary "10000100". We know that 0x84 is decimal 132 (8x16 + 4). If we divide 132 in half, we get 66. So, let's shift our binary number one place to the right—we get "01000010", which is 0x42, which is decimal 66 (4x16 + 2).

Therefore, we can see that reducing the input to 1/8 (dividing by 8) is simply shifting right three places ($8 = 2^3$). But, how do we achieve the stored 7/8 reduction? We stick to the binary powers. Think of 7/8 as seven eighths, literally—a quantity of seven of eighths, which, keeping with binary powers, is four eighths (1/2) plus two eighths (1/4) plus one eighth (1/8). Like so:

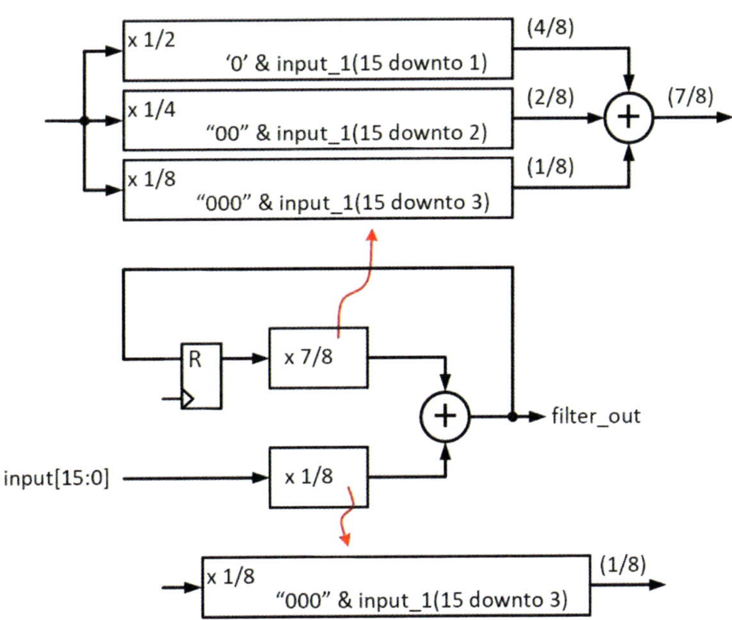

When we shift right by one place (inserting an MSB zero), we are reducing by 1/2. When we shift right by two places (inserting two MSBs zeros), we reduce by 1/4, etc..

So far, so good. However, we run into a problem if we try to operate this structure at high speeds, the kind of clock rates that push the operational envelope. We're asking the vendor's synthesis software to lay down four shifts (easy), and four additions simultaneously (not so easy—remember, the three stored-and-shifted feedback values are also added to the shifted input value).

Instead of performing the first three additions simultaneously (the ones associated with the stored feedback), we can break them up, so that only two additions are done at a time:

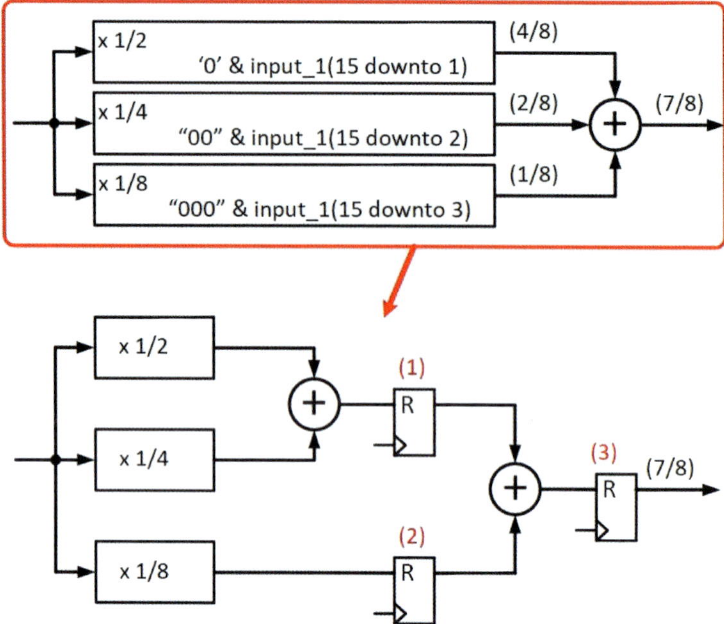

The vendor's placement and routing tools now only need to get one addition completed within one clock period prior to register (1), and another two-party addition then completed within one clock period prior to register (3). Register (2) is necessary for pipeline alignment. Without it, we'd be adding the 1/2 + 1/4 result a clock period behind the 1/8 shift—we'd be scrambling the data. Adding extra register stages for timing alignment in parallel paths is a common consequence when applying pipeline registers, often forgotten until simulation reveals gibberish results.

Here's the final design:

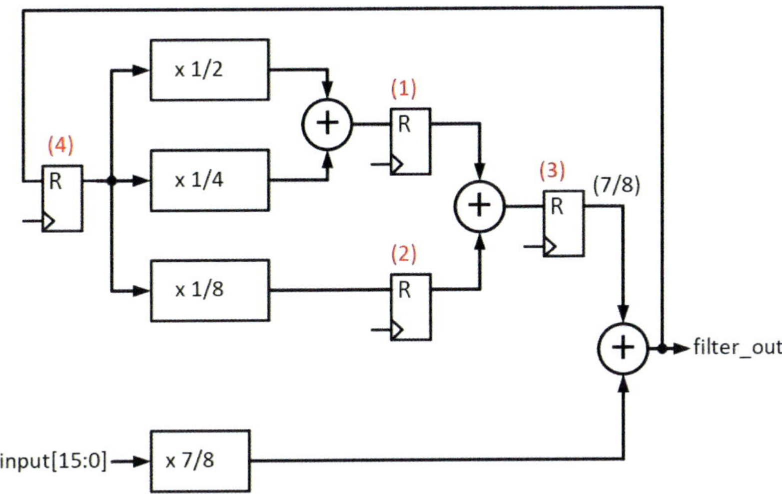

Register (4) is the original feedback storage stage, and is not algorithmically required, since both the (1)/(2) register pair and register (3) serve to store running results, but we're leaving it in place as another timing pipeline register (providing timing segregation between the output addition and the addition prior to register (1)).

We pause to note that adding timing pipeline stages requires caution beyond ensuring proper alignment (as we did with register (2)). By adding timing pipeline registers in the stored feedback path, we have altered the filter's operation somewhat.

The following figure shows the response of both the original and the pipeline-modified filters to an input that suddenly goes from zero to a large amplitude, what we call a step function, and which is often used to characterize filters. You can see that, although both filters have the same general response (a somewhat smooth climb from zero), and both eventually reach the new input amplitude, the pipeline version is visibly the ugly duckling brother. This is an example of why system engineers developing algorithms must work closely with FPGA design engineers.

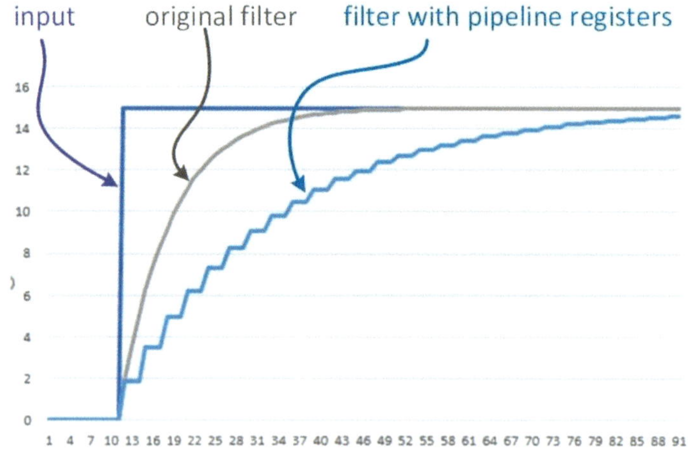

Exercises, Chapter 8

Exercise 8-1:

Modify this cross-clock domain circuit such that only pulses that are two clock_1 clocks wide on the left side, cross the domains, and that the result is then a pulse two clock_2 clocks wide on the right side.

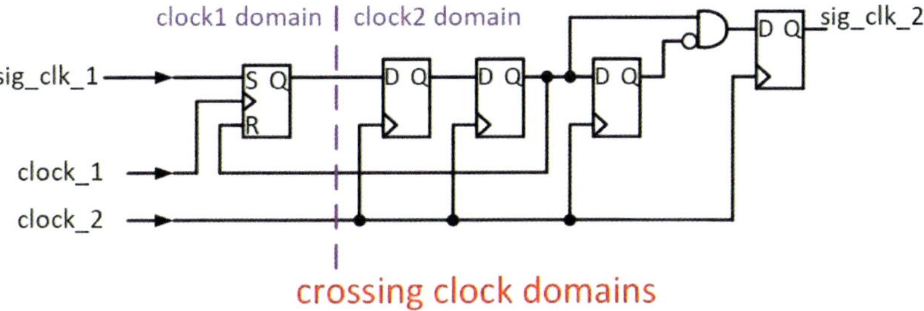

Exercise 8-2:

Assume that this counter is running at such a high clock rate that the vendor's placement/router tools cannot accommodate more than three inputs on any AND gate feeding each stage's XOR gate. Add a pipeline register to allow the tool to build this counter.

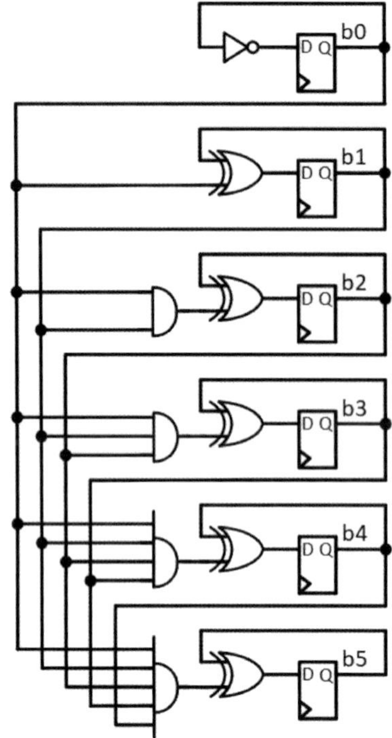

Exercise Answers, Chapter 1

A reminder that the labeling of the files in these exercises was made for the combined, single-volume, edition. Since volume 1 ended with chapter five, the exercises of this chapter (in that context) would be chapter six.

Exercise 1-1:
Correcting coding errors in Modelsim.

1) download Modelsim and a text editor (if you haven't already);
2) set up a folder for your simulation exercises;
3) download the "sim_exercise_6_1.vhdl" testbench file into your project area;
4) download the "run_6_1.do" file into your project area, and modify the "CD" line per your folder structure. Copy and paste that modified line into Modelsim;
5) run the simulation by typing "do run_6_1.do" in Modelsim, and correct the resulting errors.

The first exercise gets you set up and running with Modelsim, and introduces you to its error-reporting features, which in itself is extremely useful. Modelsim doesn't show all the errors at once, but rather doles them out.

The first pass at simulation should result in this error, at line 34.

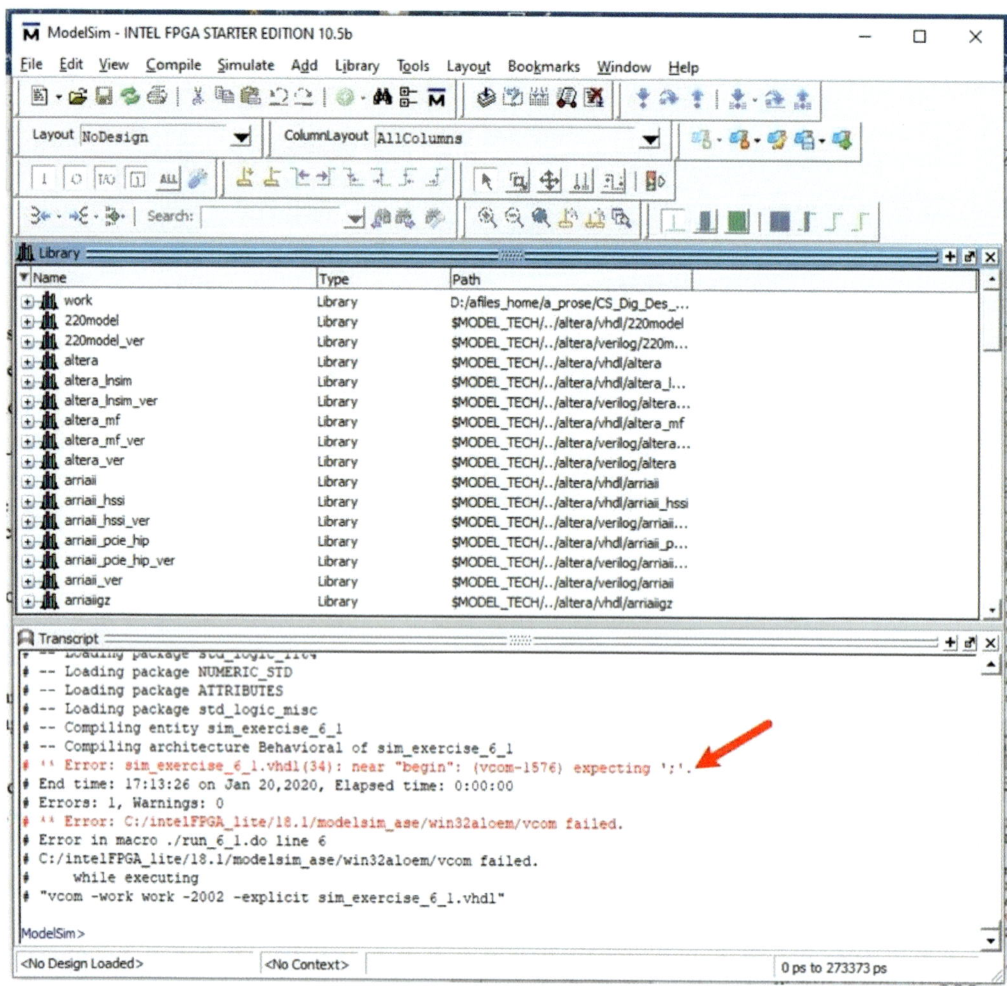

Looking at line 34, we see that it is just "begin," and that looks fine. When this happens, it often means that something tripped up the tool earlier, and, indeed, this line (32) is missing the pesky semi-colon.

```
1  -- ---------------------------------------------------------------------------
2  --                    C O P Y R I G H T   N O T I C E                       --
3  --                                                                          --
4  --                COPYRIGHT (c) 2020 Blaine C. Readler                      --
5  -- This program is provided as part of text book, and, as such, permission  --
6  -- is granted for use thereof.                                              --
7  -- ---------------------------------------------------------------------------
8  --
9  -- Create Date:     Jan. 22, 2020
10 -- Design Name:     Simulation testbench for chapter 6
11 -- Project Name:    Digital Design text
12 -- Target Devices:  unspecified
13 -- Tool versions:   unspecified
14 -- Description:     Exercise 3.6.1
15 --
16 -- Revisions: 1/22/20 -- initial version.
17 ------------------------------------------------------------------------------
18
19 library IEEE;
20 use IEEE.STD_LOGIC_1164.all;
21 use IEEE.NUMERIC_STD.all;
22 use IEEE.STD_LOGIC_MISC.all;
23
24
25 entity sim_exercise_6_1 is
26 end entity sim_exercise_6_1;
27
28 architecture Behavioral of sim_exercise_6_1 is
29
30    constant STOP         : std_logic := '1';
31    signal clock          : std_logic; -- clock
32    signal combo_sig      : std_logic_vector(7 downto 0)
33
34 begin
35
36      clock <= '0' after 10 ns when clock = '1' else
37               '1' after 10 ns;
38
39      nested_for  process
40         variable j  : integer;
41         variable k  : integer;
42      begin
43         for j in 1 to 2 loop
44            for k in 2 to 5 loop
45               combo_sig <= std_logic_vector(to_unsigned(j,4)) & std_logic_vector(to_unsigned(k,4));
46               wait until rising_edge(clock);
47            end loop;
48         end loop;
49         --
50         wait until STOP = 0;
51      end process;
52
53 end architecture Behavioral;
```

We fix that, run the simulation again, and the next error is at line 39, and Modelsim is apparently having a problem with an operator.

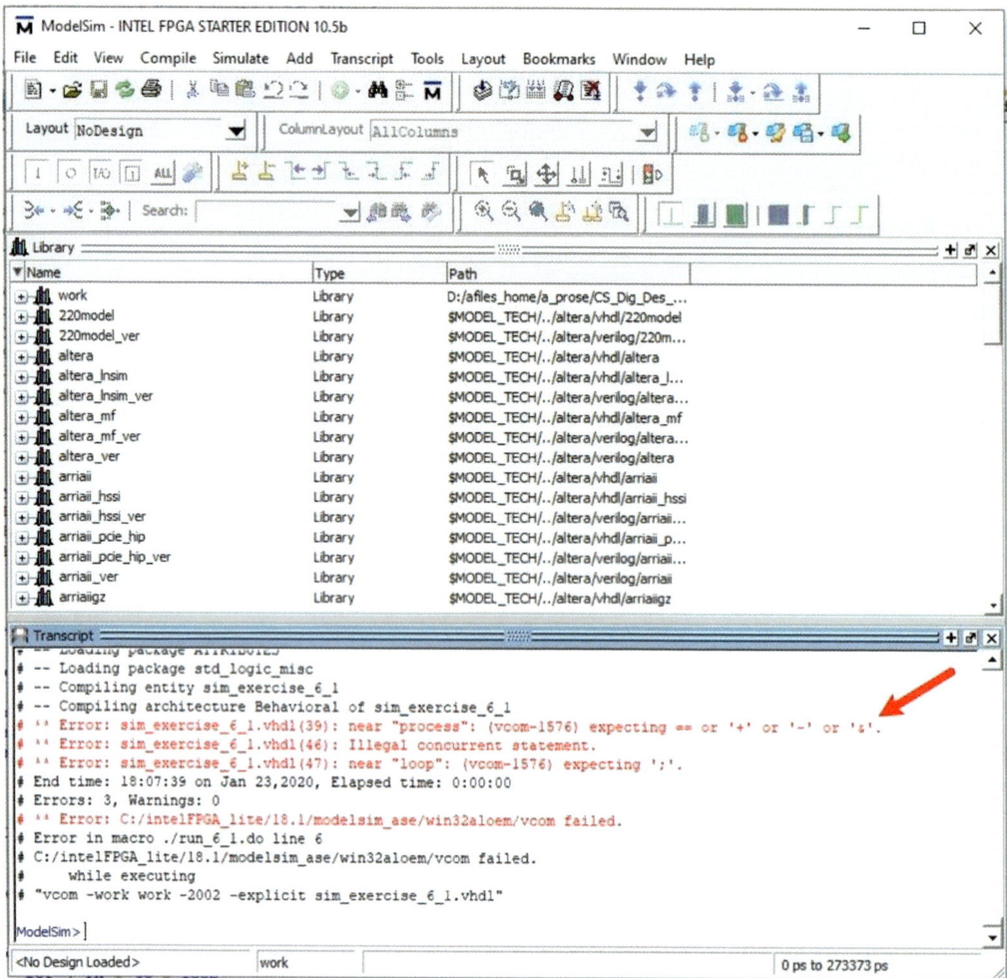

Here's line 39, the first line of a process statement.

```
1  -- ------------------------------------------------------------------------
2  --                    C O P Y R I G H T   N O T I C E                    --
3  --                                                                       --
4  --               COPYRIGHT (c) 2020 Blaine C. Readler                    --
5  -- This program is provided as part of text book, and, as such, permission --
6  -- is granted for use thereof.                                           --
7  -- ------------------------------------------------------------------------
8  --
9  -- Create Date:     Jan. 22, 2020
10 -- Design Name:     Simulation testbench for chapter 6
11 -- Project Name:    Digital Design text
12 -- Target Devices:  unspecified
13 -- Tool versions:   unspecified
14 -- Description:     Exercise 3.6.1
15 --
16 -- Revisions: 1/22/20 -- initial version.
17 ----------------------------------------------------------------------------
18
19 library IEEE;
20 use IEEE.STD_LOGIC_1164.all;
21 use IEEE.NUMERIC_STD.all;
22 use IEEE.STD_LOGIC_MISC.all;
23
24
25 entity sim_exercise_6_1 is
26 end entity sim_exercise_6_1;
27
28 architecture Behavioral of sim_exercise_6_1 is
29
30    constant STOP        : std_logic := '1';
31    signal clock         : std_logic; -- clock
32    signal combo_sig     : std_logic_vector(7 downto 0);
33
34 begin
35
36       clock <= '0' after 10 ns when clock = '1' else
37                '1' after 10 ns;
38
39       nested_for  process
40          variable j  : integer;
41          variable k  : integer;
42       begin
43          for j in 1 to 2 loop
44             for k in 2 to 5 loop
45                combo_sig <= std_logic_vector(to_unsigned(j,4)) & std_logic_vector(to_unsigned(k,4));
46                wait until rising_edge(clock);
47             end loop;
48          end loop;
49          --
50          wait until STOP = 0;
51       end process;
52
53 end architecture Behavioral;
```

Modelsim was taking a guess that line 39 needed an operator somehow, but we see immediately that it's missing the colon. We add that, and see that Modelsim is indicating that the next error is at line 46, an illegal concurrent statement …

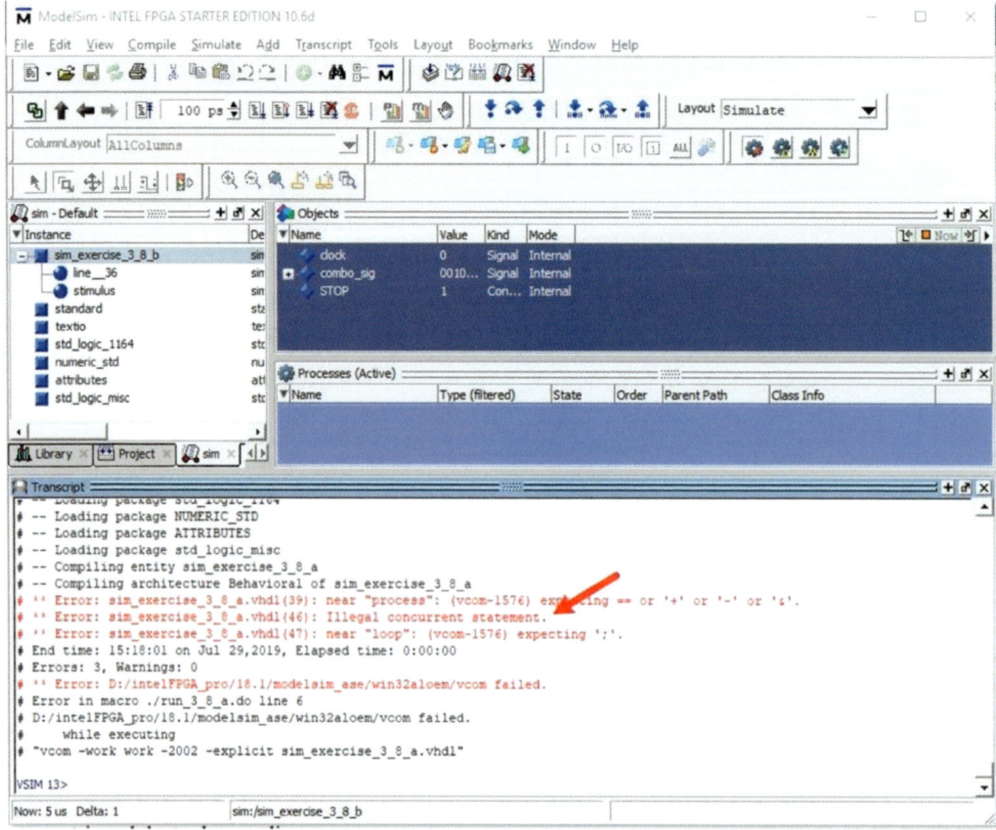

... but there's nothing apparently wrong with the "wait until" statement at that line.

We pause, and notice that the previous error was in the establishment of the process statement. If Modelsim doesn't realize that we're inside a process statement here, then this "wait until" is indeed not legal.

The lesson here is that sometimes we have to correct earlier errors before taking the next ones too seriously.

So, we run the simulation again with the fixes we've made so far, and, sure enough, the complaint about line 46 disappears, and we're left with an error at line 50.

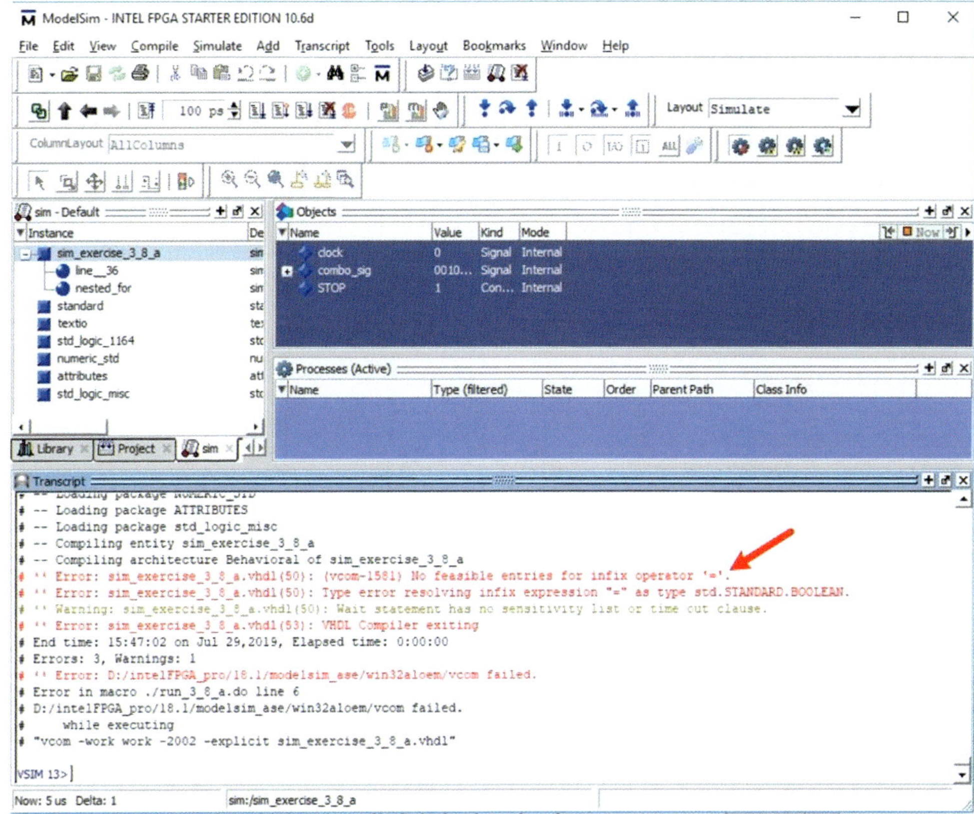

Here, Modelsim tells us exactly what the problem is. Ignoring the "infix" jargon, we see that it's complaining that we're using an equal sign to compare something that has the wrong type.

Sure enough, we've declared the constant "STOP" as std_logic, and we need quotes around the zero for that. Like so:

```
-- --------------------------------------------------------------------------
--                    C O P Y R I G H T   N O T I C E                      --
--                                                                          --
--             COPYRIGHT (c) 2020 Blaine C. Readler                         --
-- This program is provided as part of text book, and, as such, permission --
-- is granted for use thereof.                                              --
-- --------------------------------------------------------------------------
--
-- Create Date:     Jan. 22, 2020
-- Design Name:     Simulation testbench for chapter 6
-- Project Name:    Digital Design text
-- Target Devices:  unspecified
-- Tool versions:   unspecified
-- Description:     Exercise 3.6.1
--
-- Revisions: 1/22/20 -- initial version.
-----------------------------------------------------------------------------

library IEEE;
use IEEE.STD_LOGIC_1164.all;
use IEEE.NUMERIC_STD.all;
use IEEE.STD_LOGIC_MISC.all;


entity sim_exercise_6_1 is
end entity sim_exercise_6_1;

architecture Behavioral of sim_exercise_6_1 is

   constant STOP          : std_logic := '1';
   signal clock           : std_logic; -- clock
   signal combo_sig       : std_logic_vector(7 downto 0);

begin

     clock <= '0' after 10 ns when clock = '1' else
              '1' after 10 ns;

     nested_for : process
        variable j  : integer;
        variable k  : integer;
     begin
        for j in 1 to 2 loop
           for k in 2 to 5 loop
              combo_sig <= std_logic_vector(to_unsigned(j,4)) & std_logic_vector(to_unsigned(k,4));
              wait until rising_edge(clock);
           end loop;
        end loop;
        --
        wait until STOP = '0';
     end process;

end architecture Behavioral;
```

Now, when we run a simulation, Modelsim should complete with no errors.

==

Exercise 1-2:

1) download the "sim_exercise_6_2.vhdl" testbench file into your project area. Your "sim_exercise_6_1.vhdl" file from the previous step should now look like this one (after your corrections).

2) download the "run_6_2.do" into your project area, modify the first line, and run it (type "do run_6_2.do" – note that you'll have to type "quit -sim" first to exit the previous simulation).

3) open the waveform window in Modelsim, and add the "clock" and "combo_sig" signals. Save the "wave.do" file!

4) using a text editor, remove the "#" comment before the "do wave.do".

5) re-run the simulation (run the "do run_6_2.do" script).

6) expand the "combo_sig" signal in the waveform window via this "+":

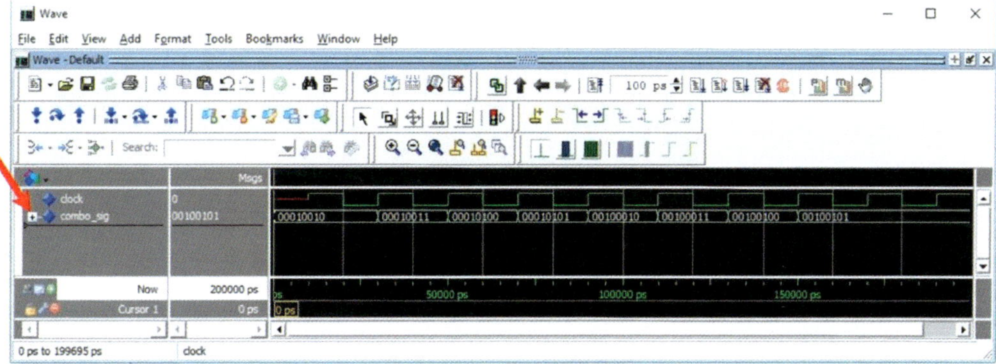

7) your waveform should look like this:

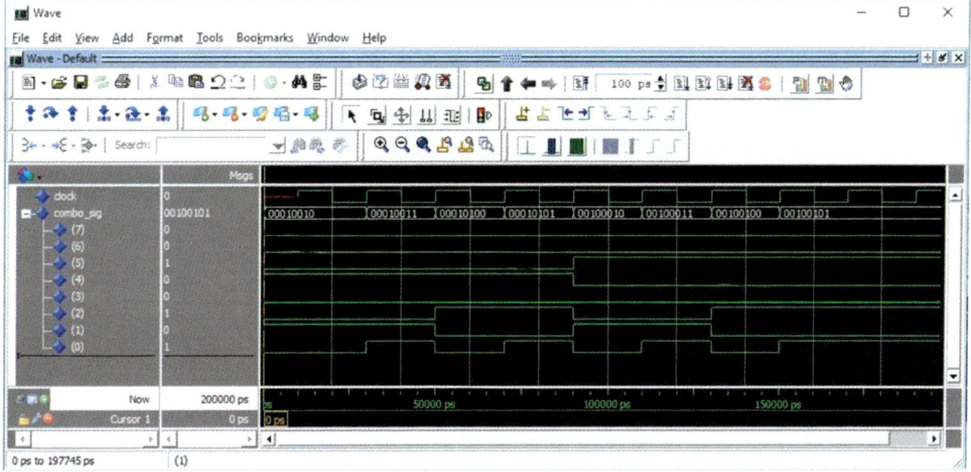

In this exercise, we use Modlesim's waveform window to observe the results of the previous corrected code, once we've corrected the original mistakes.

There are a few new points in the code to review. First, we'll look at this long assignment statement that introduces two new operations.

```
-- ---------------------------------------------------------------------------
--                    C O P Y R I G H T   N O T I C E                       --
--                                                                           --
--              COPYRIGHT (c) 2020 Blaine C. Readler                        --
-- This program is provided as part of text book, and, as such, permission --
-- is granted for use thereof.                                              --
-- ---------------------------------------------------------------------------
--
-- Create Date:      Jan. 22, 2020
-- Design Name:      Simulation testbench for chapter 6
-- Project Name:     Digital Design text
-- Target Devices:   unspecified
-- Tool versions:    unspecified
-- Description:      Exercise 3.6.2
--
-- Revisions: 1/22/20 -- initial version.
------------------------------------------------------------------------------


library IEEE;
use IEEE.STD_LOGIC_1164.all;
use IEEE.NUMERIC_STD.all;
use IEEE.STD_LOGIC_MISC.all;


entity sim_exercise_6_2 is
end entity sim_exercise_6_2;

architecture Behavioral of sim_exercise_6_2 is

   constant STOP        : std_logic := '1';
   signal clock         : std_logic; -- clock
   signal combo_sig     : std_logic_vector(7 downto 0);

begin

      clock <= '0' after 10 ns when clock = '1' else
               '1' after 10 ns;

      nested_for : process
         variable j  : integer;
         variable k  : integer;
      begin
         for j in 1 to 2 loop
            for k in 2 to 5 loop
               combo_sig <= std_logic_vector(to_unsigned(j,4)) & std_logic_vector(to_unsigned(k,4));
               wait until rising_edge(clock);
            end loop;
         end loop;
         --
         wait until STOP = '0';
      end process;

end architecture Behavioral;
```

The first of these simply converts the variables "j" and "k" to standard logic vector values. It's a two-step process. The first step is to convert the variable to an unsigned number. In order to do that, we have to tell the tool how many bits to use.

```
combo_sig <= std_logic_vector(to_unsigned(j,4)) & std_logic_vector(to_unsigned(k,4));
```

In this case, we use 4, since neither variable is larger than 15—the maximum value possible with four bits—and 4 is a convenient amount, since it can be represented as a single hex number.

The resulting 4-bit unsigned numbers are then converted to standard-logic-vectors. This leaves us with two 4-bit numbers.

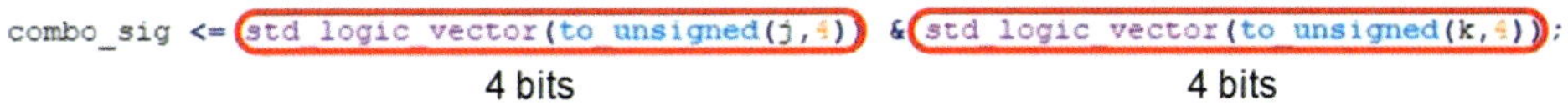

```
combo_sig <= std_logic_vector(to_unsigned(j,4)) &std_logic_vector(to_unsigned(k,4));
```

4 bits   4 bits

Notice that the "combo_sig" signal is 8 bits, as it must be, since this ampersand symbol means concatenation, which in turn means simply that we join the two 4-bit values into a single 8-bit byte.

```
-- -----------------------------------------------------------------------
--                   C O P Y R I G H T   N O T I C E                    --
--                                                                      --
--              COPYRIGHT (c) 2020 Blaine C. Readler                    --
-- This program is provided as part of text book, and, as such, permission --
-- is granted for use thereof.                                          --
-- -----------------------------------------------------------------------
--
-- Create Date:      Jan. 22, 2020
-- Design Name:      Simulation testbench for chapter 6
-- Project Name:     Digital Design text
-- Target Devices:   unspecified
-- Tool versions:    unspecified
-- Description:      Exercise 3.6.2
--
-- Revisions: 1/22/20 -- initial version.
--------------------------------------------------------------------------

library IEEE;
use IEEE.STD_LOGIC_1164.all;
use IEEE.NUMERIC_STD.all;
use IEEE.STD_LOGIC_MISC.all;


entity sim_exercise_6_2 is
end entity sim_exercise_6_2;

architecture Behavioral of sim_exercise_6_2 is

   constant STOP        : std_logic := '1';
   signal clock         : std_logic; -- clock
   signal combo_sig     : std_logic_vector(7 downto 0);

begin

     clock <= '0' after 10 ns when clock = '1' else
              '1' after 10 ns;

     nested_for : process
        variable j  : integer;
        variable k  : integer;
     begin
        for j in 1 to 2 loop
           for k in 2 to 5 loop
              combo_sig <= std_logic_vector(to_unsigned(j,4)) & std_logic_vector(to_unsigned(k,4));
              wait until rising_edge(clock);
           end loop;
        end loop;
        --
        wait until STOP = '0';
     end process;

end architecture Behavioral;
```

concatenation

For example, going back to the waveform, the Modelsim simulator indeed shows “combo_sig” as an 8-bit value.

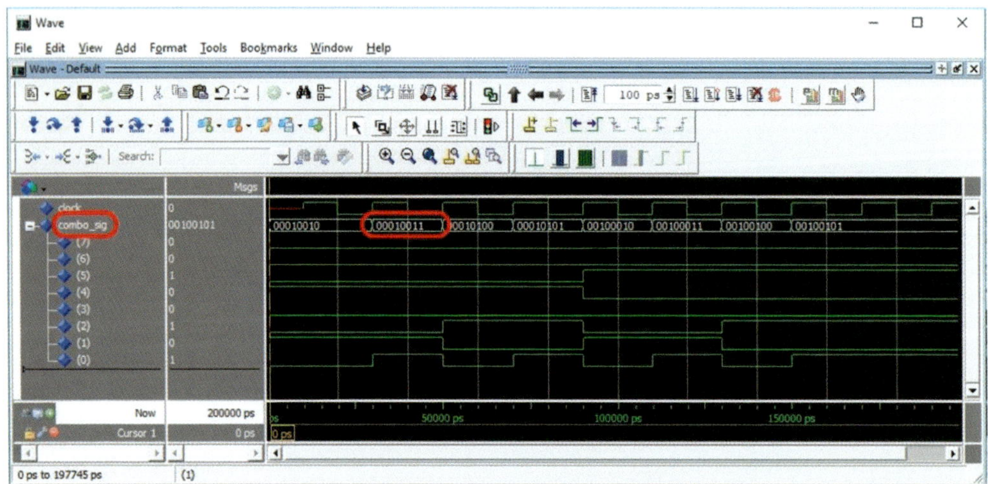

Taking a closer look at one of the values, this is “j”:

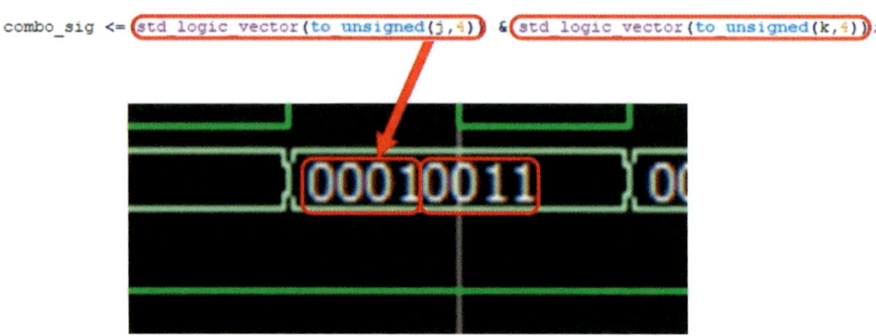

and this is “k”:

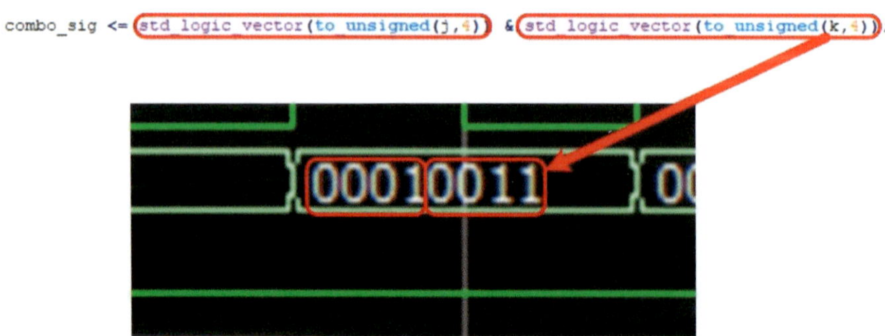

We’ve combined them together. Note that we have not added them. If we had, the result would have been one plus three, but by concatenating them, the result is nineteen—sixteen plus three.

Next, we introduce nested for-loops.

```
-- ----------------------------------------------------------------------------
--                    C O P Y R I G H T   N O T I C E                        --
--                                                                           --
--                COPYRIGHT (c) 2020 Blaine C. Readler                       --
-- This program is provided as part of text book, and, as such, permission  --
-- is granted for use thereof.                                               --
-- ----------------------------------------------------------------------------
--
-- Create Date:     Jan. 22, 2020
-- Design Name:     Simulation testbench for chapter 6
-- Project Name:    Digital Design text
-- Target Devices:  unspecified
-- Tool versions:   unspecified
-- Description:     Exercise 3.6.2
--
-- Revisions: 1/22/20 -- initial version.
-------------------------------------------------------------------------------

library IEEE;
use IEEE.STD_LOGIC_1164.all;
use IEEE.NUMERIC_STD.all;
use IEEE.STD_LOGIC_MISC.all;


entity sim_exercise_6_2 is
end entity sim_exercise_6_2;

architecture Behavioral of sim_exercise_6_2 is

   constant STOP         : std_logic := '1';
   signal clock          : std_logic; -- clock
   signal combo_sig      : std_logic_vector(7 downto 0);

begin

      clock <= '0' after 10 ns when clock = '1' else
               '1' after 10 ns;

      nested_for : process
         variable j  : integer;
         variable k  : integer;
      begin
         for j in 1 to 2 loop
            for k in 2 to 5 loop
               combo_sig <= std_logic_vector(to_unsigned(j,4)) & std_logic_vector(to_unsigned(k,4));
               wait until rising_edge(clock);
            end loop;
         end loop;

         wait until STOP = '0';
      end process;

end architecture Behavioral;
```

The idea is straight-forward. For every increment of "J", "k" increments from 2 through 5.

```
nested_for : process
   variable j  : integer;
   variable k  : integer;
begin
   for j in 1 to 2 loop
      for k in 2 to 5 loop
         combo_sig <= std_logic_vector(to_unsign
         wait until rising_edge(clock);
      end loop;
   end loop;
   --
```

Once "k" is done with 5, "J" increments, and "K" starts over.

The values for “j” and “k” look like this.

j	k
1	2
1	3
1	4
1	5
2	2
2	3
2	4
2	5

Once “j” reaches 2, and “k” reaches 5, the nested loop is done, and the code proceeds, in this case, it halts at the “stop” constant.

Here’s how the concatenated nested for-loop variables appear in the waveform.

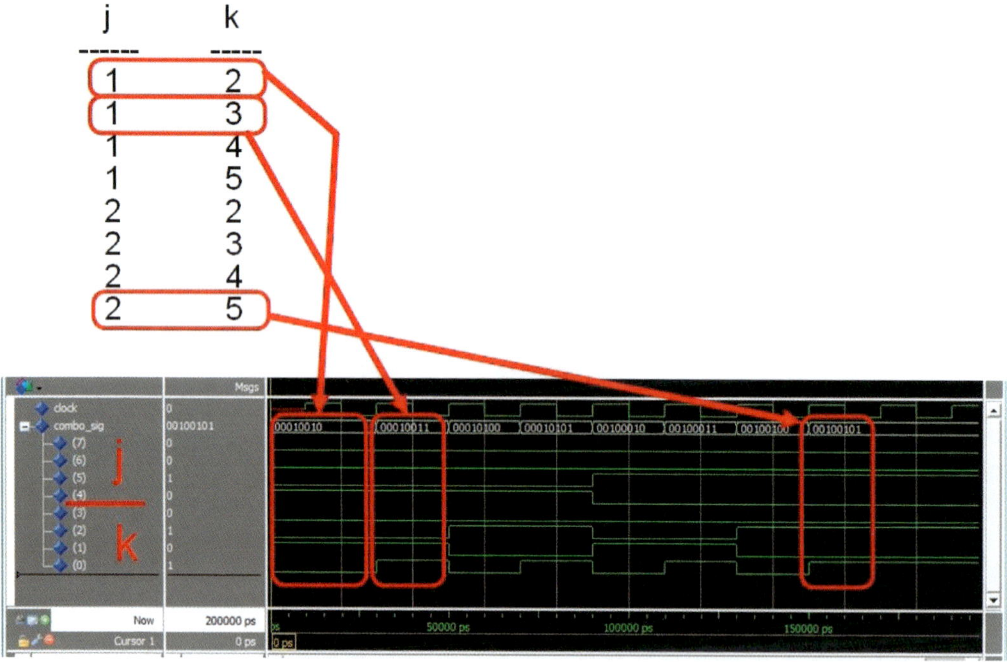

==

Exercise 1-3:

1) download the “pulse_gen_6_3.vhdl” file into your project area. This module creates a pulse every three clocks, but only when enabled.

2) copy the “sim_exercise_6_2.vhdl” file in your project area, and rename it “sim_exercise_6_3.vhdl”.

3) copy the “do run_6_2.do” file in your project area, rename it “run_6_3.do”, and modify it to include both the renamed “sim_exercise_6_3.vhdl” testbench, as well as the downloaded “pulse_gen_6_3.vhdl” file. Set it to run for 400 nanoseconds.

4) instantiate the downloaded "pulse_gen_6_3.vhdl" file in the new testbench, and modify the testbench to create an enable signal for the instantiated module that goes active for 7 clocks, and inactive for 2 clocks, repeatedly. The testbench should be able to run for at least two full cycles of the 7-clock/2-clock repetition.

5) your resulting waveform should look something like this.

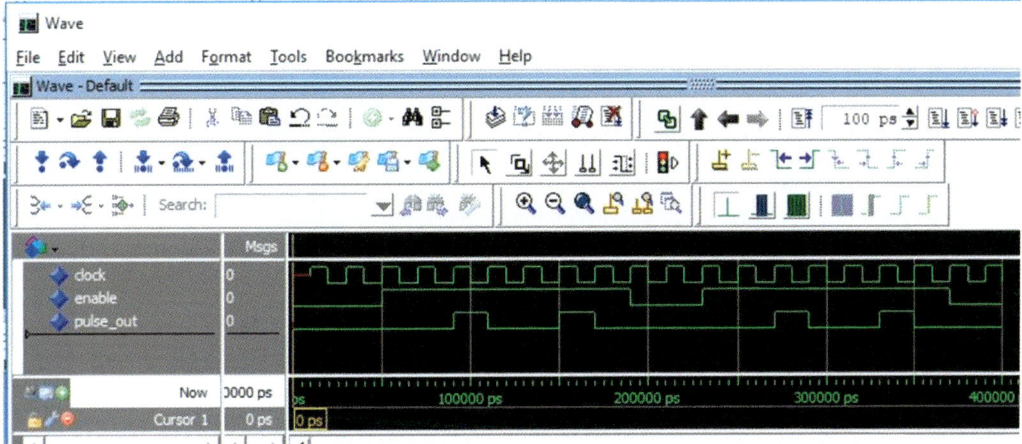

For the third exercise, we'll explore a number of ways to build the testbench that create something like this waveform.

The first method implemented in the file "sim_exercise_6_3_a" may be the most obvious approach based on the testbench construction we learned in this chapter. Let's take a look at the process statement.

```
entity sim_exercise_6_3_a is
end entity sim_exercise_6_3_a;

architecture Behavioral of sim_exercise_6_3_a is

component pulse_gen_6_3 is
port (
        clock       : in   std_logic;
        enable      : in   std_logic;
        pulse_out   : out  std_logic
    );
end component;

    constant STOP         : std_logic := '1';
    signal clock          : std_logic;
    signal enable         : std_logic := '0';
    signal pulse_out      : std_logic;

begin

    clock <= '0' after 10 ns when clock = '1' else
             '1' after 10 ns;

    exercise_6_3_a : process
        variable m  : integer;
        variable n  : integer;
    begin
        enable <= '0';
        wait until rising_edge(clock);
        wait until rising_edge(clock);
        for m in 1 to 2 loop
            enable <= '1';
            for n in 1 to 7 loop
                wait until rising_edge(clock);
            end loop;
            --
            enable <= '0';
            for n in 1 to 2 loop
                wait until rising_edge(clock);
            end loop;
        end loop;
        --
        wait until STOP = '0';
    end process;

    pulse_gen_3_8_i :  pulse_gen_6_3
    port map
      (
        clock       => clock,
        enable      => enable,
        pulse_out   => pulse_out
      );

end architecture Behavioral;
```

This outer for-loop,

```
exercise_6_3_a : process
   variable m  : integer;
   variable n  : integer;
begin
   enable <= '0';
   wait until rising_edge(clock);
   wait until rising_edge(clock);
   for m in 1 to 2 loop
      enable <= '1';
      for n in 1 to 7 loop
         wait until rising_edge(clock);
      end loop;
      --
      enable <= '0';
      for n in 1 to 2 loop
         wait until rising_edge(clock);
      end loop;
   end loop;
   --
   wait until STOP = '0';
end process;
```

corresponds with this part of the waveform.

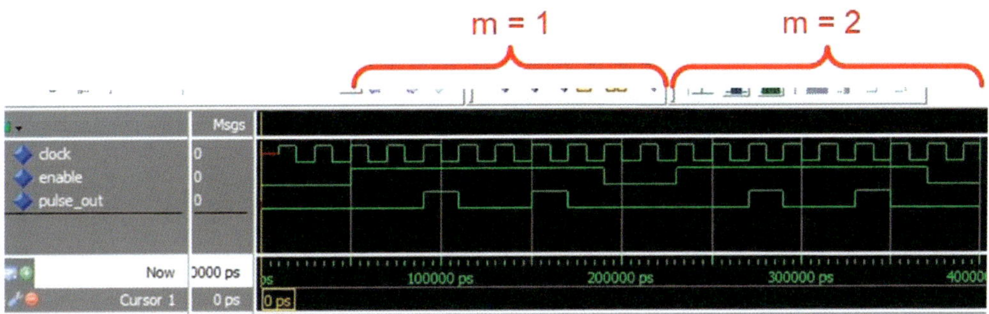

and these inner loops, to these parts of the waveform.

```
exercise_6_3_a : process
   variable m  : integer;
   variable n  : integer;
begin
   enable <= '0';
   wait until rising_edge(clock);
   wait until rising_edge(clock);
   for m in 1 to 2 loop
      enable <= '1';
      for n in 1 to 7 loop
         wait until rising_edge(clock);
      end loop;
      --
      enable <= '0';
      for n in 1 to 2 loop
         wait until rising_edge(clock);
      end loop;
   end loop;
   --
   wait until STOP = '0';
end process;
```

I added these two initial clocks just so that the outer loop would end at 400 ns.

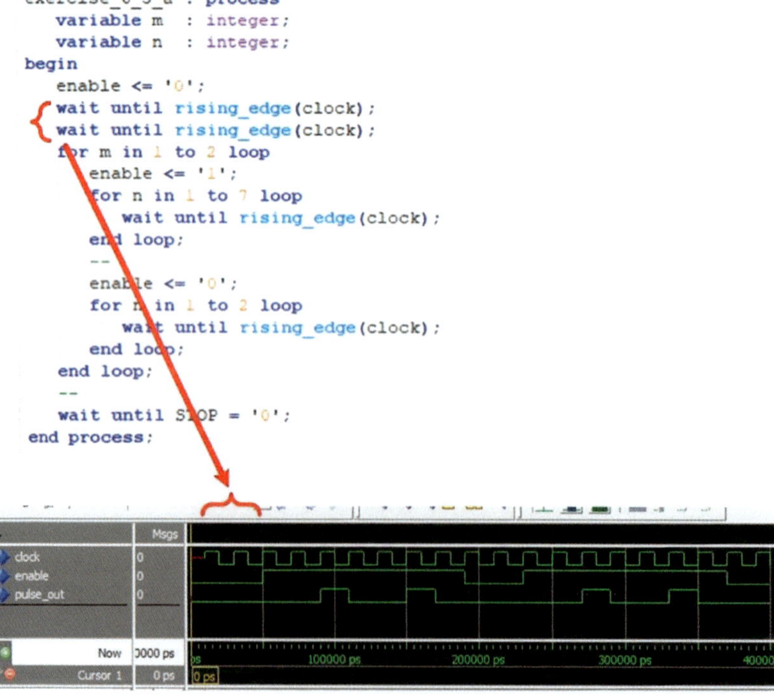

```
exercise_6_3_a : process
   variable m  : integer;
   variable n  : integer;
begin
   enable <= '0';
   wait until rising_edge(clock);
   wait until rising_edge(clock);
   for m in 1 to 2 loop
      enable <= '1';
      for n in 1 to 7 loop
         wait until rising_edge(clock);
      end loop;
      --
      enable <= '0';
      for n in 1 to 2 loop
         wait until rising_edge(clock);
      end loop;
   end loop;
   --
   wait until STOP = '0';
end process;
```

But, here's an alternative way to create the repeating cycles of the enable signal, implemented in the file "sim_exercise_6_3_b".

```
entity sim_exercise_6_3_b is
end entity sim_exercise_6_3_b;

architecture Behavioral of sim_exercise_6_3_b is

component pulse_gen_6_3 is
port (
        clock       : in    std_logic;
        enable      : in    std_logic;
        pulse_out   : out   std_logic
     );
end component;

   constant STOP          : std_logic := '1';
   signal clock           : std_logic;
   signal enable          : std_logic := '0';
   signal pulse_out       : std_logic;

begin

     clock <= '0' after 10 ns when clock = '1' else
              '1' after 10 ns;

     exercise_6_3_b : process
        variable n   : integer;
     begin
        enable <= '0';
        for n in 1 to 2 loop
           wait until rising_edge(clock);
        end loop;
        --
        enable <= '1';
        for n in 1 to 7 loop
           wait until rising_edge(clock);
        end loop;
        --
     end process;

   pulse_gen_3_8_i :  pulse_gen_6_3
   port map
     (
        clock       => clock,
        enable      => enable,
        pulse_out   => pulse_out
     );

end architecture Behavioral;
```

The changes are all inside the process statement. First, I've removed the process stall via the STOP constant.

```
exercise_6_3_b : process
   variable n   : integer;
begin
   enable <= '0';
   for n in 1 to 2 loop
      wait until rising_edge(clock);
   end loop;
   --
   enable <= '1';
   for n in 1 to 7 loop
      wait until rising_edge(clock);
   end loop;
   --
end process;
```

You will remember that without that stall, the process statement starts over again each time it finishes. This achieves our requirement to repeat the enable on/off cycles.

The other change is that I swapped the "on" and "off" periods of the enable.

```
exercise_6_3_b : process
   variable n   : integer;
begin
   enable <= '0';
   for n in 1 to 2 loop
      wait until rising_edge(clock);
   end loop;
   --
   enable <= '1';
   for n in 1 to 7 loop
      wait until rising_edge(clock);
   end loop;
   --
end process;
```

This was just a convenience so that the waveform would start off the same as with the other method.

Let's look at what I did in the pulse generator module—the module under test that we instantiated in the testbench.

```
entity pulse_gen_6_3 is
port (
            clock       : in   std_logic;
            enable      : in   std_logic;
            pulse_out   : out  std_logic
          );
end entity;

architecture behavioral of pulse_gen_6_3 is

   signal pulse_cnt      : unsigned(3 downto 0);
   signal pulse_out_lcl  : std_logic := '0';

begin

   ex_3_8 : process(clock)
   begin
      if rising_edge(clock) then
         if (enable = '1') then
            if ( pulse_cnt = X"3" ) then
               pulse_cnt      <= X"1";
               pulse_out_lcl  <= '1';
            else
               pulse_cnt      <= pulse_cnt + 1;
               pulse_out_lcl  <= '0';
            end if;
         else
            pulse_cnt      <= X"2";
            pulse_out_lcl  <= '0';
         end if;
      end if;
   end process;

   pulse_out <= pulse_out_lcl;

end architecture behavioral;
```

This is the enabled operation of the counter. It rolls over to 1 when it reaches 3, creating a pulse every 3 clock cycles (when the count is 3).

```
entity pulse_gen_6_3 is
port (
            clock       : in   std_logic;
            enable      : in   std_logic;
            pulse_out   : out  std_logic
          );
end entity;

architecture behavioral of pulse_gen_6_3 is

    signal pulse_cnt     : unsigned(3 downto 0);
    signal pulse_out_lcl : std_logic := '0';

begin

    ex_3_8 : process(clock)
    begin
        if rising_edge(clock) then
            if (enable = '1') then
                if ( pulse_cnt = X"3" ) then
                    pulse_cnt       <= X"1";
                    pulse_out_lcl   <= '1';
                else
                    pulse_cnt       <= pulse_cnt + 1;
                    pulse_out_lcl   <= '0';
                end if;
            else
                pulse_cnt       <= X"2";
                pulse_out_lcl   <= '0';
            end if;
        end if;
    end process;

    pulse_out <= pulse_out_lcl;

end architecture behavioral;
```

roll over

pulse

Notice what happens when the enable is low:

```
entity pulse_gen_6_3 is
port (
                clock          : in   std_logic;
                enable         : in   std_logic;
                pulse_out      : out  std_logic
            );
end entity;

architecture behavioral of pulse_gen_6_3 is

    signal pulse_cnt       : unsigned(3 downto 0);
    signal pulse_out_lcl : std_logic := '0';

begin

    ex_3_8 : process(clock)
    begin
        if rising_edge(clock) then
            if (enable = '1') then
                if ( pulse_cnt = X"3" ) then
                    pulse_cnt       <= X"1";
                    pulse_out_lcl   <= '1';
                else
                    pulse_cnt       <= pulse_cnt + 1;
                    pulse_out_lcl   <= '0';
                end if;
            else
                pulse_cnt       <= X"2";
                pulse_out_lcl   <= '0';
            end if;
        end if;
    end process;

    pulse_out <= pulse_out_lcl;

end architecture behavioral;
```

We force the pulse low. This way, if the enable happens to go low when the count is at three, the pulse would not be stuck "on." We also force the counter to two. That way, the next time the enable goes active again, we won't have to wait until the count reaches three for the first pulse to occur.

Finally, notice that I created a local version of the pulse_out signal so that I can force the signal to zero at the very beginning—this is just for simulation.

```
entity pulse_gen_6_3 is
port (
            clock        : in   std_logic;
            enable       : in   std_logic;
            pulse_out    : out  std_logic
        );
end entity;

architecture behavioral of pulse_gen_6_3 is

    signal pulse_cnt      : unsigned(3 downto 0);
    signal pulse_out_lcl  : std_logic := '0';

begin

    ex_3_8 : process(clock)
    begin
        if rising_edge(clock) then
            if (enable = '1') then
                if ( pulse_cnt = X"3" ) then
                    pulse_cnt       <= X"1";
                    pulse_out_lcl   <= '1';
                else
                    pulse_cnt       <= pulse_cnt + 1;
                    pulse_out_lcl   <= '0';
                end if;
            else
                pulse_cnt       <= X"2";
                pulse_out_lcl   <= '0';
            end if;
        end if;
    end process;

    pulse_out <= pulse_out_lcl;

end architecture behavioral;
```

==

Exercise 1-4:

1) copy the “sim_exercise_6_3.vhdl” file in your project area, and rename it “sim_exercise_6_4.vhdl”.

2) redesign the new “sim_exercise_6_4.vhdl” to use a clocked process statement instead of the existing one.

3) the resulting waveform should look like that from exercise 6-3.

The last exercise for this chapter demonstrates that a testbench is not limited to a linear-progressing process statement with “wait until” statements, but can be constructed with clocked process statements like we’ve been using. As in the previous exercise, there are a number of ways for implementing this. Here’s one, using a counter and a case statement.

```
architecture Behavioral of sim_exercise_6_4 is

component pulse_gen_6_3 is
port (
         clock        : in    std_logic;
         enable       : in    std_logic;
         pulse_out    : out   std_logic
      );
end component;

      signal clock              : std_logic;
      signal counter            : unsigned(3 downto 0) := X"0";
      signal enable             : std_logic := '0';
      signal pulse_out          : std_logic;

begin

      clock <= '0' after 10 ns when clock = '1' else
               '1' after 10 ns;

      exercise_3_8c : process(clock)
      begin
         if rising_edge(clock) then

            if (counter = X"9") then
               counter <= X"1";
            else
               counter <= counter + 1;
            end if;

            case (counter) is
               when X"1"      => enable <= '1';
               when X"2"      => enable <= '1';
               when X"3"      => enable <= '1';
               when X"4"      => enable <= '1';
               when X"5"      => enable <= '1';
               when X"6"      => enable <= '1';
               when X"7"      => enable <= '1';
               when X"8"      => enable <= '0';
               when X"9"      => enable <= '0';
               when others => null;
            end case;

         end if;
      end process;

      pulse_gen_3_8_i :  pulse_gen_6_3
      port map
         (
            clock        => clock,
            enable       => enable,
            pulse_out    => pulse_out
         );

end architecture Behavioral;
```

We're already familiar with the counter, where the per-clock incrementing rolls back to 1 when it reaches 9. We're also familiar with using a case statement to decode each count of the counter. Seven of the counts set the enable to one, and two counts set the enable to zero. The resulting waveform is the same as exercise three.

Here's another approach.

```
architecture Behavioral of sim_exercise_6_4 is

component pulse_gen_6_3 is
port (
        clock       : in   std_logic;
        enable      : in   std_logic;
        pulse_out   : out  std_logic
     );
end component;

     signal clock           : std_logic;
     signal counter         : unsigned(3 downto 0) := X"0";
     signal enable          : std_logic := '0';
     signal pulse_out       : std_logic;

begin

        clock <= '0' after 10 ns when clock = '1' else
                 '1' after 10 ns;

        exercise_3_8c : process(clock)
        begin
           if rising_edge(clock) then
              if (counter = X"9") then
                 counter <= X"1";
              else
                 counter <= counter + 1;
              end if;

              if (     counter >= X"1"
                   and counter <= X"7"
                 ) then
                 enable <= '1';
              else
                 enable <= '0';
              end if;

           end if;
        end process;

     pulse_gen_3_8_i :  pulse_gen_6_3
     port map
        (
         clock       => clock,
         enable      => enable,
         pulse_out   => pulse_out
        );

end architecture Behavioral;
```

counter

bounded IFs

It has the same counter—this would be common to most approaches—and instead of a case statement, we have one IF statement, where we set the enable to one when the counter is greater-than-or-equal-to one, and also less-than-or-equal-to seven. Otherwise, the enable is zero. This works the same as the case statement, and produces the same waveform.

Notice how we indicate both the "greater-than-or-equal-to" and "less-than-or-equal-to" conditions—just as they read.

greater-than-or-equal-to

```
if (     counter >= X"1"
    and counter <= X"7"
   ) then
   enable <= '1';
else
   enable <= '0';
end if;
```

less-than-or-equal-to

bounded IFs

Even though the "less-than-or-equal-to" looks just like the assignment indication ("less-than" followed by "equal"), the simulator and compiler understand that this is inside an IF condition test.

In summary, control with bounded IFs are convenient for simple conditions, and case statements are practical for more complex ones.

Exercise Answers, Chapter 2

Exercise 2-1:
Given this read-before-write dual-port memory, fill in the rd_data_a and rd_data_b signals in the timing diagram.

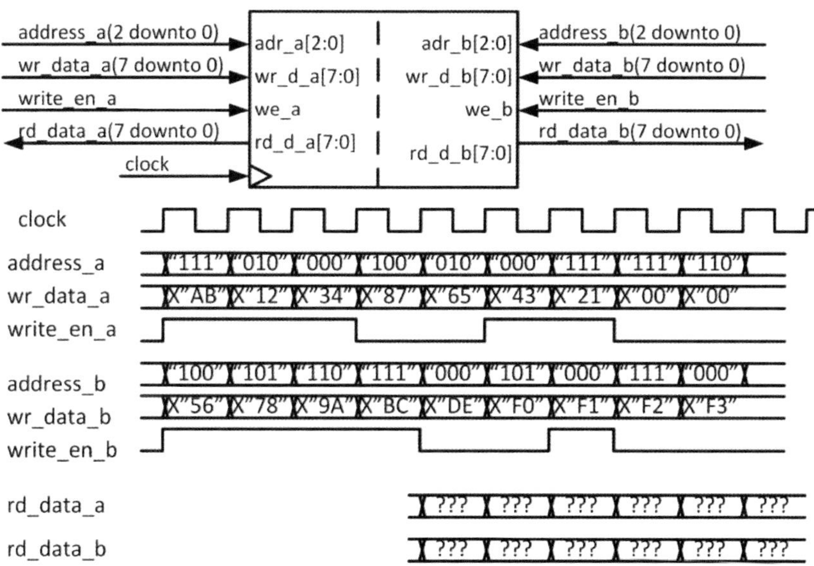

In the first exercise, we explore how a dual-port memory works. At the beginning of the write sequence, we write a hex 56 to the B port address "100", which we then read out from the A port three clocks later.

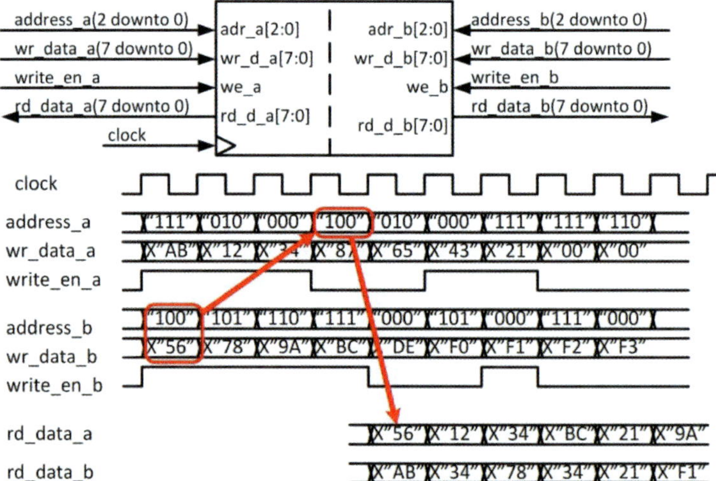

On the A port, we write a hex AB to address "111", and three clocks later, read it out on port B.

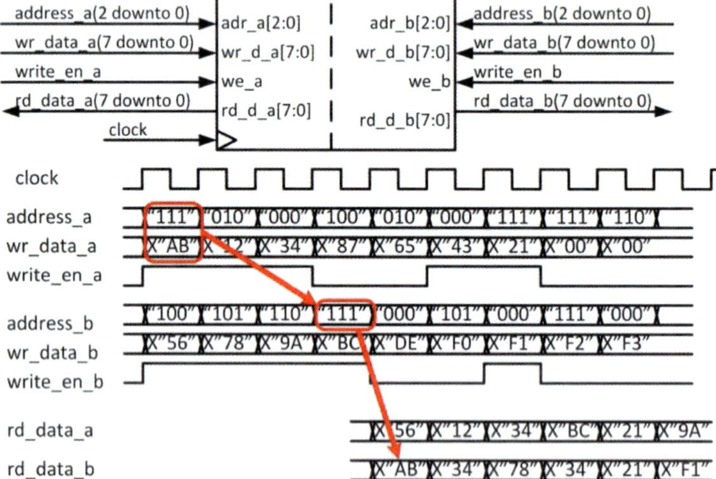

Moving along, we write hex 12 to Port A address “010”, and read it out on port A, write hex 34 to port A address “000”, and read it out on Port B.

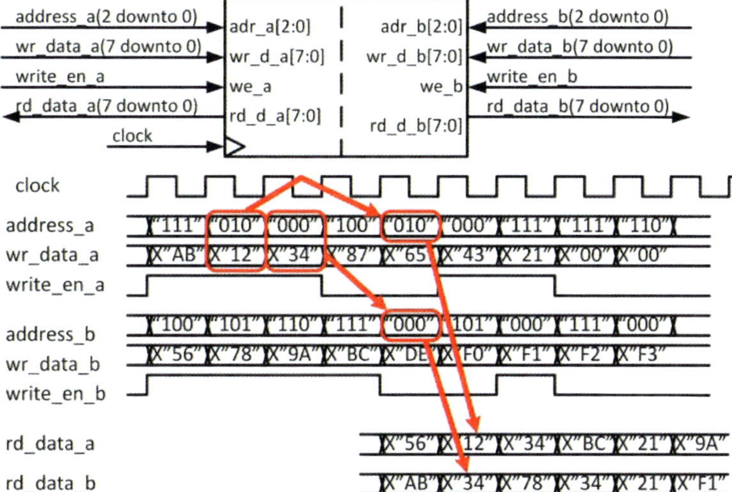

The hex 34 that we wrote to port A address “000”, we also read out on port A, and the hex 78 that we wrote to port B address “101” we read out on port B.

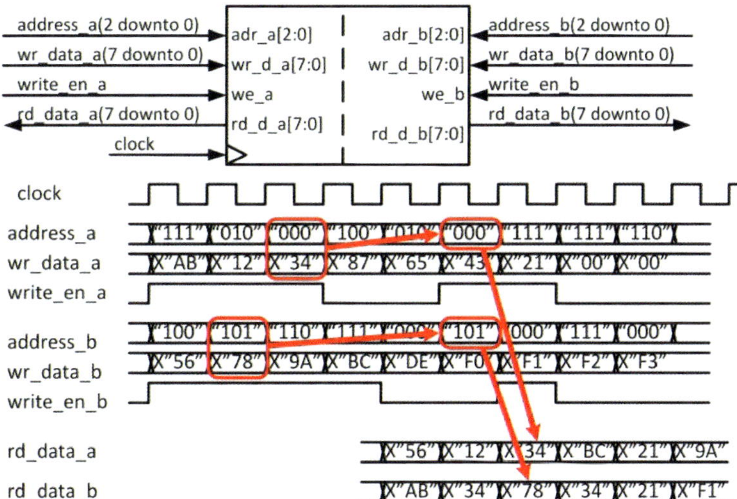

Here we see how the hex AB we first wrote to port A address "111", we later overwrite with hex BC on port B. We then read the new hex BC out on the A port.

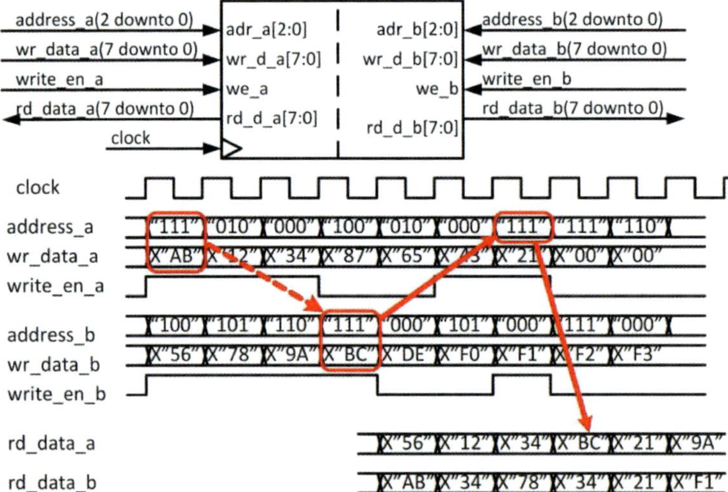

We read the hex 34 on port B that we wrote on port A address "000".

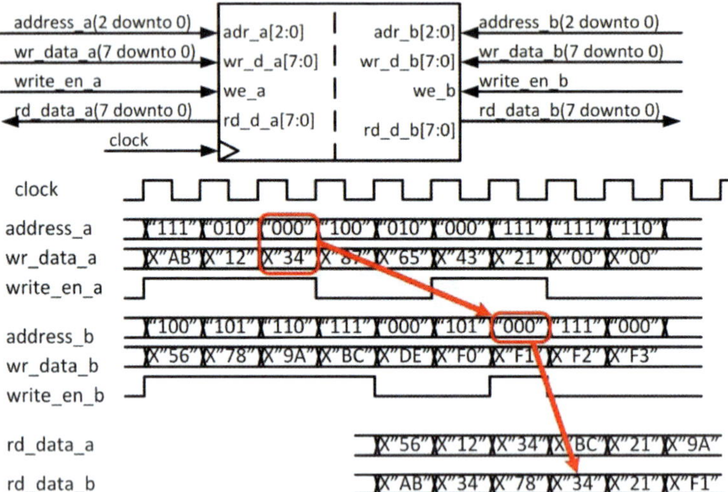

Note that the write enable was not active when port B used address "000".

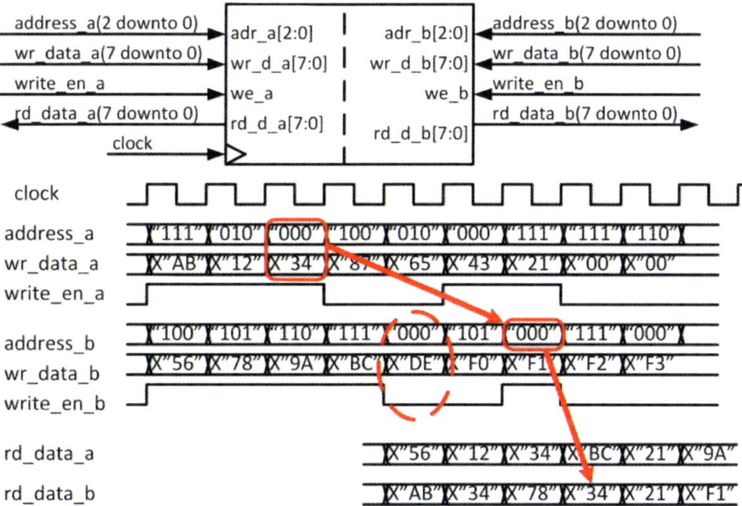

Here we write a hex 21 to port A address "111", and then read it out one clock period later on both ports A and B.

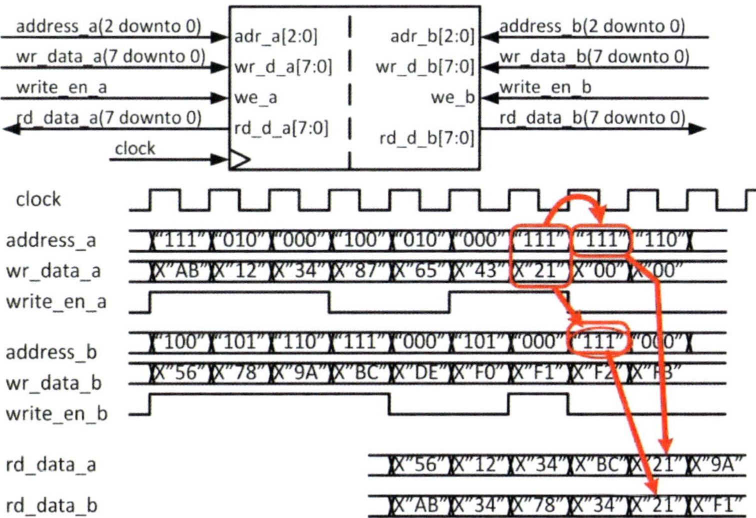

Note that we overwrote the hex BC from port B, which had overwritten the hex AB from port A.

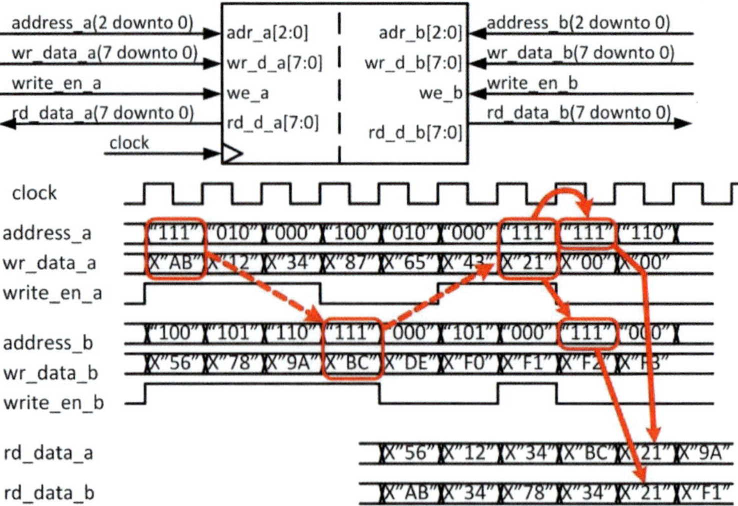

Here, we simply read hex 9A.

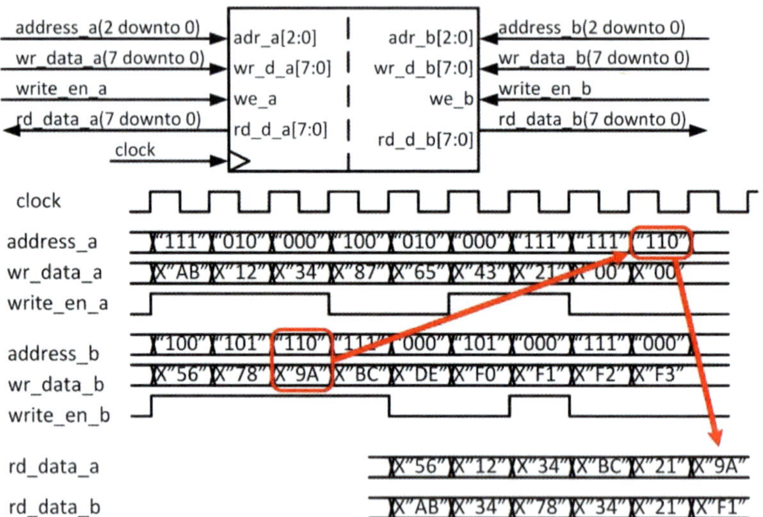

And finally, we see how address "000" was overwritten twice before reading as hex F1.

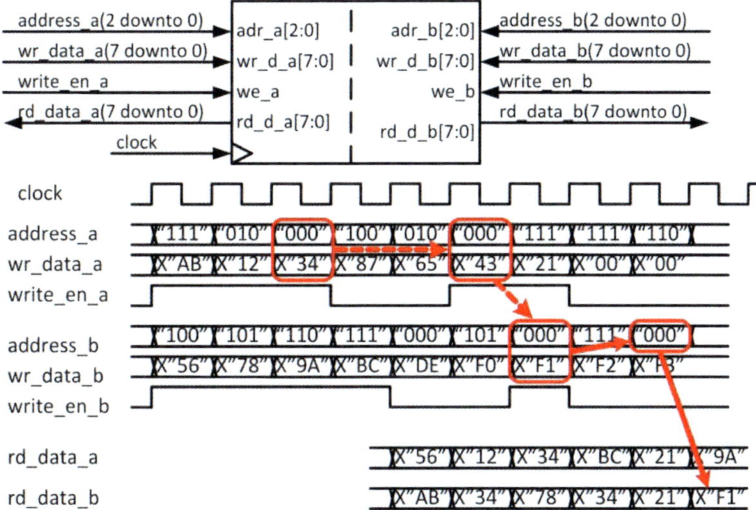

==

Exercise 2-2:

Implement these two registered muxes in VHDL code.

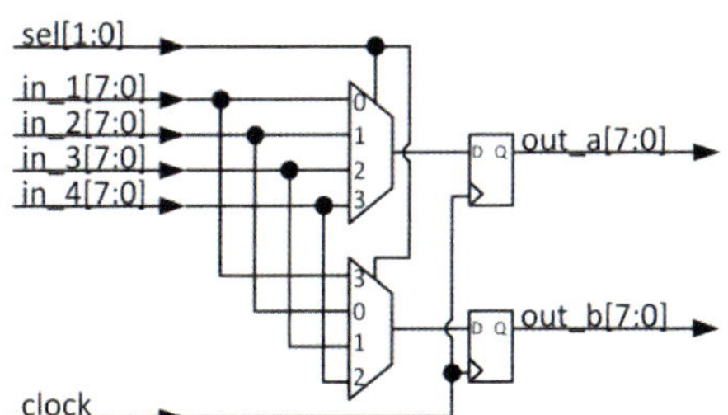

In exercise 2, we have some practice with multiplexers. Here's the code.

```
entity prob_4_4_p2 is
port (
        clock       : in   std_logic;
        sel         : in   std_logic_vector(1 downto 0);
        in_1        : in   std_logic_vector(7 downto 0);
        in_2        : in   std_logic_vector(7 downto 0);
        in_3        : in   std_logic_vector(7 downto 0);
        in_4        : in   std_logic_vector(7 downto 0);
        --
        out_a       : out  std_logic_vector(7 downto 0);
        out_b       : out  std_logic_vector(7 downto 0)
     );
end entity;

architecture behavioral of prob_4_4_p2 is

begin

   dual_mux : process(clock)
   begin
      if rising_edge(clock) then
         case(sel) is
            when "00" => out_a <= in_1;
            when "01" => out_a <= in_2;
            when "10" => out_a <= in_3;
            when "11" => out_a <= in_4;
            when others => null;
         end case;
         --
         case(sel) is
            when "00" => out_b <= in_2;
            when "01" => out_b <= in_3;
            when "10" => out_b <= in_4;
            when "11" => out_b <= in_1;
            when others => null;
         end case;
      end if;
   end process;

end architecture behavioral;
```

I'm using a case statement to implement each mux.

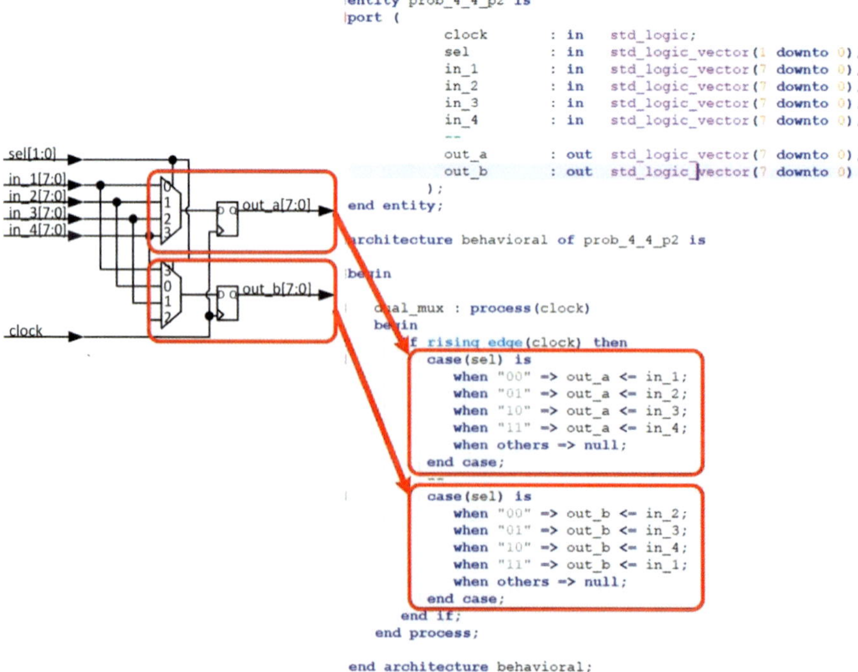

Note that I've arranged the inputs to implement the non-sequential ordering of this mux.

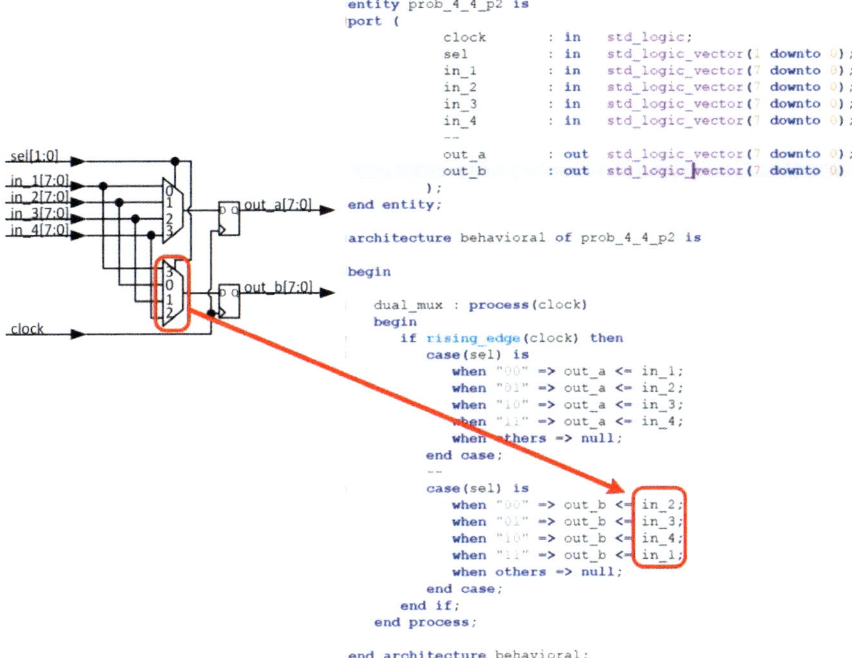


```
entity prob_4_4_p2 is
port (
            clock        : in   std_logic;
            sel          : in   std_logic_vector(1 downto 0);
            in_1         : in   std_logic_vector(7 downto 0);
            in_2         : in   std_logic_vector(7 downto 0);
            in_3         : in   std_logic_vector(7 downto 0);
            in_4         : in   std_logic_vector(7 downto 0);
            --
            out_a        : out  std_logic_vector(7 downto 0);
            out_b        : out  std_logic_vector(7 downto 0)
          );
end entity;

architecture behavioral of prob_4_4_p2 is

begin

   dual_mux : process(clock)
   begin
      if rising_edge(clock) then
         case(sel) is
            when "00" => out_a <= in_1;
            when "01" => out_a <= in_2;
            when "10" => out_a <= in_3;
            when "11" => out_a <= in_4;
            when others => null;
         end case;
         --
         case(sel) is
            when "00" => out_b <= in_2;
            when "01" => out_b <= in_3;
            when "10" => out_b <= in_4;
            when "11" => out_b <= in_1;
            when others => null;
         end case;
      end if;
   end process;

end architecture behavioral;
```

I could have also done it by changing the assignment order instead. Both methods work the same.

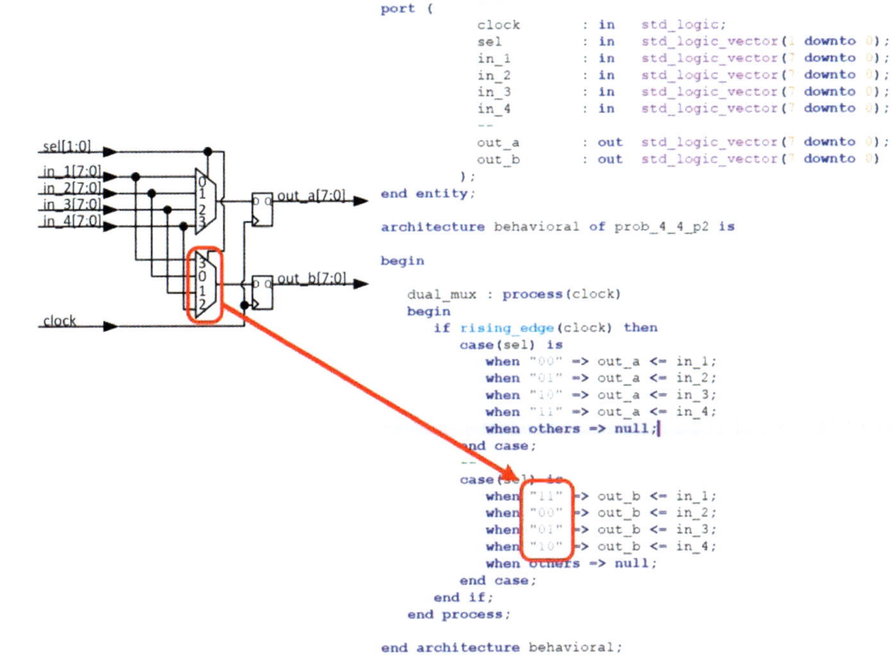


```
entity prob_4_4_p2 is
port (
            clock        : in   std_logic;
            sel          : in   std_logic_vector(1 downto 0);
            in_1         : in   std_logic_vector(7 downto 0);
            in_2         : in   std_logic_vector(7 downto 0);
            in_3         : in   std_logic_vector(7 downto 0);
            in_4         : in   std_logic_vector(7 downto 0);
            --
            out_a        : out  std_logic_vector(7 downto 0);
            out_b        : out  std_logic_vector(7 downto 0)
          );
end entity;

architecture behavioral of prob_4_4_p2 is

begin

   dual_mux : process(clock)
   begin
      if rising_edge(clock) then
         case(sel) is
            when "00" => out_a <= in_1;
            when "01" => out_a <= in_2;
            when "10" => out_a <= in_3;
            when "11" => out_a <= in_4;
            when others => null;
         end case;
         --
         case(sel) is
            when "11" => out_b <= in_1;
            when "00" => out_b <= in_2;
            when "01" => out_b <= in_3;
            when "10" => out_b <= in_4;
            when others => null;
         end case;
      end if;
   end process;

end architecture behavioral;
```

==

Exercise 2-3:

Implement these registers as an array in VHDL code.

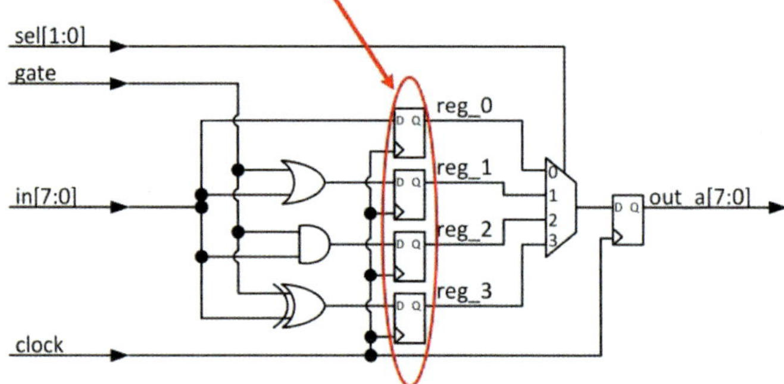

Now some practice with arrays. Here's my code, and we'll zoom in on just the architecture.

```
entity prob_4_4_p3 is
port (
            clock       : in    std_logic;
            sel         : in    std_logic_vector(1 downto 0);
            gate        : in    std_logic;
            sig_in      : in    std_logic_vector(7 downto 0);
            --
            out_a       : out   std_logic_vector(7 downto 0)
        );
end entity;

architecture behavioral of prob_4_4_p3 is

   type array_4x8 is array(3 downto 0) of std_logic_vector(7 downto 0);
   signal reg   : array_4x8;
   signal gate_vector : std_logic_vector(7 downto 0);

begin

   reg_array : process(clock)
      variable i : integer := 0;
   begin
      for i in 0 to 7 loop
         gate_vector(i) <= gate;
      end loop;
      --
      if rising_edge(clock) then
         reg(0) <= sig_in;
         reg(1) <= gate_vector OR  sig_in;
         reg(2) <= gate_vector AND sig_in;
         reg(3) <= gate_vector XOR sig_in;
         --
         case(sel) is
            when "00" => out_a <= reg(0);
            when "01" => out_a <= reg(1);
            when "10" => out_a <= reg(2);
            when "11" => out_a <= reg(3);
            when others => null;
         end case;
      end if;
   end process;

end architecture behavioral;
```

We declare a 4-by-eight array, meaning an array of four elements, where each element is 8 bits wide, a byte. We then declare that the signal "reg" is of that array type.

```
architecture behavioral of prob_4_4_p3 is

   type array_4x8 is array(3 downto 0) of std_logic_vector(7 downto 0);
   signal reg   : array_4x8;
   signal gate_vector : std_logic_vector(7 downto 0);

begin

   reg_array : process(clock)
      variable i : integer := 0;
   begin
      for i in 0 to 7 loop
         gate_vector(i) <= gate;
      end loop;
      --
      if rising_edge(clock) then
         reg(0) <= sig_in;
         reg(1) <= gate_vector OR  sig_in;
         reg(2) <= gate_vector AND sig_in;
         reg(3) <= gate_vector XOR sig_in;
         --
         case(sel) is
            when "00" => out_a <= reg(0);
            when "01" => out_a <= reg(1);
            when "10" => out_a <= reg(2);
            when "11" => out_a <= reg(3);
            when others => null;
         end case;
      end if;
   end process;

end architecture behavioral;
```

We have a bit of a dilemma here. The "gate" input is scalar, meaning it's just one bit, but we're performing logical operations on a vector signal, the 8-bit "sig_in" signal.

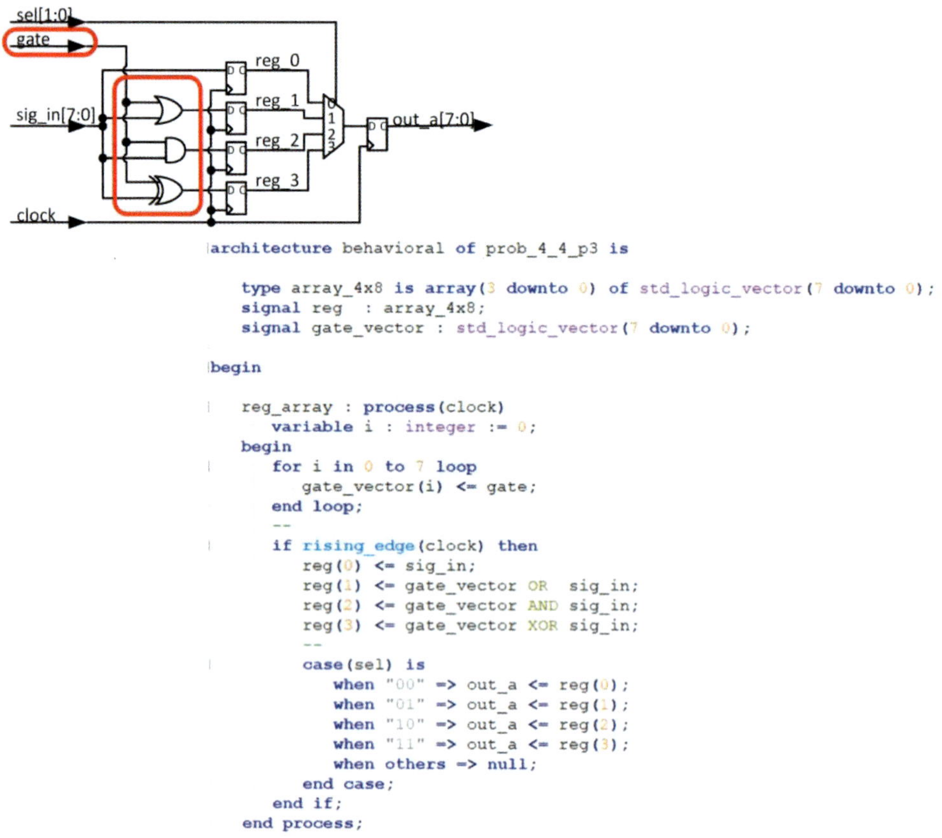


```
architecture behavioral of prob_4_4_p3 is

   type array_4x8 is array(3 downto 0) of std_logic_vector(7 downto 0);
   signal reg  : array_4x8;
   signal gate_vector : std_logic_vector(7 downto 0);

begin

   reg_array : process(clock)
      variable i : integer := 0;
   begin
      for i in 0 to 7 loop
         gate_vector(i) <= gate;
      end loop;
      --
      if rising_edge(clock) then
         reg(0) <= sig_in;
         reg(1) <= gate_vector OR  sig_in;
         reg(2) <= gate_vector AND sig_in;
         reg(3) <= gate_vector XOR sig_in;
         --
         case(sel) is
            when "00" => out_a <= reg(0);
            when "01" => out_a <= reg(1);
            when "10" => out_a <= reg(2);
            when "11" => out_a <= reg(3);
            when others => null;
         end case;
      end if;
   end process;

end architecture behavioral;
```

Verilog allows this, but not VHDL. So we create an intermediate signal that *is* an 8-bit vector, and use a FOR loop to assign the gate signal to every bit.

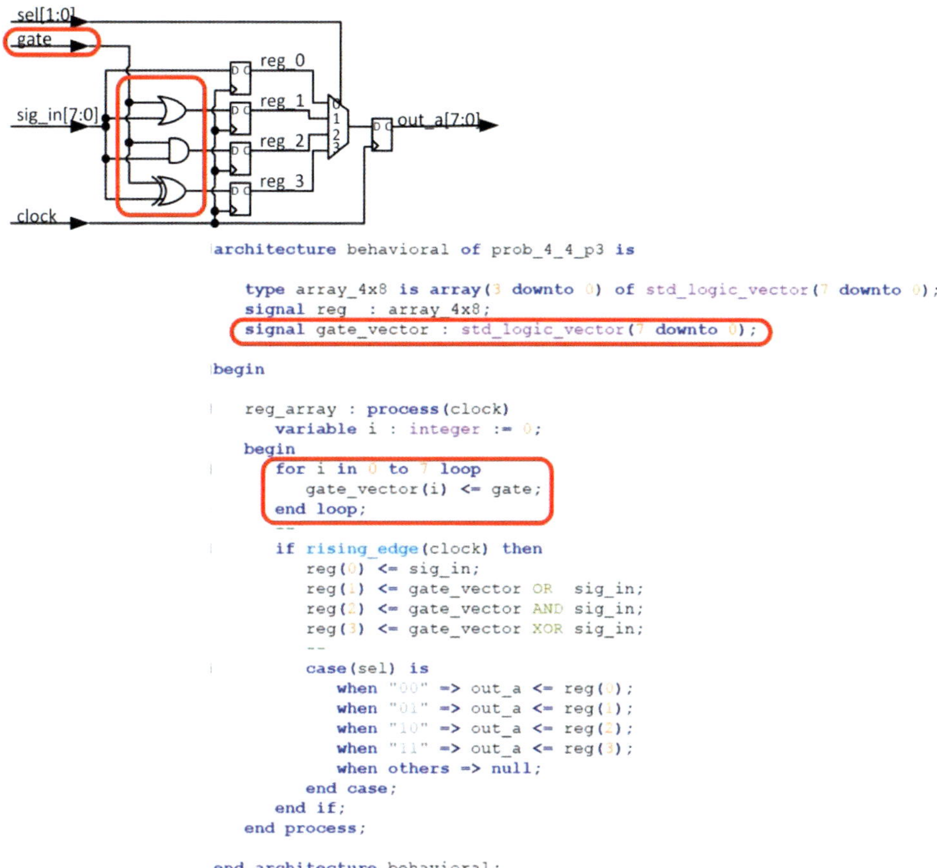


```
architecture behavioral of prob_4_4_p3 is

   type array_4x8 is array(3 downto 0) of std_logic_vector(7 downto 0);
   signal reg   : array_4x8;
   signal gate_vector : std_logic_vector(7 downto 0);

begin

   reg_array : process(clock)
      variable i : integer := 0;
   begin
      for i in 0 to 7 loop
         gate_vector(i) <= gate;
      end loop;
      --
      if rising_edge(clock) then
         reg(0) <= sig_in;
         reg(1) <= gate_vector OR  sig_in;
         reg(2) <= gate_vector AND sig_in;
         reg(3) <= gate_vector XOR sig_in;
         --
         case(sel) is
            when "00" => out_a <= reg(0);
            when "01" => out_a <= reg(1);
            when "10" => out_a <= reg(2);
            when "11" => out_a <= reg(3);
            when others => null;
         end case;
      end if;
   end process;

end architecture behavioral;
```

I had indicated earlier that we rarely use FOR loops in implemented code, and I'm using it here to show that there are some applications where it can be handy.

Here we:

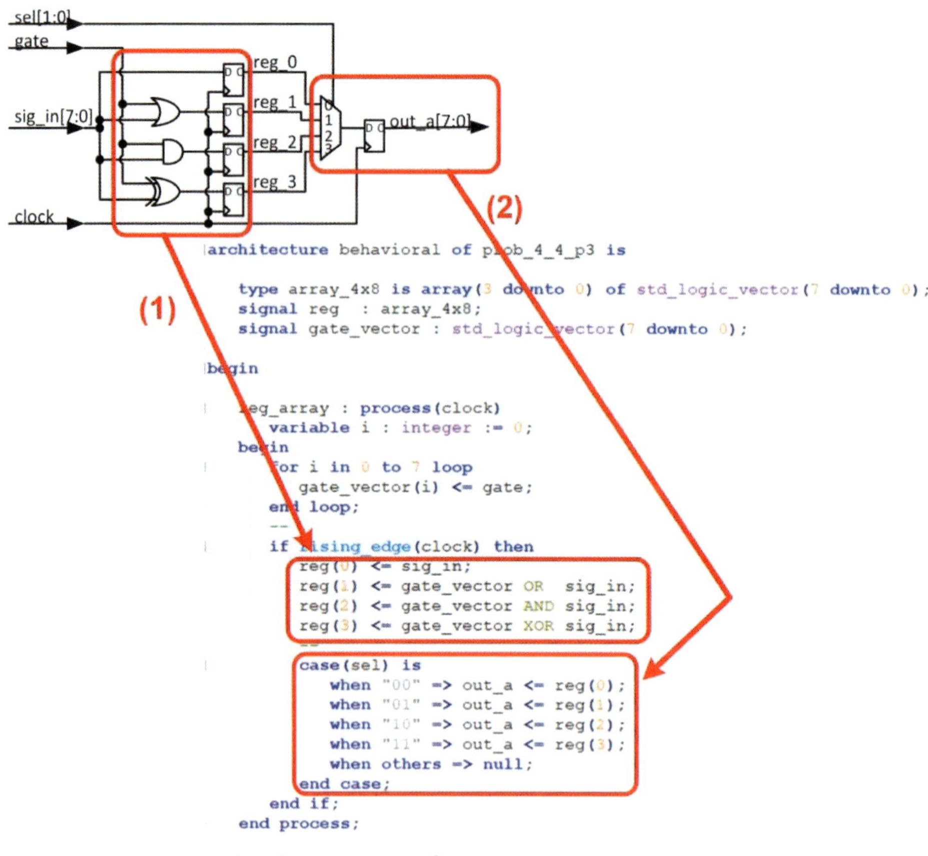

(1) assign the logical operations on the array; and
(2) implement the mux with a case statement.

Here's the simulation (using testbench prob_4_4_p3_tb).

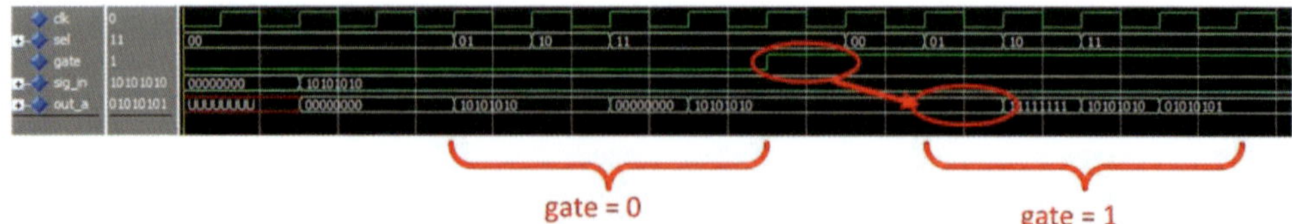

The gate signal input is effectively low and then high during these sets of four clocks, remembering that there's a two-clock delay from changes in the gate signal to resulting effects on the output.

There's a one-clock delay from the select to the output.

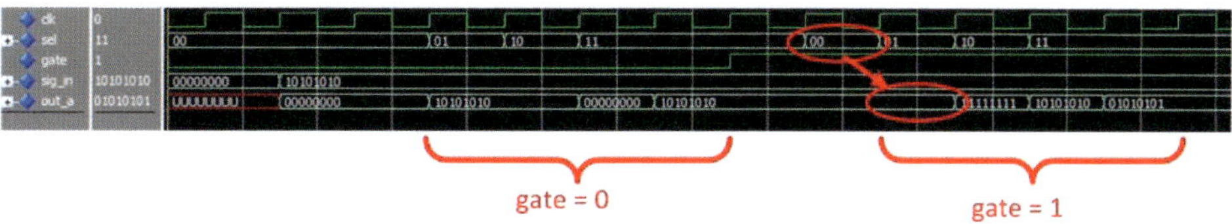

This clock period selects the sig_in signal straight through with no logical operation.

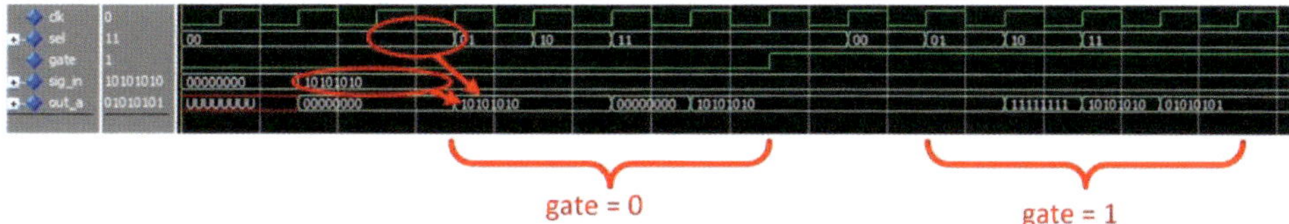

This clock period is the OR operation, and so the sig_in signal again is unchanged.

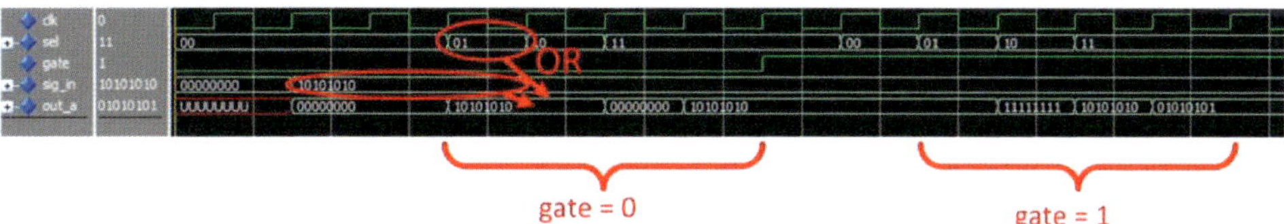

However, when we get to the AND operation, the low gate forces the output to all zeros.

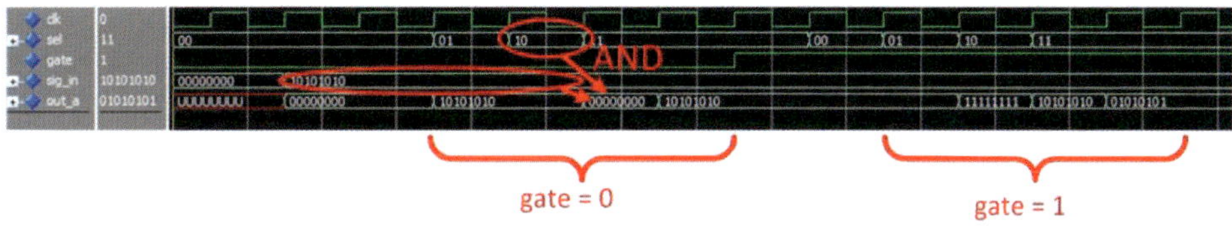

The last operation is the XOR, and since XORing a zero does nothing, again, the output is essentially a pass-through.

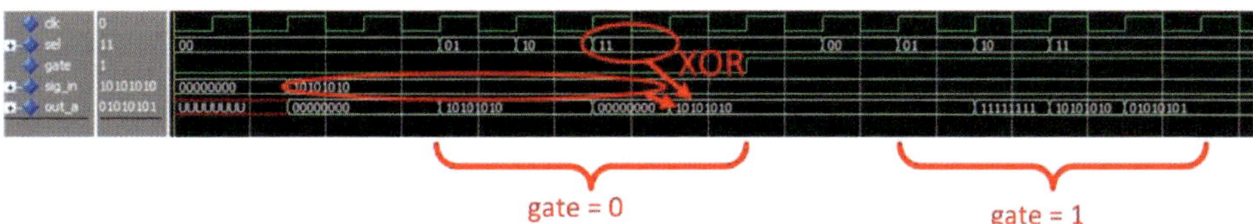

We move on to a high gate signal input …

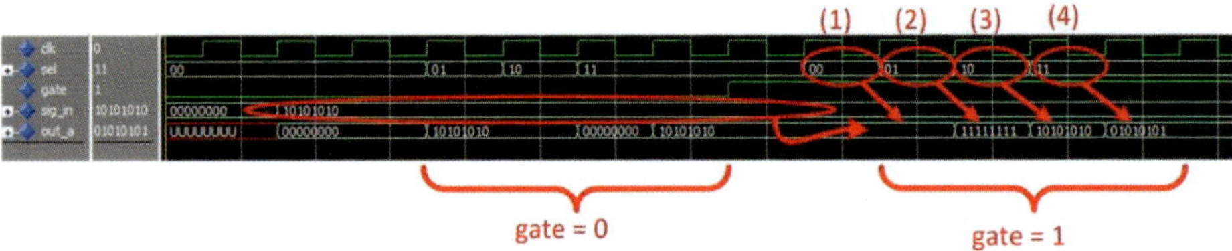

… where:
(1) it has no affect on the first clock period;
(2) but now the OR operation results in all ones;
(3) the AND is a pass-thru, and;
(4) the XOR inverts the signal, from the original hex AA to a hex 55.

===

Exercise 2-4:
Code a dual-port memory with separate read and write clocks.

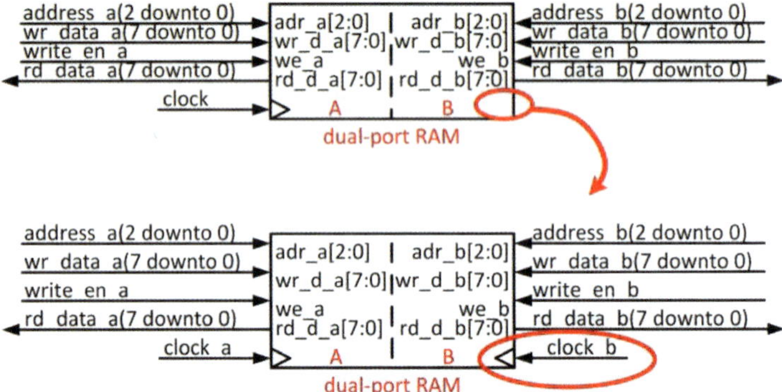

We learned about dual-port memories, but those were single-clock configurations. In this exercise, we want to have each port use its own clock. This is a very useful configuration, for example in systems where a micro-processor is operating in one clock domain, and the RTL in another.

Here's one version of the two-clock dual-port memory, and we'll zoom in on the meat of the architecture.

```
entity prob_4_4_p4_1 is
port (
            clock_a        : in   std_logic;
            wr_d_a         : in   std_logic_vector(7 downto 0);
            adr_a          : in   std_logic_vector(2 downto 0);
            we_a           : in   std_logic;
            rd_d_a         : out  std_logic_vector(7 downto 0);
            --
            clock_b        : in   std_logic;
            wr_d_b         : in   std_logic_vector(7 downto 0);
            adr_b          : in   std_logic_vector(2 downto 0);
            we_b           : in   std_logic;
            rd_d_b         : out  std_logic_vector(7 downto 0)
      );
end entity;

architecture behavioral of prob_4_4_p4_1 is
   type array_8x8 is array (7 downto 0) of std_logic_vector(7 downto 0);
   signal mem_dat  : array_8x8;
begin
   inferred_dp_memory_a : process(clock_a)
   begin
      if rising_edge(clock_a) then
         -- writes ---
         if ( we_a = '1' ) then
            mem_dat(to_integer(unsigned(adr_a))) <= wr_d_a;
         end if;
         -- reads ---
         rd_d_a <= mem_dat(to_integer(unsigned(adr_a)));
      end if;
   end process;

   inferred_dp_memory_b : process(clock_b)
   begin
      if rising_edge(clock_b) then
         -- writes ---
         if ( we_b = '1' ) then
            mem_dat(to_integer(unsigned(adr_a))) <= wr_d_b;
         end if;
         -- reads ---
         rd_d_b <= mem_dat(to_integer(unsigned(adr_b)));
      end if;
   end process;

end architecture behavioral;
```

This is the A port.

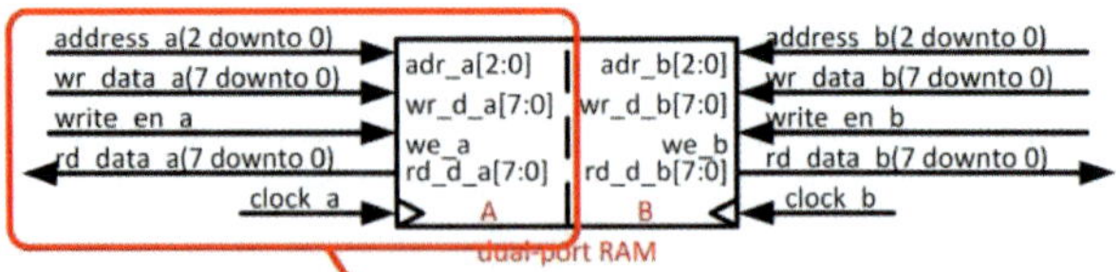

```
inferred_dp_memory_a : process(clock_a)
begin
   if rising_edge(clock_a) then
      -- writes ---
      if ( we_a = '1' ) then
         mem_dat(to_integer(unsigned(adr_a))) <= wr_d_a;
      end if;
      -- reads ---
      rd_d_a <= mem_dat(to_integer(unsigned(adr_a)));
   end if;
end process;
```

```
inferred_dp_memory_b : process(clock_b)
begin
   if rising_edge(clock_b) then
      -- writes ---
      if ( we_b = '1' ) then
         mem_dat(to_integer(unsigned(adr_a))) <= wr_d_b;
      end if;
      -- reads ---
      rd_d_b <= mem_dat(to_integer(unsigned(adr_b)));
   end if;
end process;
```

It has its own dedicated clocked process statement.

And this is the B port with its own dedicated clocked process statement.

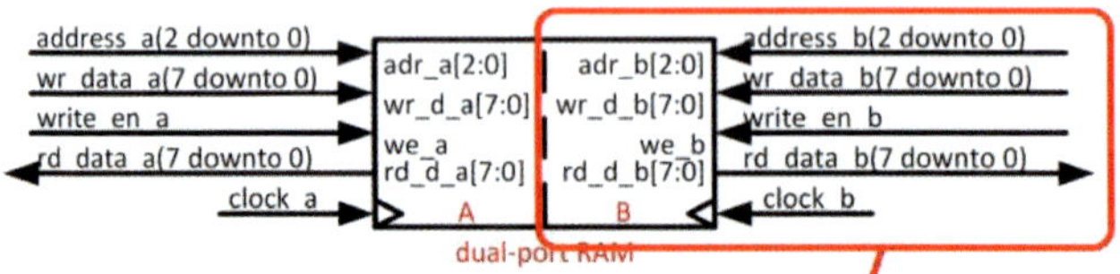

```
inferred_dp_memory_a : process(clock_a)
begin
   if rising_edge(clock_a) then
      -- writes ---
      if ( we_a = '1' ) then
         mem_dat(to_integer(unsigned(adr_a))) <= wr_d_a;
      end if;
      -- reads ---
      rd_d_a <= mem_dat(to_integer(unsigned(adr_a)));
   end if;
end process;
```

```
inferred_dp_memory_b : process(clock_b)
begin
   if rising_edge(clock_b) then
      -- writes ---
      if ( we_b = '1' ) then
         mem_dat(to_integer(unsigned(adr_a))) <= wr_d_b;
      end if;
      -- reads ---
      rd_d_b <= mem_dat(to_integer(unsigned(adr_b)));
   end if;
end process;
```

It's as though we simply combined two single-port memories into the same architecture.

The key, of course, is that they share the same memory core, the same memory array.

```
inferred_dp_memory_a : process(clock_a)
begin
   if rising_edge(clock_a) then
      -- writes ---
      if ( we_a = '1' ) then
        mem_dat(to_integer(unsigned(adr_a))) <= wr_d_a;
      end if;
      -- reads ---
      rd_d_a <= mem_dat(to_integer(unsigned(adr_a)));
   end if;
end process;

inferred_dp_memory_b : process(clock_b)
begin
   if rising_edge(clock_b) then
      -- writes ---
      if ( we_b = '1' ) then
        mem_dat(to_integer(unsigned(adr_a))) <= wr_d_b;
      end if;
      -- reads ---
      rd_d_b <= mem_dat(to_integer(unsigned(adr_b)));
   end if;
end process;
```

Normally, this would be a compiler no-no, having the same signal assigned in different process statements. The question is whether the compiler recognizes this structure as an inferred memory.

Here's an alternative approach.

```
entity prob_4_4_p4_2 is
port (
            clock_a         : in    std_logic;
            wr_d_a          : in    std_logic_vector(7 downto 0);
            adr_a           : in    std_logic_vector(2 downto 0);
            we_a            : in    std_logic;
            rd_d_a          : out   std_logic_vector(7 downto 0);
            --
            clock_b         : in    std_logic;
            wr_d_b          : in    std_logic_vector(7 downto 0);
            adr_b           : in    std_logic_vector(2 downto 0);
            we_b            : in    std_logic;
            rd_d_b          : out   std_logic_vector(7 downto 0)
    );
end entity;

architecture behavioral of prob_4_4_p4_2 is
   type array_8x8 is array (7 downto 0) of std_logic_vector(7 downto 0);
   signal mem_dat   : array_8x8;
begin

   inferred_dp_memory : process(clock_a, clock_b)
   begin
     if rising_edge(clock_a) then
        -- writes ---
        if ( we_a = '1' ) then
           mem_dat(to_integer(unsigned(adr_a))) <= wr_d_a;
        end if;
        -- reads ---
        rd_d_a <= mem_dat(to_integer(unsigned(adr_a)));
     elsif rising_edge(clock_b) then
        -- writes ---
        if ( we_b = '1' ) then
           mem_dat(to_integer(unsigned(adr_a))) <= wr_d_b;
        end if;
        -- reads ---
        rd_d_b <= mem_dat(to_integer(unsigned(adr_b)));
     end if;
   end process;

end architecture behavioral;
```

The one process statement includes both clocks. Although more correct in a VHDL coding point of view, most (probably all) FPGA logic fabric does not accommodate two clocks on the same registered target. Here again, the question is whether the compiler recognizes this structure as an inferred memory.

We should pause a moment to understand that if you use memory in an FPGA design, you will almost assuredly use the vendor's built-in RAM blocks, and there will be no need for scratching your head over the proper inferred memory structure.

Exercise Answers, Chapter 3

Exercise 3-1:
Add "empty" and "full" flags to this FIFO from the chapter.

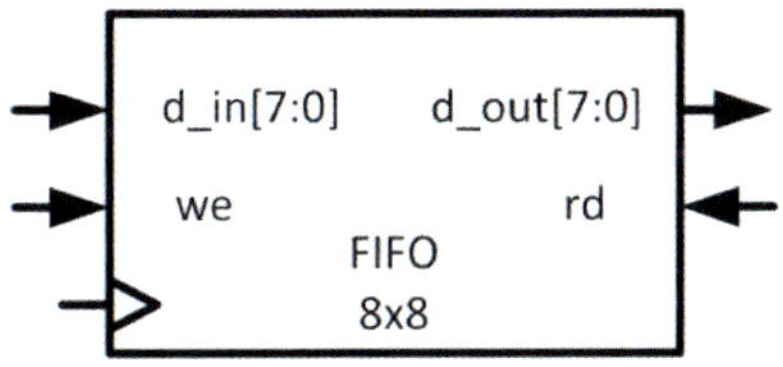
d_in[7:0]
d_out[7:0]
we
rd
FIFO
8x8

```
architecture behavioral of fifo is

   type array_8x8 is array (7 downto 0) of std_logic_vector(7 downto 0);
   signal mem_dat  : array_8x8;
   signal addr_write    : unsigned(2 downto 0) := "000";
   signal addr_read     : unsigned(2 downto 0) := "000";

begin

   fifo_memory : process(clock)
   begin
      if rising_edge(clock) then
         -- writes ---
         if ( we = '1' ) then
            mem_dat(to_integer(unsigned(addr_write))) <= d_in;
            addr_write <= addr_write + 1;
         end if;
         -- reads ---
         if (rd = '1') then
            d_out     <= mem_dat(to_integer(unsigned(addr_read)));
            addr_read <= addr_read + 1;
         end if;
      end if;
   end process;

end architecture behavioral;
```

Most FIFOs include these flags, indicating that the FIFO is either empty or full. When a FIFO is either empty or full, the internal write and read addresses are the same. Imagine a FIFO that has just one value entered. Further, let's say that it's located at address 5, i.e., the last FIFO write was to address 5. Since in our FIFO code we increment the write address after the write, the write address is now 6. We've been emptying the FIFO, and since the last read would have been from address 4 (the one before the last value), and since we also increment the read address after each read, the read address is now at 5. If we read that last entry—rendering the FIFO empty—the read address will increment to 6, and the write and read addresses will be the same.

Similarly, imagine the same FIFO that now has just one empty location left—it's nearly full. Say that this last empty location is address 2. The write address is now also 2 (we last wrote to address 1, and the write address then incremented). The last location that had been read was this one, 2. Since then, we've only been writing, all around circular buffer. So, the read address is now at 3. Once we write to this last location, the write address will increment to 3, and again the write and read addresses are the same.

We've established that when the write and read addresses are the same, the FIFO is either empty or full. So, if we just look at the write and read addresses, how do we know which it is? We can't know. We need to know the history as well—did the addresses become the same due to a last write, or a last read?

This sounds like a job for a state machine.

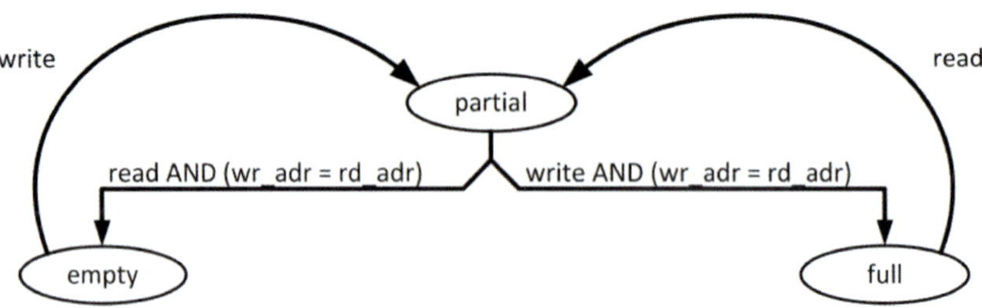

Here, we have just three states. We identify the FIFO as empty, full, or anything in between, i.e., when it's partially full.

When it is partially full, and we get a write to an address that's the same as the read address, then we deduce that the FIFO is full:

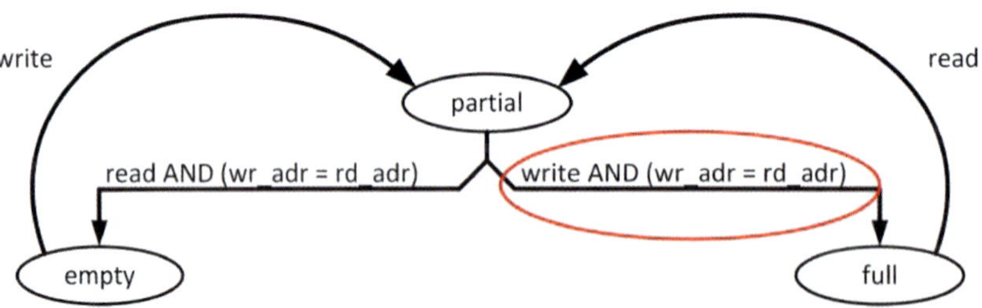

Similarly, if we are partially full, and we get a read from an address that's the same as the write address, we deduce that the FIFO is empty:

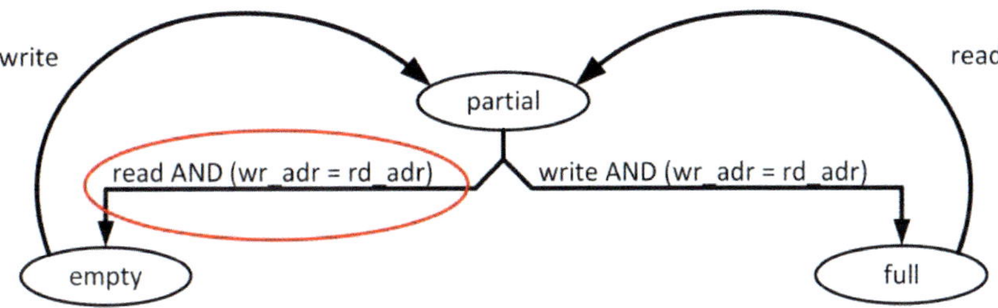

From the full and empty states, any legitimate write or read takes us back to the partially full condition:

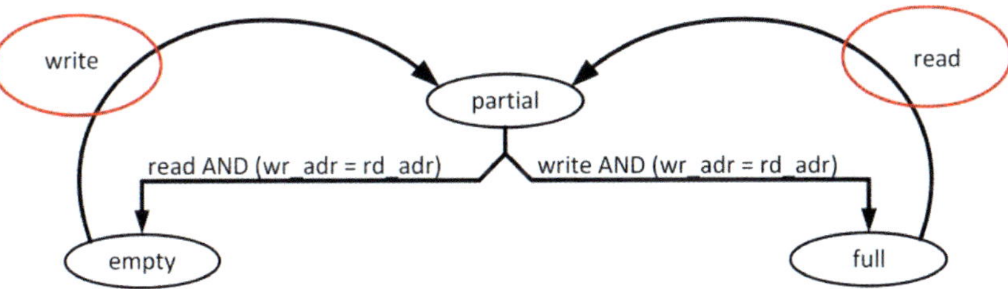

However, we understand that we increment the write and read addresses *after* we perform the write or read. Therefore, we need to compare to the operative address plus one.

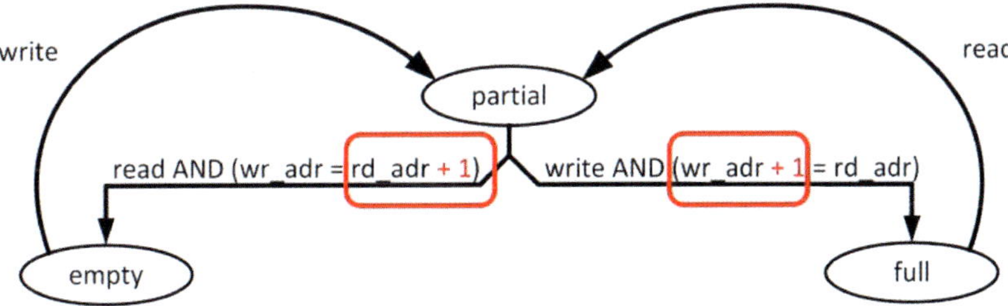

We are in effect looking ahead to see what the addresses will be after we perform the write or read. It's only after we do the write or read and increment the associated address that they're then the same.

Here's the code for my version of the FIFO—too tiny to see, but we'll zoom in.

```
library IEEE;
use IEEE.STD_LOGIC_1164.all;
use IEEE.NUMERIC_STD.all;
use IEEE.STD_LOGIC_MISC.all;
use IEEE.STD_LOGIC_UNSIGNED.all;


entity prob_8_1 is
port (
            clock          : in    std_logic;
            --
            we             : in    std_logic;
            d_in           : in    std_logic_vector(7 downto 0);
            full           : out   std_logic;
            --
            rd             : in    std_logic;
            d_out          : out   std_logic_vector(7 downto 0);
            empty          : out   std_logic
        );
end entity;

architecture behavioral of prob_8_1 is

    type array_8x8 is array (7 downto 0) of std_logic_vector(7 downto 0);
    signal mem_dat  : array_8x8;
    signal addr_write     : unsigned(2 downto 0)  := "000";
    signal addr_read      : unsigned(2 downto 0)  := "000";
    --
    type state_type is (  partial,
                          full_state,
                          empty_state
                     );
    signal state : state_type := empty_state;

begin

    fifo_memory : process(clock)
    begin
       if rising_edge(clock) then
          -- writes ---
          if (    we = '1'
              and state /= full_state
             ) then
             mem_dat(to_integer(unsigned(addr_write))) <= d_in;
             addr_write <= addr_write + 1;
          end if;
          -- reads ---
          if (    rd = '1'
              and state /= empty_state
             ) then
             d_out      <= mem_dat(to_integer(unsigned(addr_read)));
             addr_read <= addr_read + 1;
          end if;
          --
          case (state) is
             when empty_state => if (we = '1') then                                 state <= partial;  end if;
             when full_state  => if (rd = '1') then                                 state <= partial;  end if;
             when partial     => if (    rd = '1'
                                     and addr_write = addr_read + 1
                                    ) then                                          state <= empty_state;
                                 elsif (    we = '1'
                                        and addr_write + 1 = addr_read
                                       ) then                                       state <= full_state;
                                 end if;
             when others => null;
          end case;
       end if;
    end process;

    full   <= '1' when state = full_state  else '0';
    empty  <= '1' when state = empty_state else '0';

 end architecture behavioral;
```

First, the entity, with the two new flags:

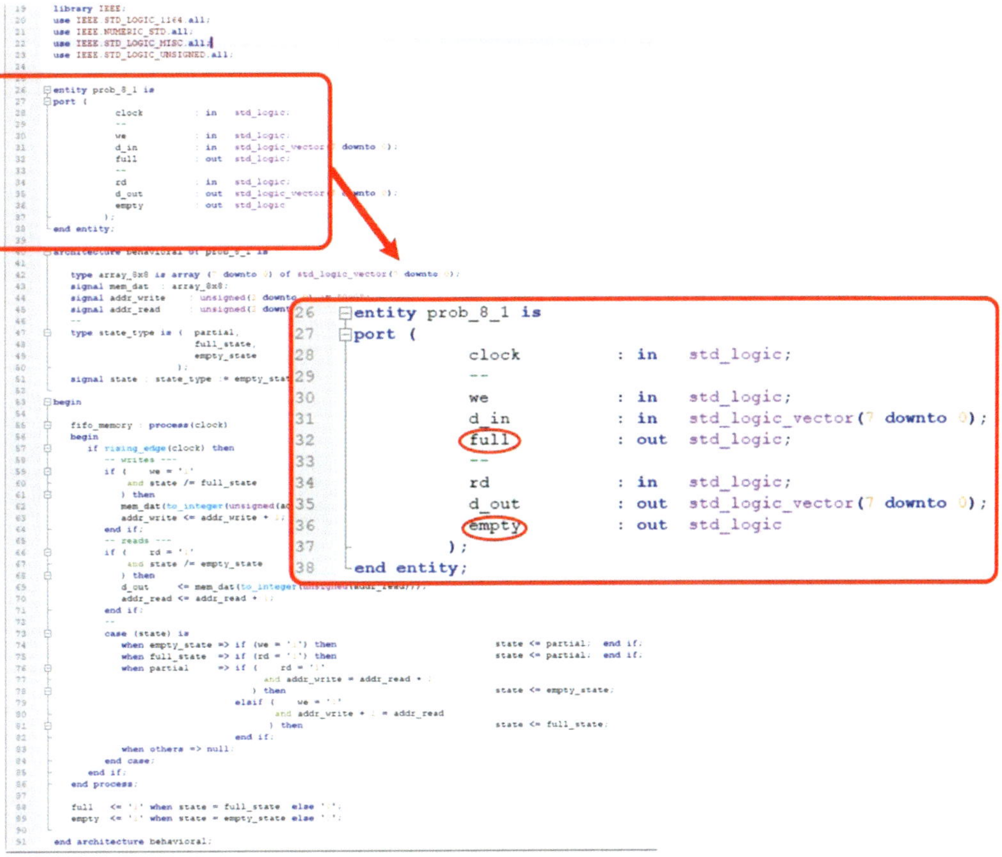

```
library IEEE;
use IEEE.STD_LOGIC_1164.all;
use IEEE.NUMERIC_STD.all;
use IEEE.STD_LOGIC_MISC.all;
use IEEE.STD_LOGIC_UNSIGNED.all;

entity prob_8_1 is
port (
            clock       : in    std_logic;
            --
            we          : in    std_logic;
            d_in        : in    std_logic_vector(7 downto 0);
            full        : out   std_logic;
            --
            rd          : in    std_logic;
            d_out       : out   std_logic_vector(7 downto 0);
            empty       : out   std_logic
        );
end entity;

architecture behavioral of prob_8_1 is

    type array_8x8 is array (7 downto 0) of std_logic_vector(7 downto 0);
    signal mem_dat  : array_8x8;
    signal addr_write     : unsigned(2 downto
    signal addr_read      : unsigned(2 down
    --
    type state_type is (  partial,
                          full_state,
                          empty_state
                       );
    signal state : state_type := empty_stat

begin

    fifo_memory : process(clock)
    begin
        if rising_edge(clock) then
            -- writes ---
            if (     we = '1'
                and state /= full_state
               ) then
                mem_dat(to_integer(unsigned(ad
                addr_write <= addr_write + 1;
            end if;
            -- reads ---
            if (     rd = '1'
                and state /= empty_state
               ) then
                d_out      <= mem_dat(to_integer(
                addr_read <= addr_read + 1;
            end if;
            --
            case (state) is
                when empty_state => if (we = '1') then                     state <= partial;  end if;
                when full_state  => if (rd = '1') then                     state <= partial;  end if;
                when partial     => if (     rd = '1'
                                         and addr_write = addr_read + 1
                                        ) then                             state <= empty_state;
                                    elsif (     we = '1'
                                            and addr_write + 1 = addr_read
                                           ) then                          state <= full_state;
                                    end if;
                when others => null;
            end case;
        end if;
    end process;

    full   <= '1' when state = full_state  else '0';
    empty  <= '1' when state = empty_state else '0';

end architecture behavioral;
```

```
entity prob_8_1 is
port (
            clock       : in    std_logic;
            --
            we          : in    std_logic;
            d_in        : in    std_logic_vector(7 downto 0);
            full        : out   std_logic;
            --
            rd          : in    std_logic;
            d_out       : out   std_logic_vector(7 downto 0);
            empty       : out   std_logic
        );
end entity;
```

... and the architecture:

```
library IEEE;
use IEEE.STD_LOGIC_1164.all;
use IEEE.NUMERIC_STD.all;
use IEEE.STD_LOGIC_MISC.all;
use IEEE.STD_LOGIC_UNSIGNED.all;


entity prob_8_1 is
port (
            clock          : in   std_logic;
            --
            we             : in   std_logic;
            d_in           : in   std_logic_vector(7 downto 0);
            full           : out  std_logic;
            --
            rd             : in   std_logic;
            d_out          : out  std_logic_vector(7 downto 0);
            empty          : out  std_logic
        );
end entity;

architecture behavioral of prob_8_1 is

    type array_8x8 is array (7 downto 0) of std_logic_vector(7 downto 0);
    signal mem_dat  : array_8x8;
    signal addr_write    : unsigned(2 downto 0) := "000";
    signal addr_read     : unsigned(2 downto 0) := "000";
    --
    type state_type is (  partial,
                          full_state,
                          empty_state
                        );
    signal state : state_type := empty_state;

begin

    fifo_memory : process(clock)
    begin
       if rising_edge(clock) then
          -- writes ---
          if (    we = '1'
              and state /= full_state
             ) then
             mem_dat(to_integer(unsigned(addr_write))) <= d_in;
             addr_write <= addr_write + 1;
          end if;
          -- reads ---
          if (    rd = '1'
              and state /= empty_state
             ) then
             d_out     <= mem_dat(to_integer(unsigned(addr_read)));
             addr_read <= addr_read + 1;
          end if;
          --
          case (state) is
             when empty_state => if (we = '1') then                state <= partial;  end if;
             when full_state  => if (rd = '1') then                state <= partial;  end if;
             when partial     => if (    rd = '1'
                                      and addr_write = addr_read + 1
                                     ) then                        state <= empty_state;
                                 elsif (    we = '1'
                                         and addr_write + 1 = addr_read
                                        ) then                     state <= full_state;
                                 end if;
             when others => null;
          end case;
       end if;
    end process;

    full   <= '1' when state = full_state  else '0';
    empty  <= '1' when state = empty_state else '0';

end architecture behavioral;
```

```
architecture behavioral of prob_8_1 is

    type array_8x8 is array (7 downto 0) of std_logic_vector(7 downto 0);
    signal mem_dat  : array_8x8;
    signal addr_write    : unsigned(2 downto 0) := "000";
    signal addr_read     : unsigned(2 downto 0) := "000";
    --
    type state_type is (  partial,
                          full_state,
                          empty_state
                       );
    signal state : state_type := empty_state;

begin

    fifo_memory : process(clock)
    begin
       if rising_edge(clock) then
          -- writes ---
          if (     we = '1'
              and state /= full_state
             ) then
             mem_dat(to_integer(unsigned(addr_write))) <= d_in;
             addr_write <= addr_write + 1;
          end if;
          -- reads ---
          if (     rd = '1'
              and state /= empty_state
             ) then
             d_out      <= mem_dat(to_integer(unsigned(addr_read)));
             addr_read <= addr_read + 1;
          end if;
          --
          case (state) is
             when empty_state => if (we = '1') then                        state <= partial;  end if;
             when full_state  => if (rd = '1') then                        state <= partial;  end if;
             when partial     => if (     rd = '1'
                                      and addr_write = addr_read + 1
                                    ) then                                  state <= empty_state;
                                 elsif (     we = '1'
                                         and addr_write + 1 = addr_read
                                       ) then                               state <= full_state;
                                 end if;
             when others => null;
          end case;
       end if;
    end process;

    full   <= '1' when state = full_state  else '0';
    empty  <= '1' when state = empty_state else '0';

 end architecture behavioral;
```

We'll zoom in more …

… first with the declarations, including those for the state machine:

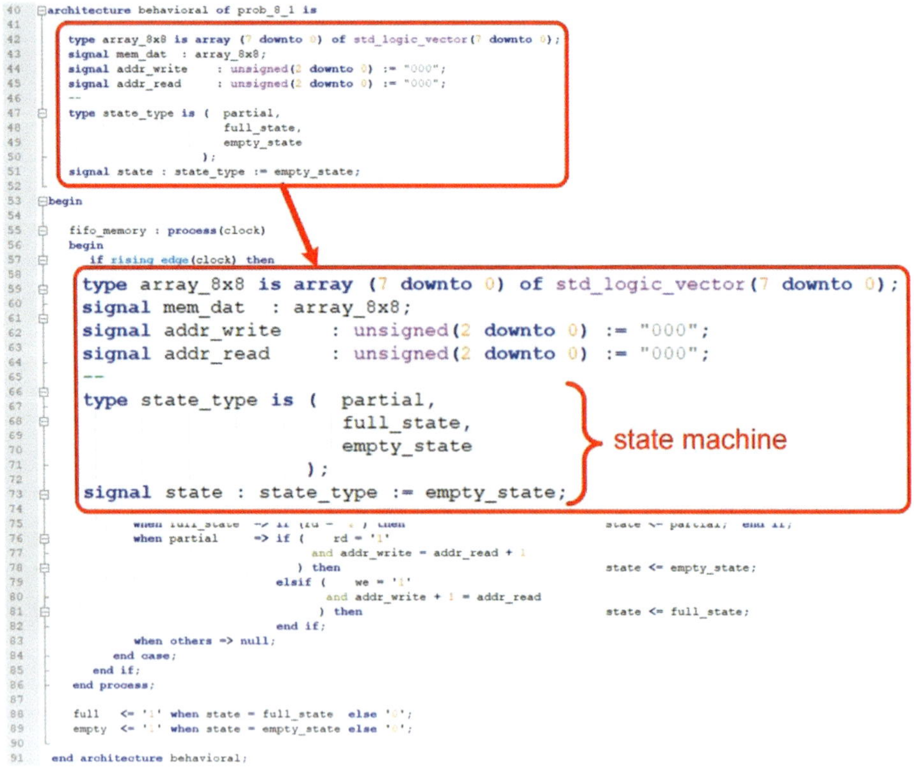

```
architecture behavioral of prob_8_1 is

  type array_8x8 is array (7 downto 0) of std_logic_vector(7 downto 0);
  signal mem_dat   : array_8x8;
  signal addr_write    : unsigned(2 downto 0) := "000";
  signal addr_read     : unsigned(2 downto 0) := "000";
  --
  type state_type is (  partial,
                        full_state,
                        empty_state
                     );
  signal state : state_type := empty_state;

begin

  fifo_memory : process(clock)
  begin
    if rising_edge(clock) then
```


```
type array_8x8 is array (7 downto 0) of std_logic_vector(7 downto 0);
signal mem_dat   : array_8x8;
signal addr_write    : unsigned(2 downto 0) := "000";
signal addr_read     : unsigned(2 downto 0) := "000";
--
type state_type is (  partial,
                      full_state,
                      empty_state
                   );
signal state : state_type := empty_state;
```


state machine

```
          when partial     => if (    rd = '1'
                                  and addr_write = addr_read + 1
                                 ) then                          state <= empty_state;
                              elsif (    we = '1'
                                     and addr_write + 1 = addr_read
                                    ) then                       state <= full_state;
                              end if;
          when others => null;
        end case;
      end if;
  end process;

  full   <= '1' when state = full_state  else '0';
  empty  <= '1' when state = empty_state else '0';

end architecture behavioral;
```

… and the clocked process statement:

```
architecture behavioral of prob_8_1 is

    type array_8x8 is array (7 downto 0) of std_logic_vector(7 downto 0);
    signal mem_dat  : array_8x8;
    signal addr_write    : unsigned(2 downto 0) := "000";
    signal addr_read     : unsigned(2 downto 0) := "000";
    --
    type state_type is (  partial,
                          full_state,
                          empty_state
                       );
    signal state : state_type := empty_state;

begin

    fifo_memory : process(clock)
    begin
       if rising_edge(clock) then
          -- writes ---
          if (     we = '1'
              and state /= full_state
             ) then
             mem_dat(to_integer(unsigned(addr_write))) <= d_in;
             addr_write <= addr_write + 1;
          end if;
          -- reads ---
          if (     rd = '1'
              and state /= empty_state
             ) then
             d_out      <= mem_dat(to_integer(unsigned(addr_read)));
             addr_read <= addr_read + 1;
          end if;
          --
          case (state) is
             when empty_state => if (we = '1') then                    state <= partial;  end if;
             when full_state  => if (rd = '1') then                    state <= partial;  end if;
             when partial     => if (     rd = '1'
                                      and addr_write = addr_read + 1
                                    ) then                             state <= empty_state;
                                 elsif (     we = '1'
                                         and addr_write + 1 = addr_read
                                       ) then                          state <= full_state;
                                 end if;
             when others => null;
          end case;
       end if;
    end process;

    full    <= '1' when state = full_state else '0';
    empty   <= '1' when state = empty_state else '0';

end architecture behavioral;
```


```
fifo_memory : process(clock)
begin
   if rising_edge(clock) then
      -- writes ---
      if (     we = '1'
          and state /= full_state
         ) then
         mem_dat(to_integer(unsigned(addr_write))) <= d_in;
         addr_write <= addr_write + 1;
      end if;
      -- reads ---
      if (     rd = '1'
          and state /= empty_state
         ) then
         d_out      <= mem_dat(to_integer(unsigned(addr_read)));
         addr_read <= addr_read + 1;
      end if;
      --
      case (state) is
         when empty_state => if (we = '1') then                    state <= partial;  end if;
         when full_state  => if (rd = '1') then                    state <= partial;  end if;
         when partial     => if (     rd = '1'
                                  and addr_write = addr_read + 1
                                ) then                             state <= empty_state;
                             elsif (     we = '1'
                                     and addr_write + 1 = addr_read
                                   ) then                          state <= full_state;
                             end if;
         when others => null;
      end case;
   end if;
end process;
```

This section is the same as the FIFO implementation from the chapter, except that I'm now qualifying the writes and reads with the full and empty conditions. Further writes are blocked when the FIFO is full, and further reads are blocked when the FIFO is empty:

```
fifo_memory : process(clock)
begin
   if rising_edge(clock) then
      -- writes ---
      if (     we = '1'
          and state /= full_state
         ) then
         mem_dat(to_integer(unsigned(addr_write))) <= d_in;
         addr_write <= addr_write + 1;
      end if;
      -- reads ---
      if (     rd = '1'
          and state /= empty_state
         ) then
         d_out      <= mem_dat(to_integer(unsigned(addr_read)));
         addr_read <= addr_read + 1;
      end if;
      --
      case (state) is
         when empty_state => if (we = '1') then                              state <= partial;  end if;
         when full_state  => if (rd = '1') then                              state <= partial;  end if;
         when partial     => if (     rd = '1'
                                  and addr_write = addr_read + 1
                                 ) then                                      state <= empty_state;
                             elsif (     we = '1'
                                     and addr_write + 1 = addr_read
                                    ) then                                   state <= full_state;
                             end if;
         when others => null;
      end case;
   end if;
end process;
```

FIFO

This is the new state machine, a straight-forward implementation of the earlier diagram, as we learned in chapter four.

```
fifo_memory : process(clock)
begin
   if rising_edge(clock) then
      -- writes ---
      if (     we = '1'
          and state /= full_state
         ) then
         mem_dat(to_integer(unsigned(addr_write))) <= d_in;
         addr_write <= addr_write + 1;
      end if;
      -- reads ---
      if (     rd = '1'
          and state /= empty_state
         ) then
         d_out      <= mem_dat(to_integer(unsigned(addr_read)));
         addr_read <= addr_read + 1;
      end if;
      --
      case (state) is
         when empty_state => if (we = '1') then                              state <= partial;  end if;
         when full_state  => if (rd = '1') then                              state <= partial;  end if;
         when partial     => if (     rd = '1'
                                  and addr_write = addr_read + 1
                                 ) then                                      state <= empty_state;
                             elsif (     we = '1'
                                     and addr_write + 1 = addr_read
                                    ) then                                   state <= full_state;
                             end if;
         when others => null;
      end case;
   end if;
end process;
```

Finally, we translate the empty and full states into output port flag signals.

```
architecture behavioral of prob_8_1 is

    type array_8x8 is array (7 downto 0) of std_logic_vector(7 downto 0);
    signal mem_dat     : array_8x8;
    signal addr_write     : unsigned(2 downto 0) := "000";
    signal addr_read      : unsigned(2 downto 0) := "000";
    --
    type state_type is (  partial,
                          full_state,
                          empty_state
                       );
    signal state : state_type := empty_state;

begin

    fifo_memory : process(clock)
    begin
       if rising_edge(clock) then
          -- writes ---
          if (     we = '1'
             and state /= full_state
             ) then
             mem_dat(to_integer(unsigned(addr_write))) <= d_in;
             addr_write <= addr_write + 1;
          end if;
          -- reads ---
          if (     rd = '1'
             and state /= empty_state
             ) then
             d_out      <= mem_dat(to_integer(unsigned(addr_read)));
             addr_read <= addr_read + 1;
          end if;
          --
          case (state) is
             when empty_state => if (we = '1') then                          state <= partial;  end if;
             when full_state  => if (rd = '1') then                          state <= partial;  end if;
             when partial     => if (     rd = '1'
                                     and addr_write = addr_read + 1
                                     ) then                                  state <= empty_state;
                                 elsif (     we = '1'
                                        and addr_write + 1 = addr_read
                                       ) then                                state <= full_state;
                                 end if;
             when others => null;
          end case;
       end if;
    end process;

    full    <= '1' when state = full_state  else '0';
    empty   <= '1' when state = empty_state else '0';

end architecture behavioral;
```

full   <= '1' when state = full_state else '0';
empty   <= '1' when state = empty_state else '0';

==

Exercise 3-2:

Create a testbench to exercise the FIFO from the previous exercise using while loops to generate the write and read control signals. The testbench should write eight times and see the full flag go active, and then read eight times to see the empty flag go active.

This is an example result:

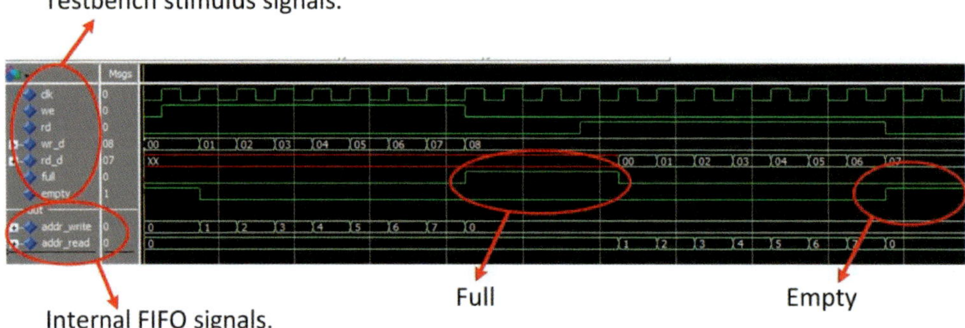

There's obviously no one testbench coding that generates this waveform. In general, however, the various versions will probably be similar. The following is my version. If yours generates something approximating this waveform—sees the full and empty flags go active—then mine may be any not better than yours.

Here's my tiny and unreadable version (we'll zoom in).

```
entity prob_8_2_tb is
end entity prob_8_2_tb;

architecture Behavioral of prob_8_2_tb is

    component prob_8_1 is
    port (
            clock       : in   std_logic;
            --
            we          : in   std_logic;
            d_in        : in   std_logic_vector(7 downto 0);
            full        : out  std_logic;
            --
            rd          : in   std_logic;
            d_out       : out  std_logic_vector(7 downto 0);
            empty       : out  std_logic
         );
    end component;

    signal clk           : std_logic := '0';
    signal we_tb         : std_logic := '0';
    signal we            : std_logic := '0';
    signal rd_tb         : std_logic := '0';
    signal rd            : std_logic := '0';
    signal wr_dat        : unsigned(7 downto 0) := X"00";
    signal wr_d          : std_logic_vector(7 downto 0);
    signal rd_d          : std_logic_vector(7 downto 0);
    signal full          : std_logic;
    signal empty         : std_logic;

begin

    clk <= '0' after 50 ns when clk = '1' else
           '1' after 50 ns;

    stimulus : process
      variable j  : integer;
    begin
      we_tb <= '0';
      --
      wait until rising_edge(clk);
      we_tb <= '1';
      wait until rising_edge(clk);
      while (full = '0') loop
         wr_dat <= wr_dat + 1;
         wait until rising_edge(clk);
      end loop;
      we_tb <= '0';
      --
      wait until rising_edge(clk);
      wait until rising_edge(clk);
      rd_tb <= '1';
      while (empty = '0') loop
         wait until rising_edge(clk);
      end loop;
      rd_tb <= '0';
      --
      for j in 0 to 50 loop
         wait until rising_edge(clk);
      end loop;
    end process;

    we   <= we_tb AND (NOT full);
    wr_d <= std_logic_vector(wr_dat);
    rd   <= rd_tb AND (NOT empty);

    fifo_1 : prob_8_1
    port map
       (
        clock   => clk,
        --
        we      => we,
        d_in    => wr_d,
        full    => full,
        --
        rd      => rd,
        d_out   => rd_d,
        empty   => empty
       );
end architecture Behavioral;
```

The FIFO component declaration and instantiation would be common to all testbench versions, other than different testbench signal names.

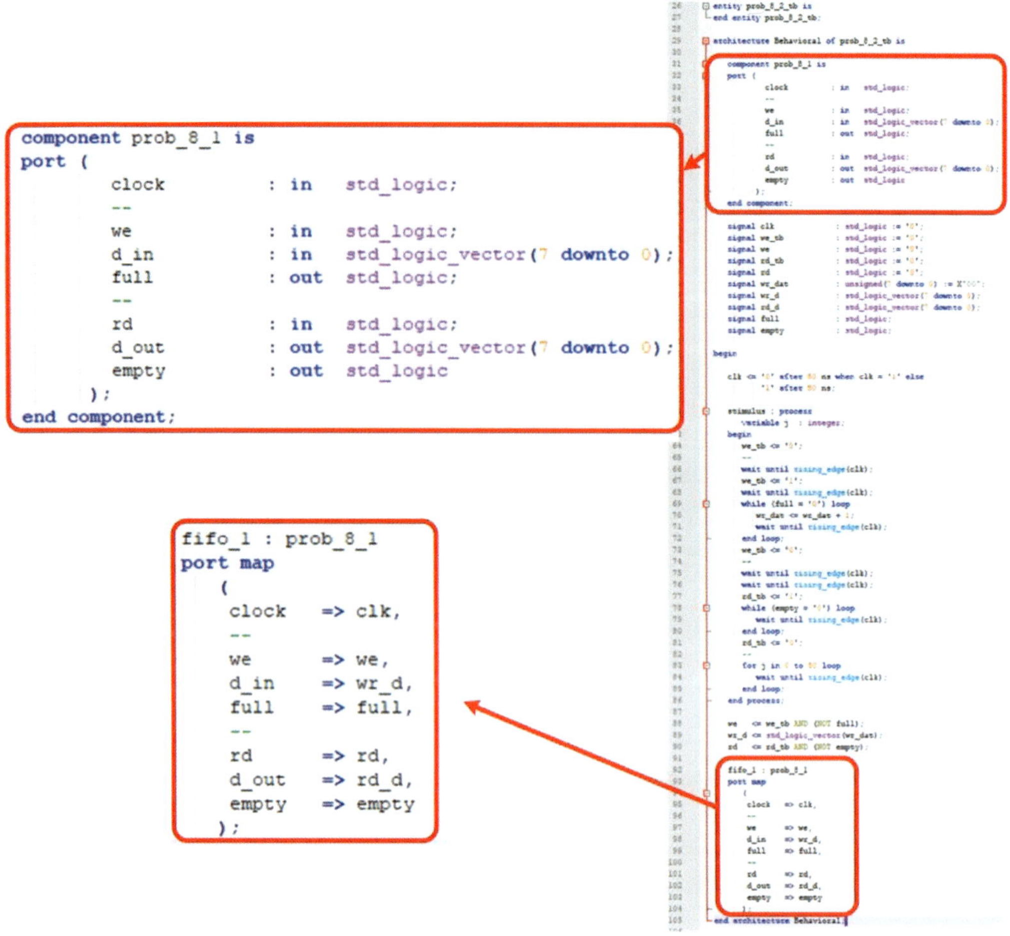

Here's the testbench signals:

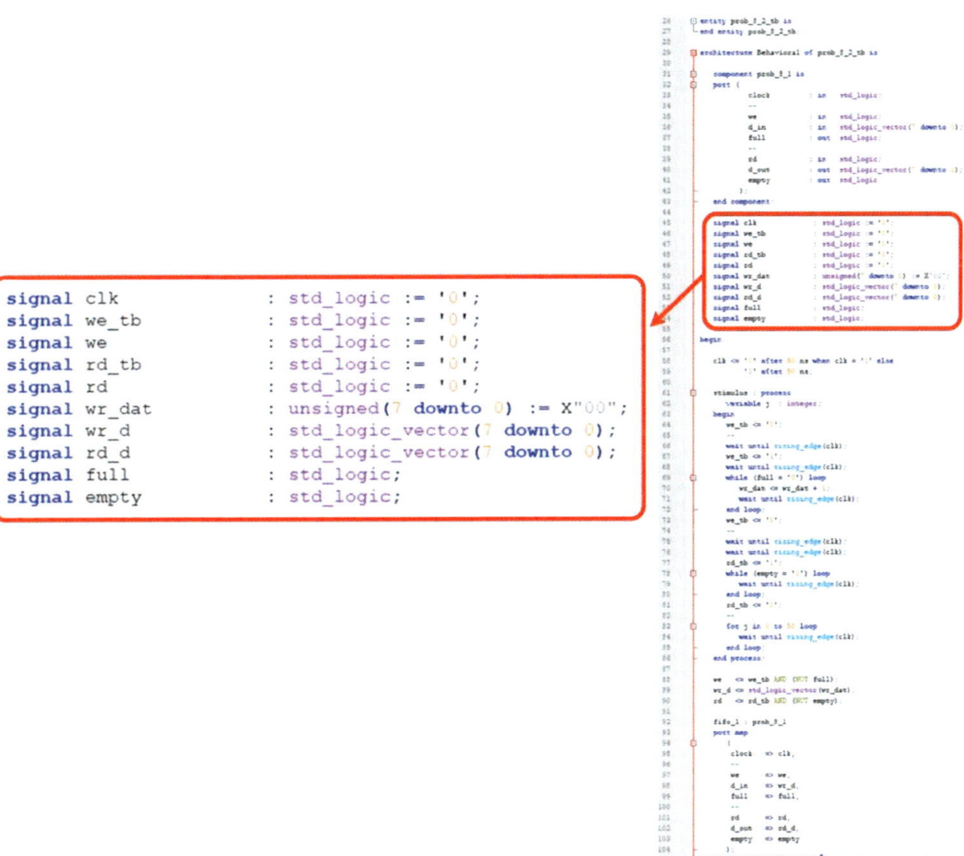

... the standard clock generation:

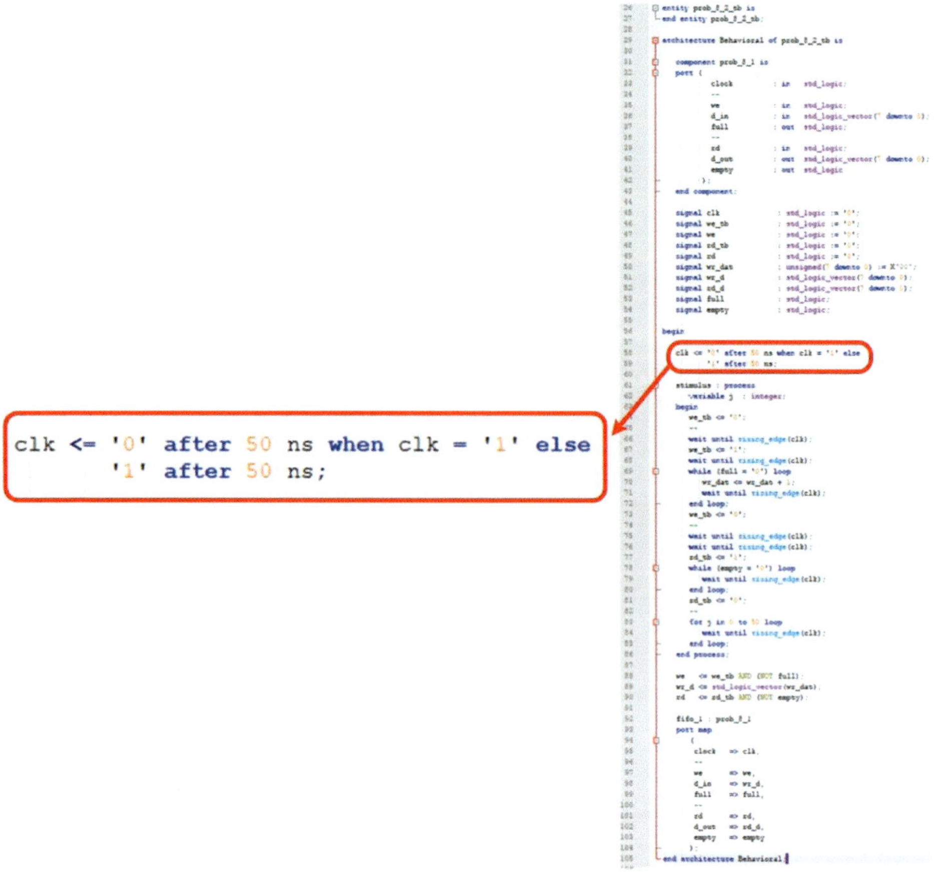

```
entity prob_8_2_tb is
end entity prob_8_2_tb;

architecture Behavioral of prob_8_2_tb is

   component prob_8_1 is
   port (
            clock        : in   std_logic;
            --
            we           : in   std_logic;
            d_in         : in   std_logic_vector(7 downto 0);
            full         : out  std_logic;
            --
            rd           : in   std_logic;
            d_out        : out  std_logic_vector(7 downto 0);
            empty        : out  std_logic
         );
   end component;

   signal clk              : std_logic := '0';
   signal we_tb            : std_logic := '0';
   signal we               : std_logic := '0';
   signal rd_tb            : std_logic := '0';
   signal rd               : std_logic := '0';
   signal wr_dat           : unsigned(7 downto 0) := X"00";
   signal wr_d             : std_logic_vector(7 downto 0);
   signal rd_d             : std_logic_vector(7 downto 0);
   signal full             : std_logic;
   signal empty            : std_logic;

begin

   clk <= '0' after 50 ns when clk = '1' else
          '1' after 50 ns;

   stimulus : process
      variable j  : integer;
   begin
      we_tb <= '0';
      --
      wait until rising_edge(clk);
      we_tb <= '1';
      wait until rising_edge(clk);
      while (full = '0') loop
         wr_dat <= wr_dat + 1;
         wait until rising_edge(clk);
      end loop;
      we_tb <= '0';
      --
      wait until rising_edge(clk);
      wait until rising_edge(clk);
      rd_tb <= '1';
      while (empty = '0') loop
         wait until rising_edge(clk);
      end loop;
      rd_tb <= '0';
      --
      for j in 0 to 50 loop
         wait until rising_edge(clk);
      end loop;
   end process;

   we   <= we_tb AND (NOT full);
   wr_d <= std_logic_vector(wr_dat);
   rd   <= rd_tb AND (NOT empty);

   fifo_1 : prob_8_1
   port map
      (
       clock   => clk,
       --
       we      => we,
       d_in    => wr_d,
       full    => full,
       --
       rd      => rd,
       d_out   => rd_d,
       empty   => empty
      );
end architecture Behavioral;
```

```
clk <= '0' after 50 ns when clk = '1' else
       '1' after 50 ns;
```

... and the meat, the process statement, and FIFO signal assignments:

```
stimulus : process
   variable j  : integer;
begin
   we_tb <= '0';
   --
   wait until rising_edge(clk);
   we_tb <= '1';
   wait until rising_edge(clk);
   while (full = '0') loop
      wr_dat <= wr_dat + 1;
      wait until rising_edge(clk);
   end loop;
   we_tb <= '0';
   --
   wait until rising_edge(clk);
   wait until rising_edge(clk);
   rd_tb <= '1';
   while (empty = '0') loop
      wait until rising_edge(clk);
   end loop;
   rd_tb <= '0';
   --
   for j in 0 to 50 loop
      wait until rising_edge(clk);
   end loop;
end process;

we   <= we_tb AND (NOT full);
wr_d <= std_logic_vector(wr_dat);
rd   <= rd_tb AND (NOT empty);
```

```
entity prob_8_2_tb is
end entity prob_8_2_tb;

architecture Behavioral of prob_8_2_tb is

   component prob_8_1 is
   port (
           clock          : in   std_logic;
           --
           we             : in   std_logic;
           d_in           : in   std_logic_vector(7 downto 0);
           full           : out  std_logic;
           --
           rd             : in   std_logic;
           d_out          : out  std_logic_vector(7 downto 0);
           empty          : out  std_logic
        );
   end component;

   signal clk              : std_logic := '0';
   signal we_tb            : std_logic := '0';
   signal we               : std_logic := '0';
   signal rd_tb            : std_logic := '0';
   signal rd               : std_logic := '0';
   signal wr_dat           : unsigned(7 downto 0) := X"00";
   signal wr_d             : std_logic_vector(7 downto 0);
   signal rd_d             : std_logic_vector(7 downto 0);
   signal full             : std_logic;
   signal empty            : std_logic;

begin

   clk <= '0' after 50 ns when clk = '1' else
          '1' after 50 ns;

   stimulus : process
      variable j  : integer;
   begin
      we_tb <= '0';
      --
      wait until rising_edge(clk);
      we_tb <= '1';
      wait until rising_edge(clk);
      while (full = '0') loop
         wr_dat <= wr_dat + 1;
         wait until rising_edge(clk);
      end loop;
      we_tb <= '0';
      --
      wait until rising_edge(clk);
      wait until rising_edge(clk);
      rd_tb <= '1';
      while (empty = '0') loop
         wait until rising_edge(clk);
      end loop;
      rd_tb <= '0';
      --
      for j in 0 to 50 loop
         wait until rising_edge(clk);
      end loop;
   end process;

   we   <= we_tb AND (NOT full);
   wr_d <= std_logic_vector(wr_dat);
   rd   <= rd_tb AND (NOT empty);

   fifo_1 : prob_8_1
   port map
      (
       clock   => clk,
       --
       we      => we,
       d_in    => wr_d,
       full    => full,
       --
       rd      => rd,
       d_out   => rd_d,
       empty   => empty
      );
end architecture Behavioral;
```

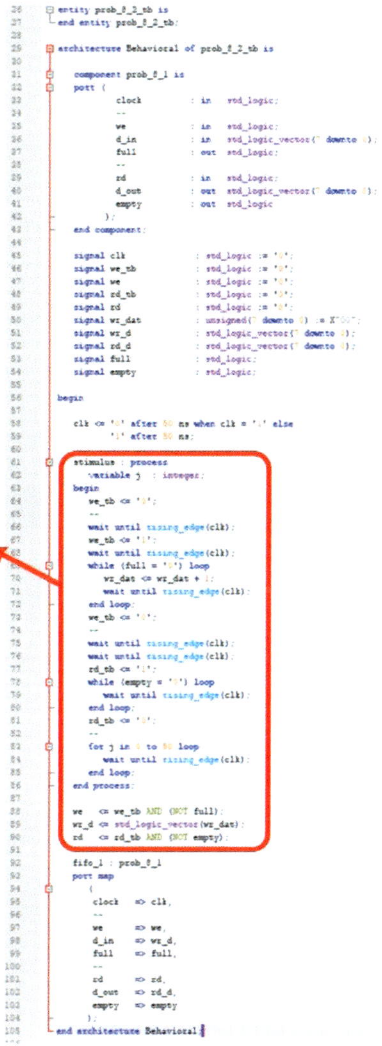

In this next figure:

```
stimulus : process
   variable j  : integer;
begin
   we_tb <= '0';
   --
   wait until rising_edge(clk);
   we_tb <= '1';
   wait until rising_edge(clk);
   while (full = '0') loop
      wr_dat <= wr_dat + 1;
      wait until rising_edge(clk);
   end loop;
   we_tb <= '0';
   --
   wait until rising_edge(clk);
   wait until rising_edge(clk);
   rd_tb <= '1';
   while (empty = '0') loop
      wait until rising_edge(clk);
   end loop;
   rd_tb <= '0';
   --
   for j in 0 to 50 loop
      wait until rising_edge(clk);
   end loop;
end process;

we   <= we_tb AND (NOT full);
wr_d <= std_logic_vector(wr_dat);
rd   <= rd_tb AND (NOT empty);
```


(1) the while loop that generates the write enable,
(2) the write enable goes active before the loop,
(3) and inactive after the loop is done,
(4) the FIFO full flag controls when we leave the loop,
(5) and we increment the write data each pass in order to create a test pattern.

After activating the write enable, I wait for a rising clock in order to catch a write to the write data zero value:

```
stimulus : process
   variable j  : integer;
begin
   we_tb <= '0';
   --
   wait until rising_edge(clk);
   we_tb <= '1';
   wait until rising_edge(clk);
   while (full = '0') loop
      wr_dat <= wr_dat + 1;
      wait until rising_edge(clk);
   end loop;
   we_tb <= '0';
   --
   wait until rising_edge(clk);
   wait until rising_edge(clk);
   rd_tb <= '1';
   while (empty = '0') loop
      wait until rising_edge(clk);
   end loop;
   rd_tb <= '0';
   --
   for j in 0 to 50 loop
      wait until rising_edge(clk);
   end loop;
end process;

we   <= we_tb AND (NOT full);
wr_d <= std_logic_vector(wr_dat);
rd   <= rd_tb AND (NOT empty);
```

… this synchronizes the internal FIFO write addresses with the write data. This is a tweak to the testbench for clarity.

This while loop creates this portion of the waveform, where the write enable signal goes active for 8 clocks, and the full flag then goes active.

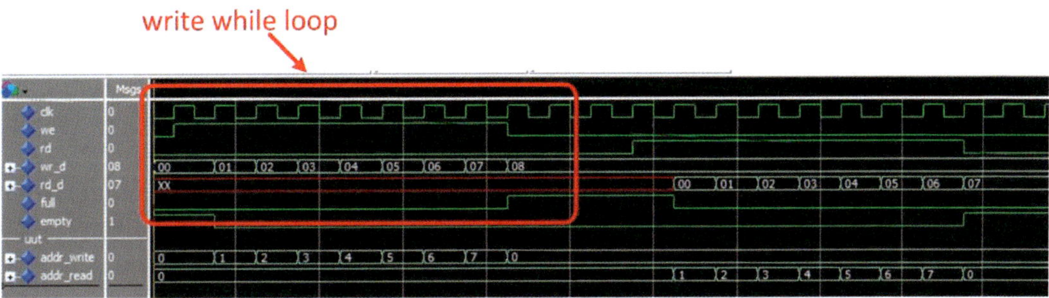

Note that in order to de-activate the write enable signal during the same clock period that the full flag goes active:

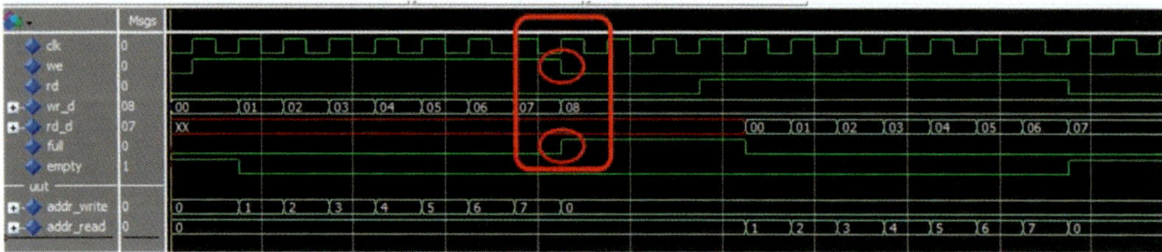

... I had to gate the FIFO write enable signal asynchronously with the full flag:

```
stimulus : process
   variable j  : integer;
begin
   we_tb <= '0';
   --
   wait until rising_edge(clk);
   we_tb <= '1';
   wait until rising_edge(clk);
   while (full = '0') loop
      wr_dat <= wr_dat + 1;
      wait until rising_edge(clk);
   end loop;
   we_tb <= '0';
   --
   wait until rising_edge(clk);
   wait until rising_edge(clk);
   rd_tb <= '1';
   while (empty = '0') loop
      wait until rising_edge(clk);
   end loop;
   rd_tb <= '0';
   --
   for j in 0 to 50 loop
      wait until rising_edge(clk);
   end loop;
end process;

we   <= we_tb AND (NOT full);
wr_d <= std_logic_vector(wr_dat);
rd   <= rd_tb AND (NOT empty);
```

Here I add two clocks just for some separation:

```
stimulus : process
   variable j  : integer;
begin
   we_tb <= '0';
   --
   wait until rising_edge(clk);
   we_tb <= '1';
   wait until rising_edge(clk);
   while (full = '0') loop
      wr_dat <= wr_dat + 1;
      wait until rising_edge(clk);
   end loop;
   we_tb <= '0';
   --
   wait until rising_edge(clk);
   wait until rising_edge(clk);
   rd_tb <= '1';
   while (empty = '0') loop
      wait until rising_edge(clk);
   end loop;
   rd_tb <= '0';
   --
   for j in 0 to 50 loop
      wait until rising_edge(clk);
   end loop;
end process;

we   <= we_tb AND (NOT full);
wr_d <= std_logic_vector(wr_dat);
rd   <= rd_tb AND (NOT empty);
```

And here's the read while loop:

```
stimulus : process
   variable j  : integer;
begin
   we_tb <= '0';
   --
   wait until rising_edge(clk);
   we_tb <= '1';
   wait until rising_edge(clk);
   while (full = '0') loop
      wr_dat <= wr_dat + 1;
      wait until rising_edge(clk);
   end loop;
   we_tb <= '0';
   --
   wait until rising_edge(clk);
   wait until rising_edge(clk);
   rd_tb <= '1';                         (1)
   while (empty = '0') loop
      wait until rising_edge(clk);       read while loop
   end loop;
   rd_tb <= '0';                         (2)
   --
   for j in 0 to 50 loop
      wait until rising_edge(clk);
   end loop;
end process;

we   <= we_tb AND (NOT full);
wr_d <= std_logic_vector(wr_dat);
rd   <= rd_tb AND (NOT empty);           (3)
```

(1) with the read enable activated before it, and

(2) de-activated after it;

(3) and, like the write enable, I asynchronously deactivate the read enable with the empty flag.

Finally, I insert a filler for loop to prevent the process statement from starting over:

```
stimulus : process
   variable j   : integer;
begin
   we_tb <= '0';
   --
   wait until rising_edge(clk);
   we_tb <= '1';
   wait until rising_edge(clk);
   while (full = '0') loop
      wr_dat <= wr_dat + 1;
      wait until rising_edge(clk);
   end loop;
   we_tb <= '0';
   --
   wait until rising_edge(clk);
   wait until rising_edge(clk);
   rd_tb <= '1';
   while (empty = '0') loop
      wait until rising_edge(clk);
   end loop;
   rd_tb <= '0';
   --
   for j in 0 to 50 loop
      wait until rising_edge(clk);
   end loop;
end process;

we   <= we_tb AND (NOT full);
wr_d <= std_logic_vector(wr_dat);
rd   <= rd_tb AND (NOT empty);
```

Looking again at the waveform …

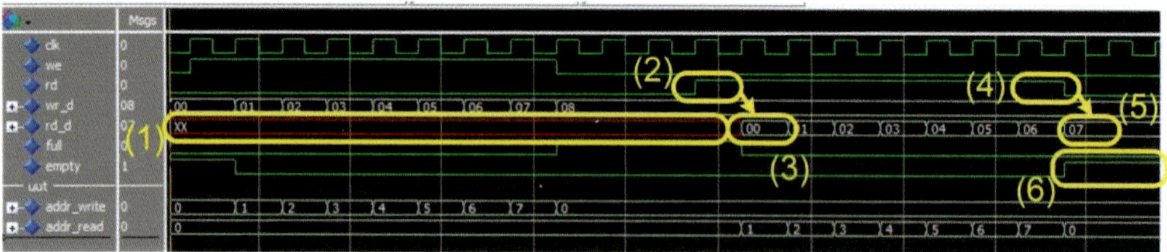

(1) we note that the output of the FIFO is indeterminate until we perform the first read. This is a simulation artifact. In an actual synthesized design, this would be whatever happened to be in the FIFO output register—either some remnant from previous writes and reads, or zero if fresh from power-up,

(2) remembering that I made sure that the write data was the same as the internal FIFO write address, we see that the first read at this clock,

(3) yields a zero value,

(4) the 8th read …

(5) … yields the last value—seven,

(6) when the empty flag then goes active.

Exercise Answers, Chapter 4

Exercise 4-1:
Fill in the locations in the memory with the data written below.

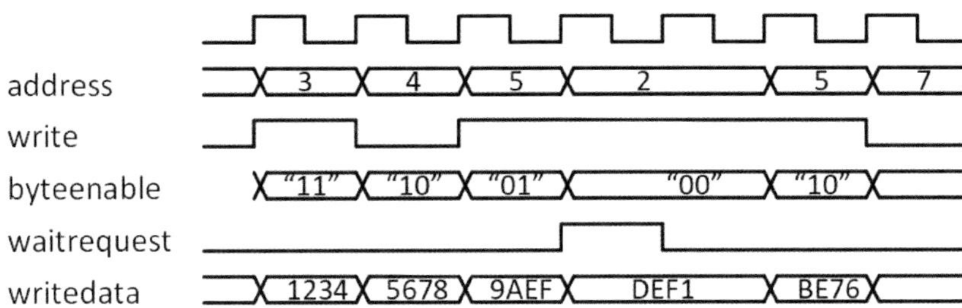

(values are in hex)

address	byte 1	byte 0
0	00	00
1	00	00
2	00	00
3	00	00
4	00	00
5	00	00
6	00	00
7	00	00

In the first exercise, we're putting a memory-mapped bus through some paces. The first clock is straight-forward, we write the hex 1234 to memory address 3:

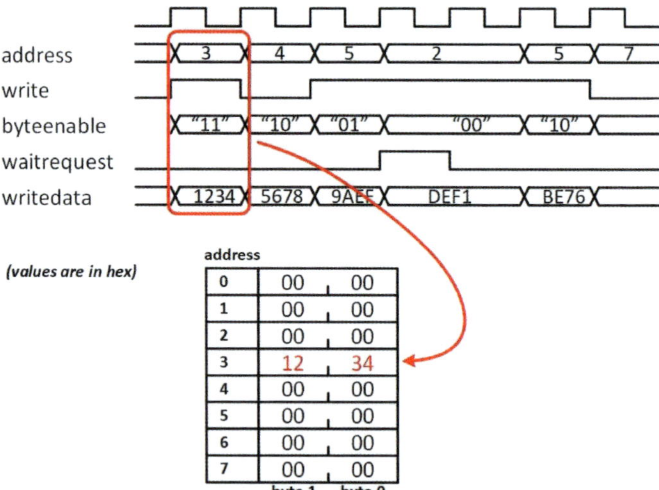

The next clock is a do-nothing, since the write signal is de-asserted:

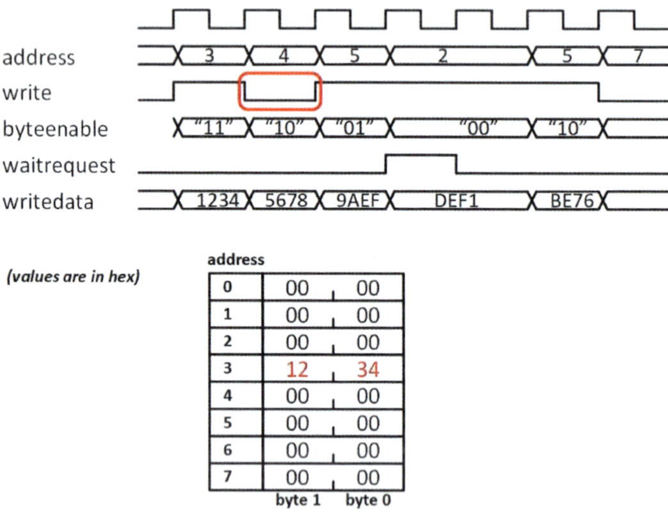

For the next clock, we write only the LS byte to address 5:

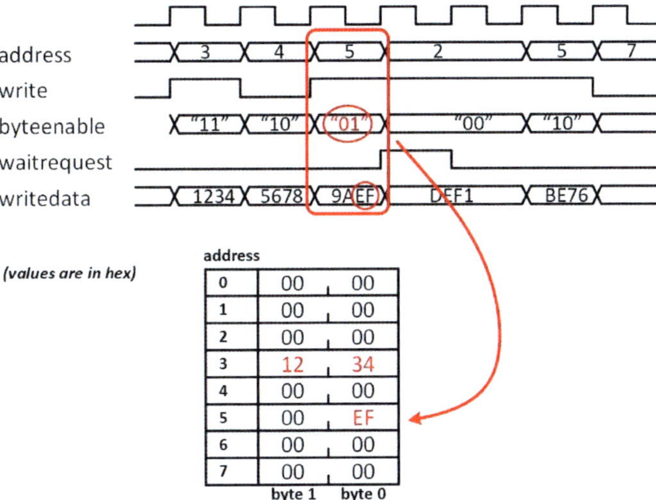

address	byte 1	byte 0
0	00	00
1	00	00
2	00	00
3	12	34
4	00	00
5	00	EF
6	00	00
7	00	00

(1) Waitrequest pauses the operation one clock, and (2) then we still don't write anything, since the byte enables are all zero:

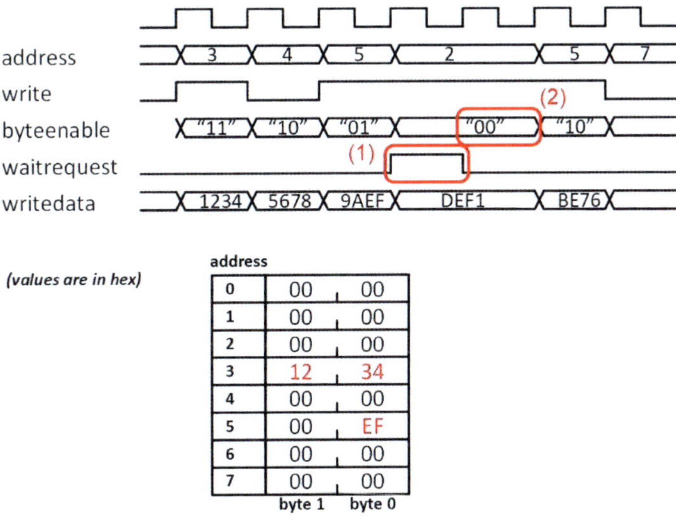

address	byte 1	byte 0
0	00	00
1	00	00
2	00	00
3	12	34
4	00	00
5	00	EF
6	00	00
7	00	00

Finally, we write just the MS byte to address 5, completing the iconic hex "beef":

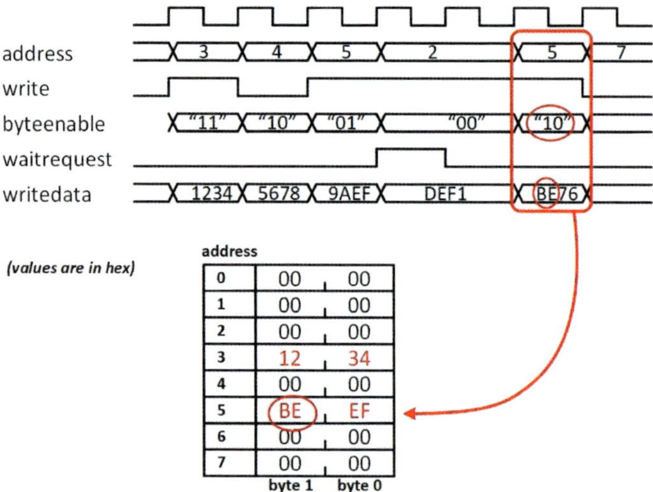

==

Exercise 4-2:

Fill in the read data from the memory below.

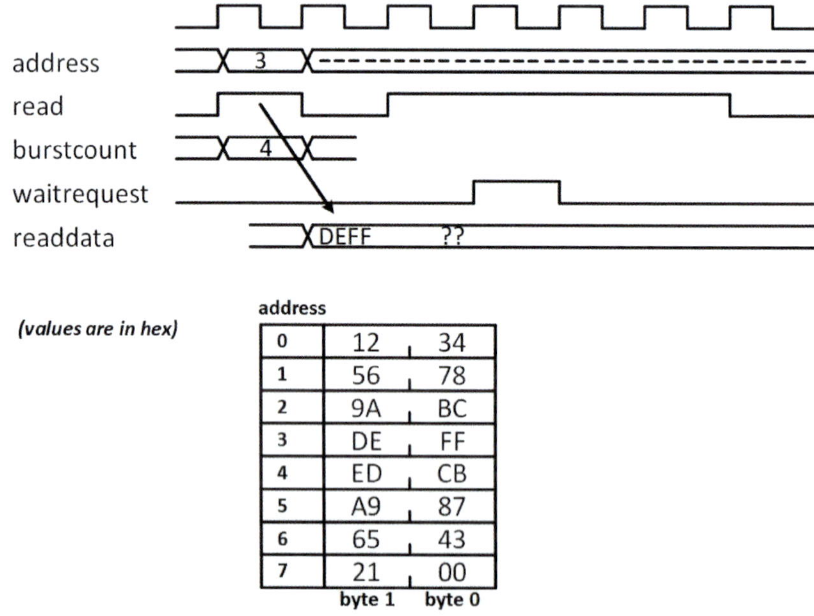

The second exercise is a burst. We kick it off by showing the first word read:

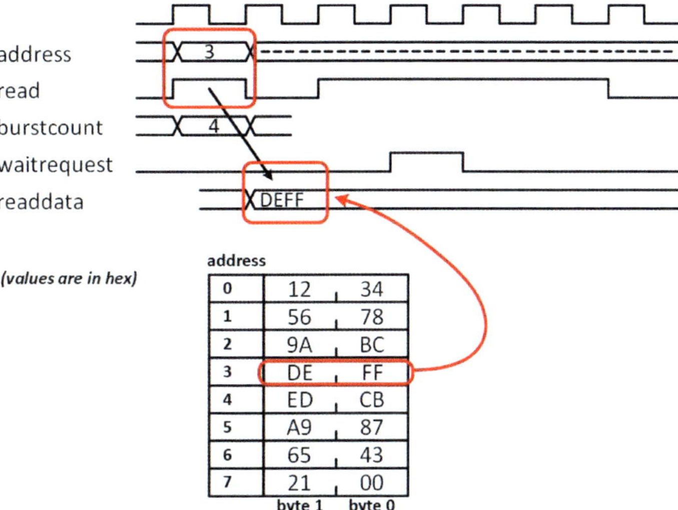

The next clock is a do-nothing:

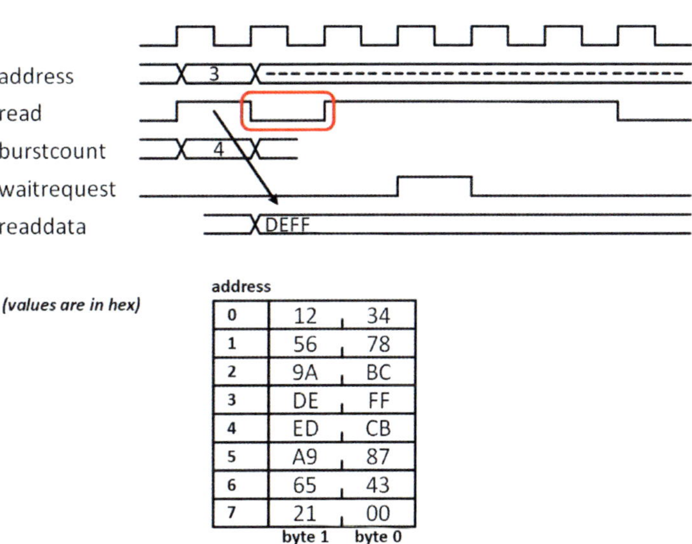

And then we (the slave side) increment our address count to 4, and present the second burst word, hex EDCB, from the memory onto the readdata bus:

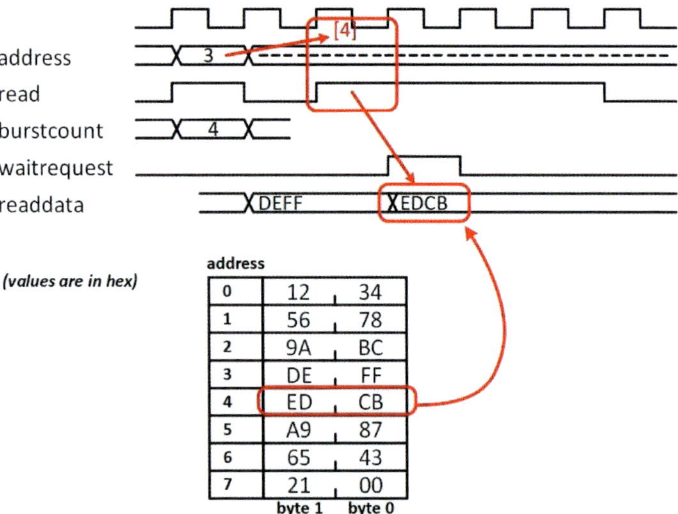

We pause for a waitrequest:

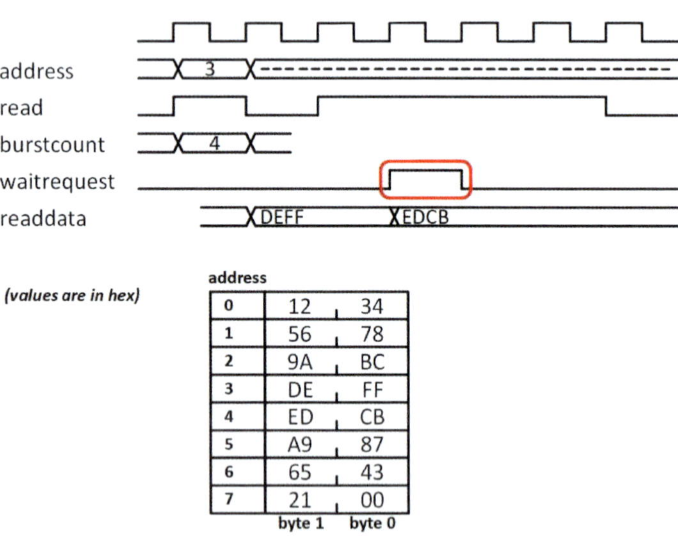

And then increment our address, and present the third word, hex A987, from address 5:

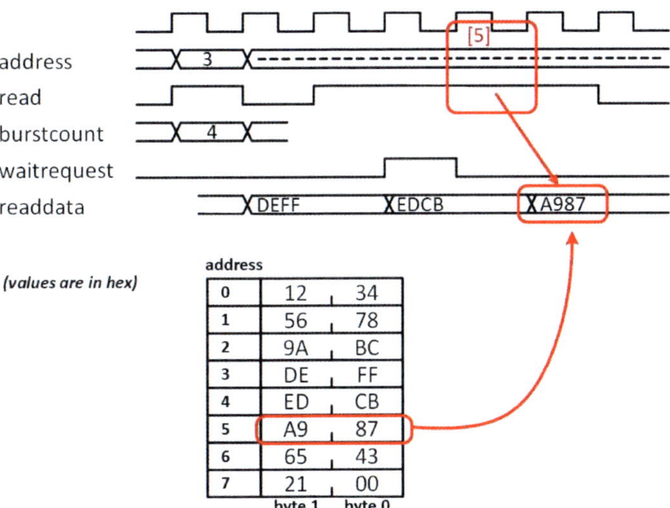

Finally, we read the last word from address 6:

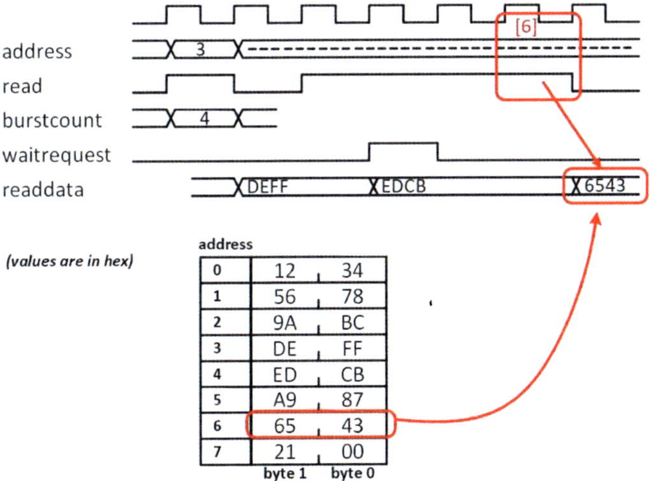

Exercise Answers, Chapter 5

Exercise 5-1:
Perform the following binary arithmetic.

i. add these two unsigned numbers

```
   0 1 1 1 1 1 1 1
 + 0 0 0 0 0 0 0 1
   ---------------
   ? ? ? ? ? ? ? ?
```

ii. add these two signed numbers

```
   S
  [0]0 1 1 1 0 1 1
 +[0]0 0 1 0 0 0 1
   ---------------
   ? ? ? ? ? ? ? ?
```

iii. add these two signed numbers

```
   S
  [0]1 0 0 0 1 0 0
 +[1]1 1 1 1 0 0 0
   ---------------
   ? ? ? ? ? ? ? ?
```

iv. subtract these two signed numbers

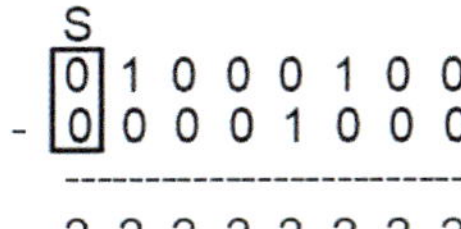

The first exercise example is just a refresher about adding binary numbers, and how we carry a one when adding two ones.

i. add these two unsigned numbers

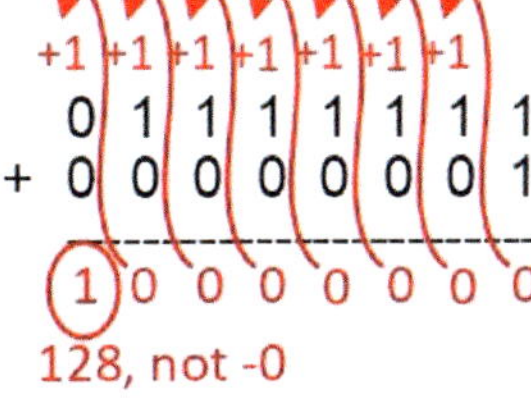

Note that since this is an unsigned number, the MS bit makes this value a positive 128, not a negative zero (whatever that is).

For the second exercise example, we move on to signed numbers, and here we simply see that adding two positive numbers is the same as adding unsigned numbers, as long as we don't roll over into the MS bit:

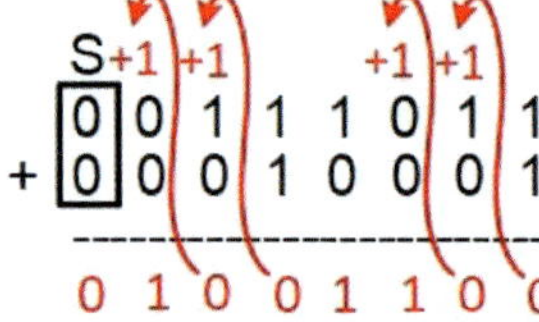

For the third exercise example, we add a signed positive number to a signed negative number, which is the same as subtracting the second from the first. But, whether positive or negative, all we ever do with binary arithmetic is add, and here, a one carries into the sign position, which makes the answer positive:

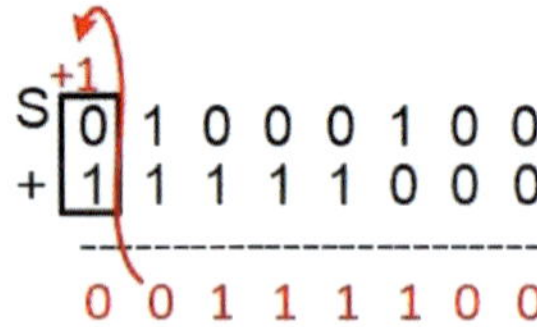

We see that the first number is a positive decimal 68, and to find the value of the second number, since it's negative, we perform a two's compliment operation to determine the negative amplitude.

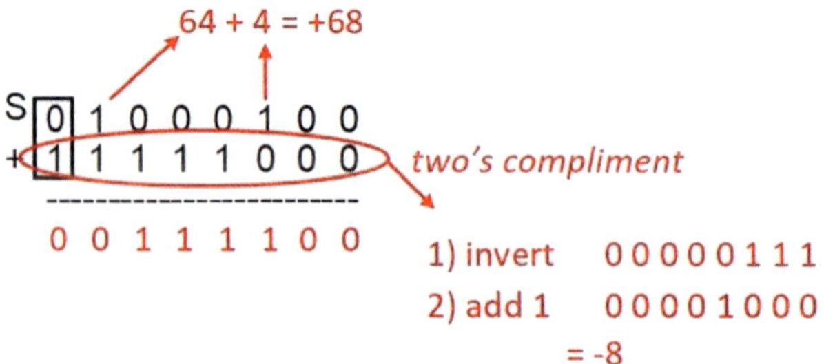

We first invert, and then add a one, which yields an amplitude of 8.

So, our addition consisted of adding a positive 68 to a negative 8, which is 60.

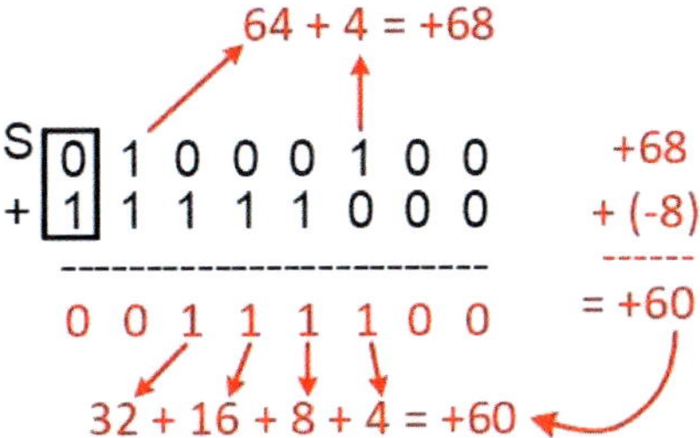

… and, sure enough the two methods agree.

What we've shown is that indeed, by using two's compliment for negative numbers, we can perform both addition and subtraction using just binary addition.

Finally, in the fourth exercise example, we're asked to do a subtraction, and we'll put the two's compliment operation to the test.

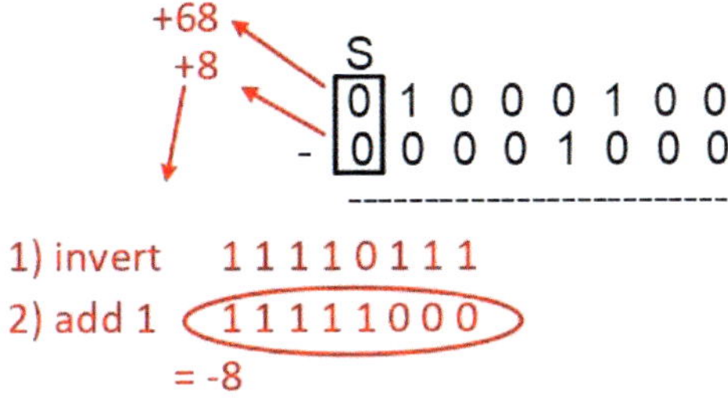

We invert and add one, to get the two's compliment of 8, i.e., a negative 8. Then we do the binary addition:

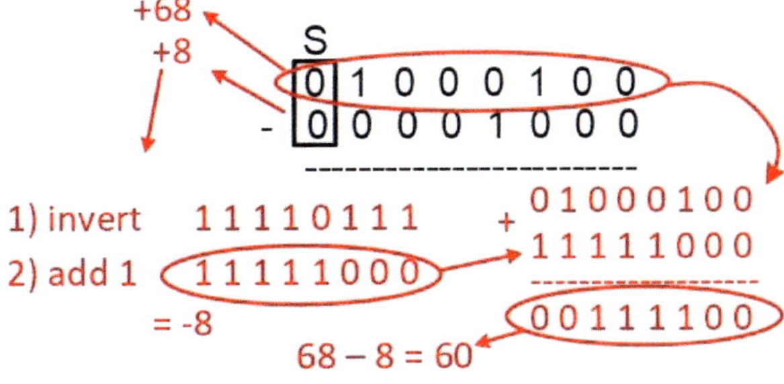

… which gives us 60. But you had no doubts.

==

Exercise 5-2:
Complete the timing diagram.

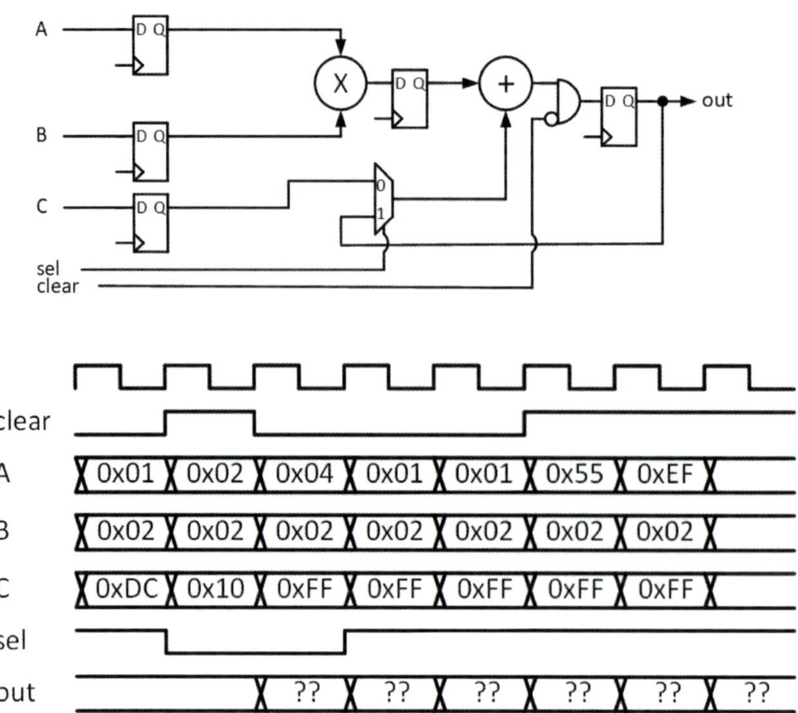

DSP IP from the FPGA vendors will always include some timing-driven pipeline registers in the path (registers that have no purpose other than to delay the signal by one clock), some of them configurable by you, the user.

The clear input at clock 2 sets the output to zero on the next clock (3).

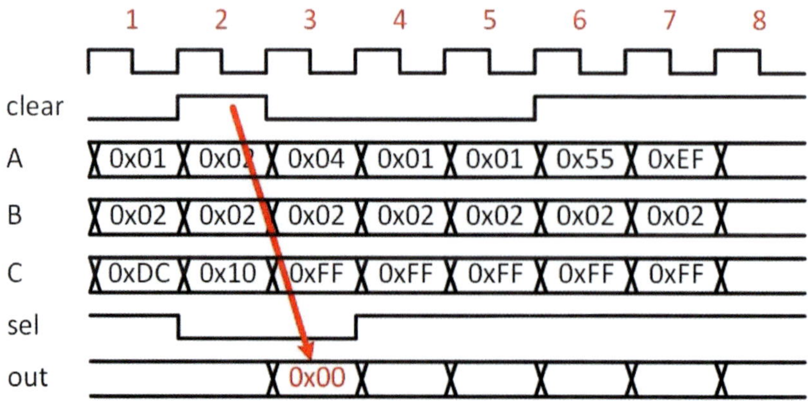

Here (clock 3), the mux select line is low, allowing the registered C input to feed the adder.

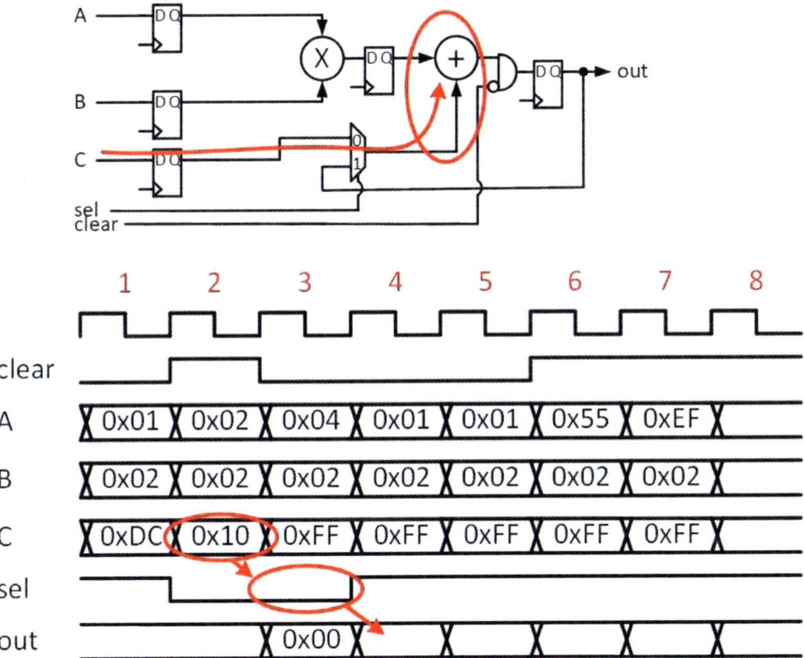

Note that because of signal C's input register, the C value at clock 2 is applied to the enabled adder at clock 3, and the result due to C's contribution is registered out at clock 4.

Since there are two registers between the A and B inputs and the adder, the values at clock 1 contribute to the output.

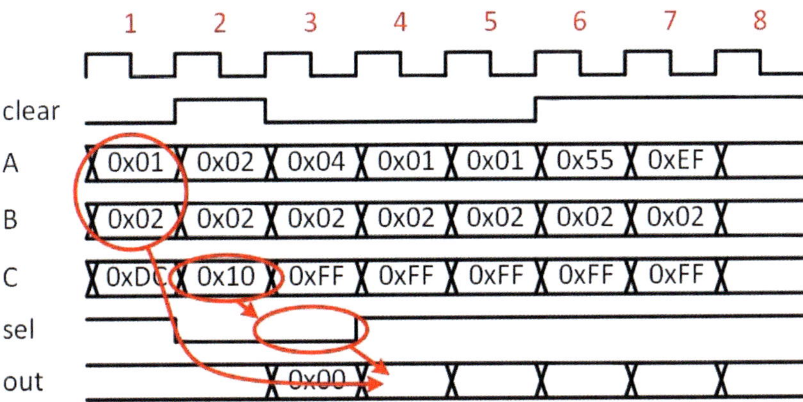

Doing the arithmetic, we get a hex 1-2 as an answer:

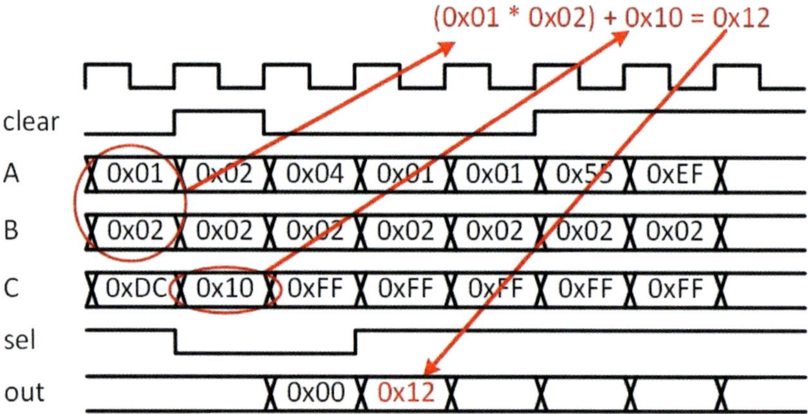

On the next clock (4), the mux select goes high, feeding the output back to the adder, for the answer at clock 5:

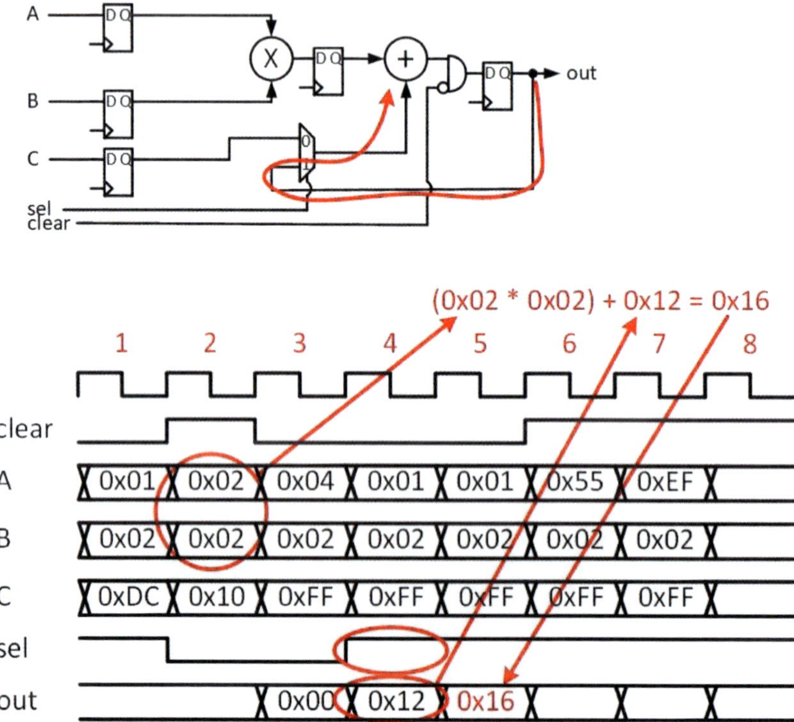

The next clock is the same operation:

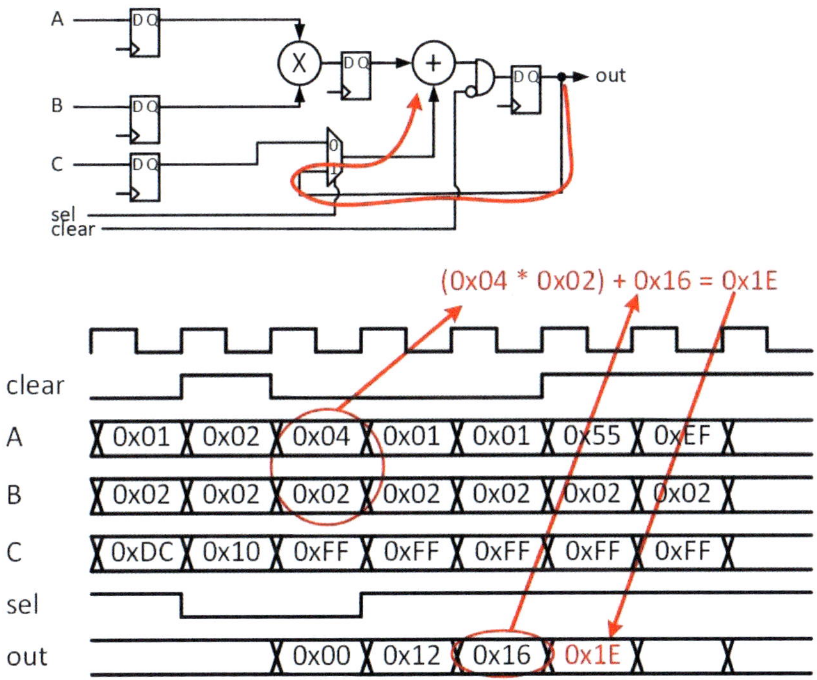

And finally, we end with the block cleared again

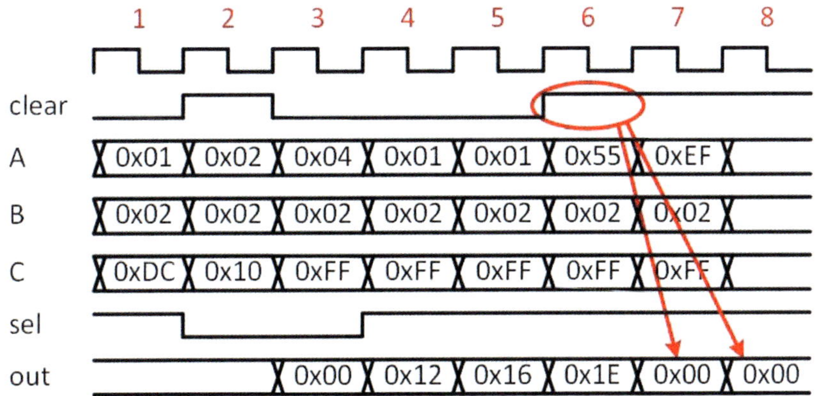

Exercise Answers, Chapter 6

Exercise 6-1:
Add parity to the Rx half of the implemented UART. The "par_err" signal goes active when parity errors are detected. The parity is even when "par_even" is asserted, otherwise it is odd.

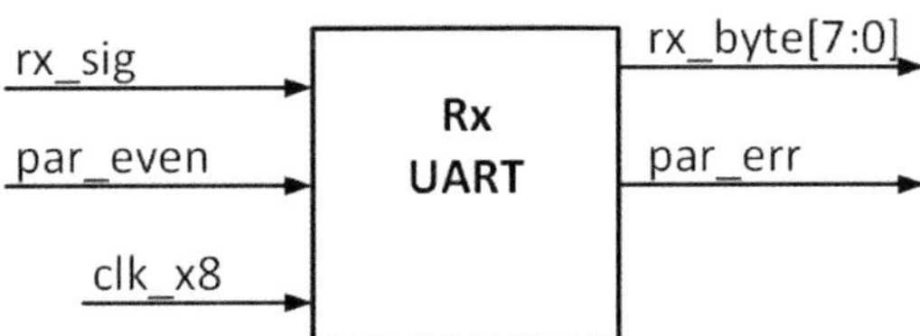

Time for some design practice.

When enabled, the parity option includes a parity bit in each transmission. The calculation on the received bits and subsequent check against the transmitted parity bit occurs in parallel with the assembly of the received character, so we can ignore the assembly logic of the original designs, and work with just the state machine.

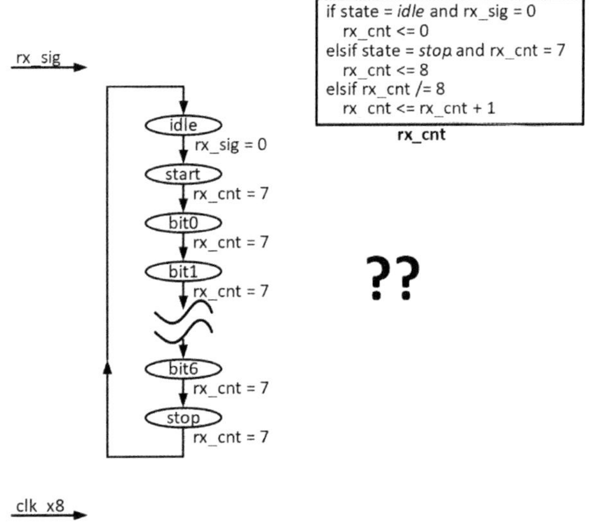

Reviewing the UART transmission, we saw that the parity bit is the eighth bit (when the UART is configured for a 7-bit character):

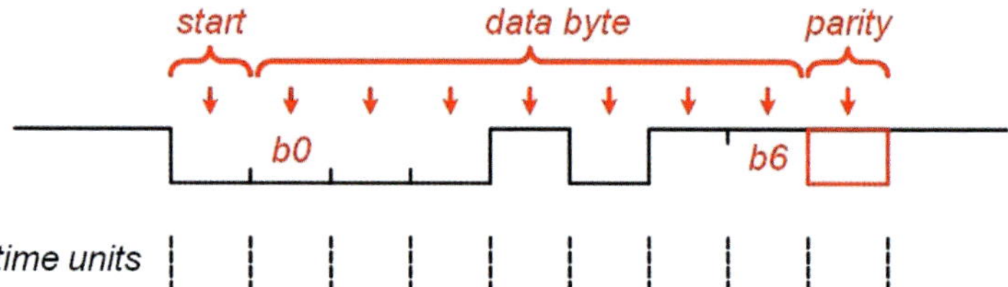

So we add a new state to the state machine.

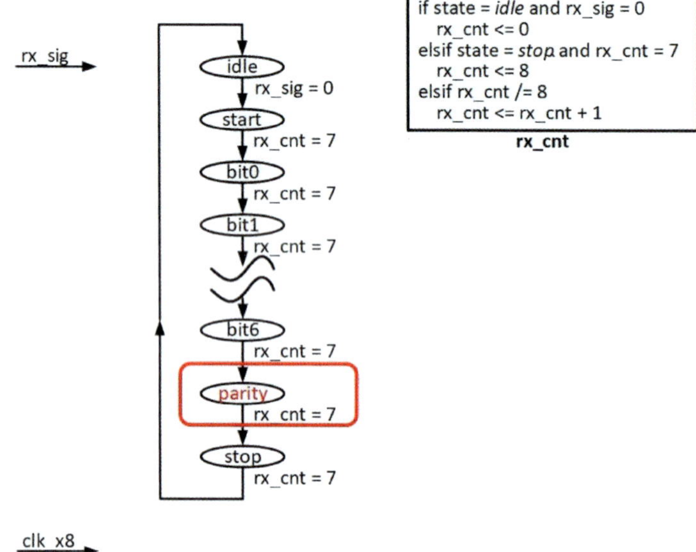

The parity calculations are performed over this span:

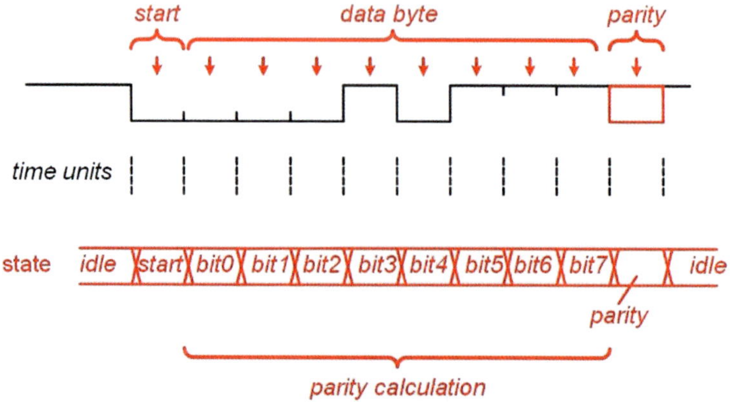

This is the pseudo-code for the parity calculation:

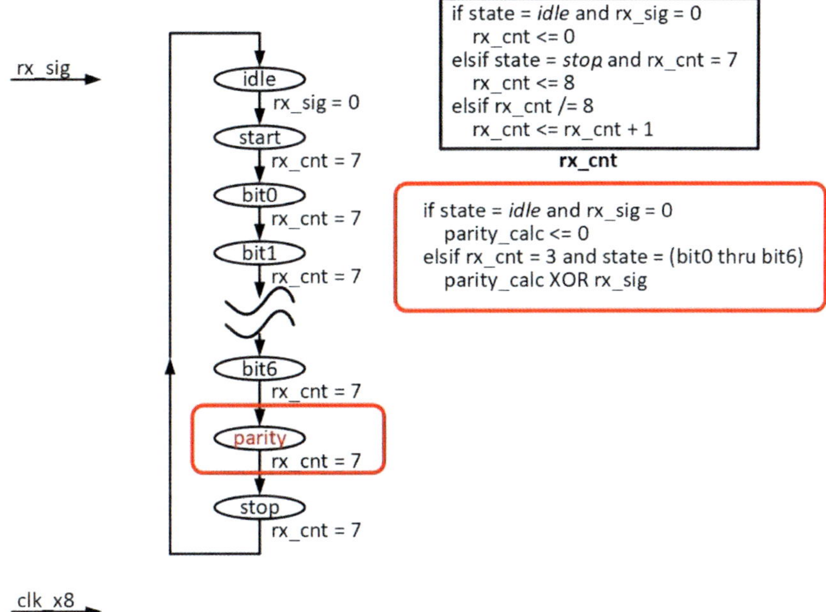

We clear the calculated parity at the beginning of each transmission.

```
if state = idle and rx_sig = 0
   parity_calc <= 0
elsif rx_cnt = 3 and state = (bit0 thru bit6)
   parity_calc XOR rx_sig
```

And, at the same time that the received bits are sampled (when the rx_cnt is 3) we XOR the current calculated parity value with that bit.

```
if state = idle and rx_sig = 0
   parity_calc <= 0
elsif rx_cnt = 3 and state = (bit0 thru bit6)
   parity_calc XOR rx_sig
```

An even parity configuration means that the transmitted parity bit makes the number of ones in the transmission an even amount.

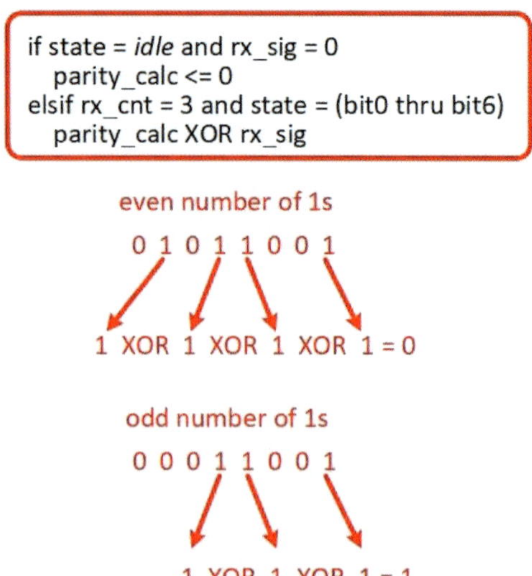

if state = *idle* and rx_sig = 0
parity_calc <= 0
elsif rx_cnt = 3 and state = (bit0 thru bit6)
parity_calc XOR rx_sig
even number of 1s
0 1 0 1 1 0 0 1
1 XOR 1 XOR 1 XOR 1 = 0
odd number of 1s
0 0 0 1 1 0 0 1
1 XOR 1 XOR 1 = 1

An XOR of two ones is zero. An XOR of another one—now three—is one. You can see that XORing an even number of ones yields zero, while XORing an odd number of ones yields a one. Imagine that a transmission has an even number of ones. That means that the transmitted parity bit is zero. If we XOR the transmitted parity—zero—with our calculated parity—also zero, the result is zero. Now imagine that a transmission has an odd number of ones. That means that the transmitted parity bit is one. If we XOR the transmitted parity—one—with our calculated parity—also one—the result is again zero. Any other combination is a one, and indicates that at least one bit was errored during transmission.

So, here we do just that.

```
if state = idle and rx_sig = 0
   parity_calc <= 0
elsif rx_cnt = 3 and state = (bit0 thru bit6)
   parity_calc XOR rx_sig

if state = parity and rx_cnt = 3
   parity_err <= rx_sig XOR parity_calc          par_err →
```

We XOR the received parity bit—rx_sig—with our calculated parity—parity_calc—and that result directly becomes our par_err output.

Here's an alternative:

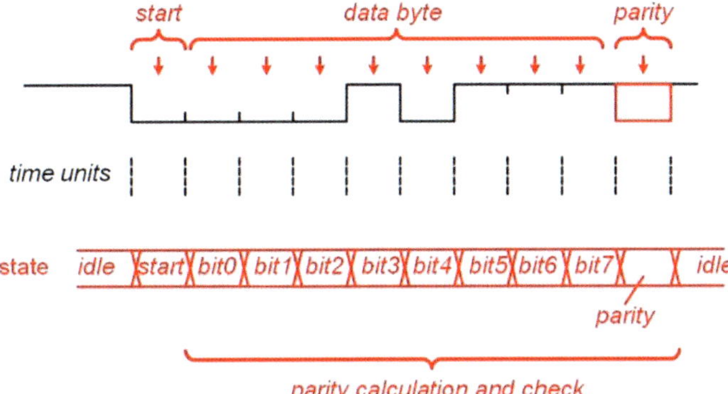

Notice that our parity calculation simply XORs all the incoming bits with the ongoing calculation. That is exactly what we just did to determine if there was an error (we XORed the parity bit with the final parity calculation of the data bits, which was just XORing them in turn).

So, the par_err logic can be slightly simplified by including the parity bit in the calculation, and latching the error indication at the very end—the stop state—after we've included the parity in the calculation.

```
if state = idle and rx_sig = 0
   parity_calc <= 0
elsif rx_cnt = 3 and state = (bit0 thru parity)
   parity_calc XOR rx_sig
```

```
if state = stop and rx_cnt = 3
   parity_err <= parity_calc
```

par_err

All of this has been for an even parity configuration. UARTs also allow a configuration for odd parity, which means that the parity bit ensures that there's an odd number if total bits in the transmission. Thus, for odd parity configurations, we simply invert the final result to become the error output:

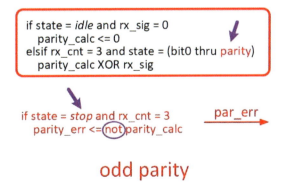

```
if state = idle and rx_sig = 0
   parity_calc <= 0
elsif rx_cnt = 3 and state = (bit0 thru parity)
   parity_calc XOR rx_sig
```

```
if state = stop and rx_cnt = 3
   parity_err <= not parity_calc
```

par_err

odd parity

Exercise Answers, Chapter 7

Exercise 7-1:
Using the I2C protocol below, show the I2C transaction when the master reads a 0x5A data value from slave device 0x52 at register address 0x03.

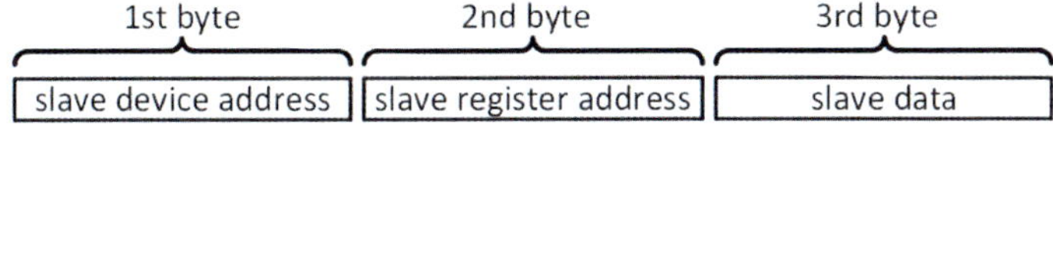

We'll start off by reviewing how the start bit is formed. In I2C, both the clock and data lines are high when idle (they only go low when the master or slave drives them). The data line is allowed to transition only when the clock is low for transaction bits, and a start bit is defined as the data line falling while the clock line is high.

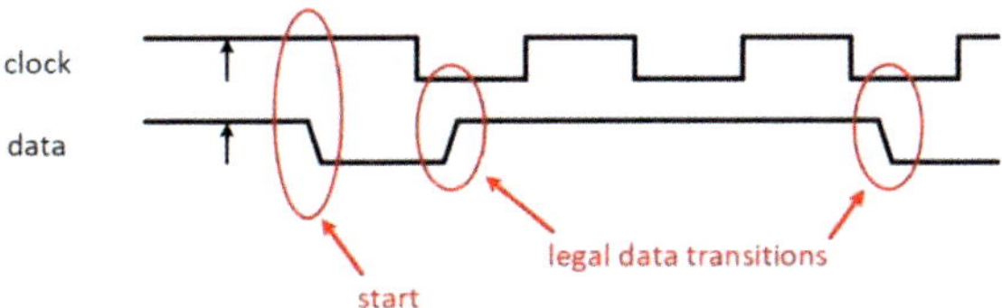

The first byte consists of the slave address, which is indicated to be a hex 52. The presents a dilemma, however. We know that the LS bit of this 8-bit field is the read/write indicator. So, is this hex 52 just from bits 6 down to 0, i.e., "1010010"? No. The standard way of indicating this is to say that the device address is the full 8 bits, from 7 down to 0. Thus, device addresses are always even, leaving the LS bit available for the read/write bit. Since this is a read transaction, we make this bit high.

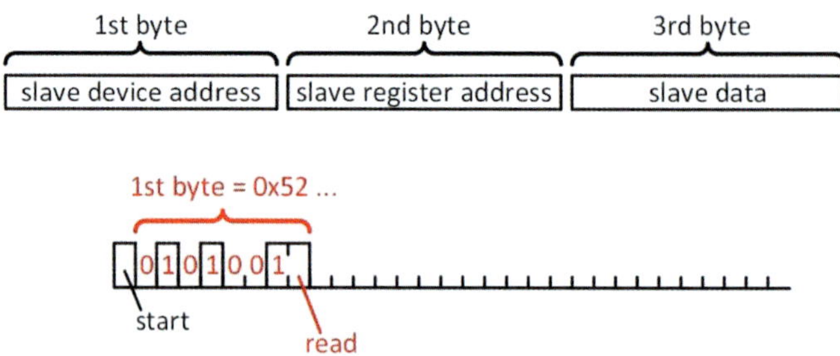

This is followed by the slave's acknowledge bit, a low, and then the slave register address:

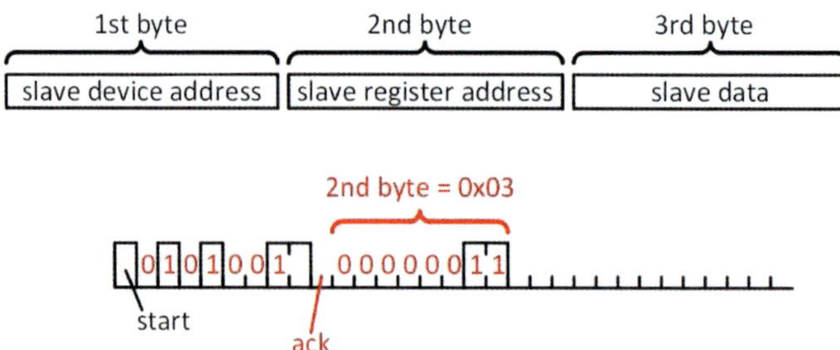

The slave acknowledges this, and the slave drives the bus with the read byte (0x5A):

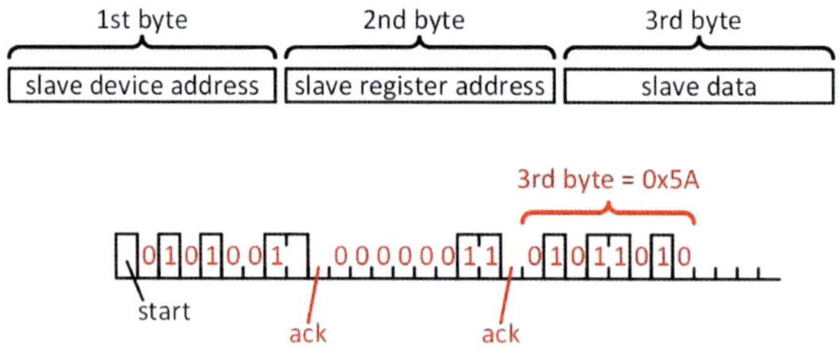

This is followed by a final slave ack, and a stop bit:

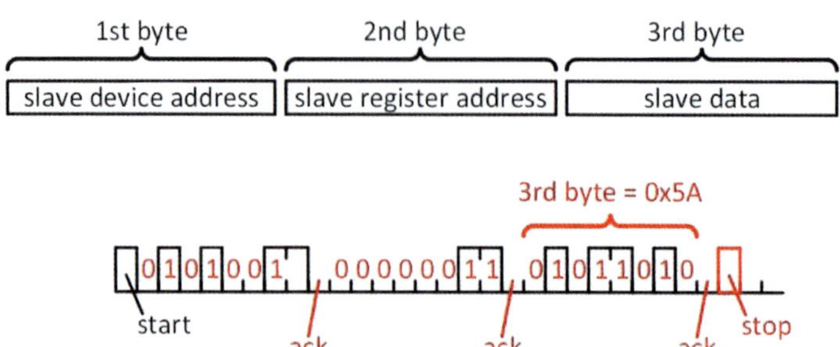

Note that this last acknowledge is driven by the master, since it is the one now on the receiving end.

Like the start bit, the stop indication consists of the data line transitioning while the clock is high, except whereas the data line goes high-to-low for a start bit, here, the data line goes low-to-high for a stop.

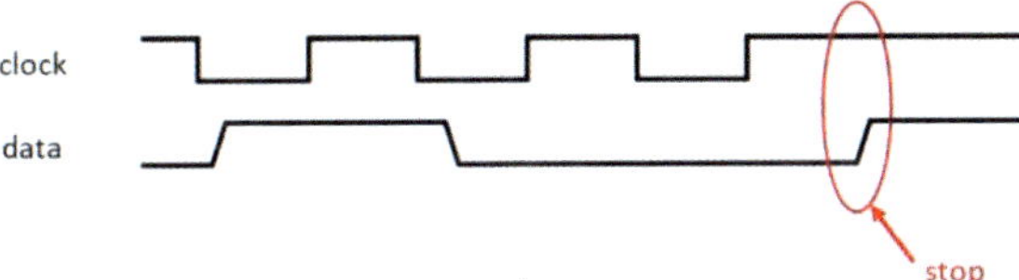

Note that, whereas UART communications are sent LSB-first, in I2C communications, it's the opposite, the MSB goes first.

Exercise Answers, Chapter 8

Exercise 8-1:
Modify this cross-clock domain circuit such that only pulses that are two clock_1 clocks wide on the left side, cross the domains, and that the result is then a pulse two clock_2 clocks wide on the right side.

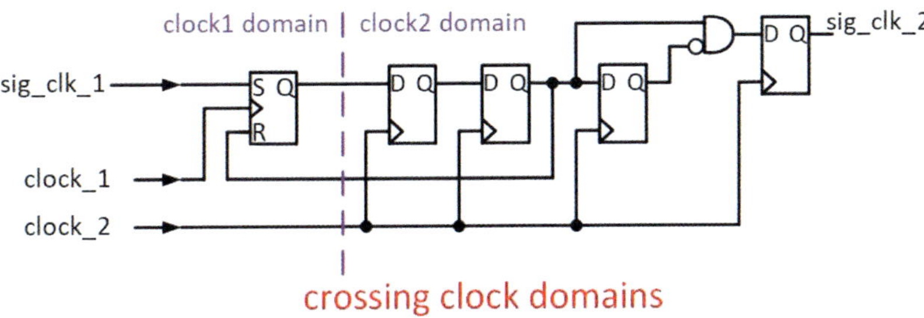

Here's one version:

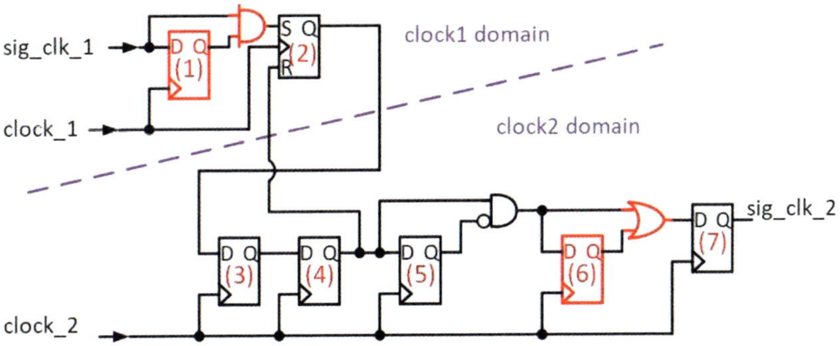

I've added the pipeline delay register (1) to the input, so that both the sig_clk_1 input and its delayed state must be high to set the set/reset flop (2). This satisfies the first part of the exercise (only two-clock long pulses get through).

The d-flop (3) isolates metastability as before, and, as before, d-flop (4) resets the set/reset d-flop, since we have now firmly captured the clock_1 double-pulse event.

Again as before, d-flop (5) and the AND gate together filter all but rising-edge events.

D-flop (6) creates the double-wide pulse out.

Exercise 8-2:
Assume that this counter is running at such a high clock rate that the vendor's placement/router tools cannot accommodate more than three inputs on any AND gate feeding each stage's XOR gate. Add a pipeline register to allow the tool to build this counter.

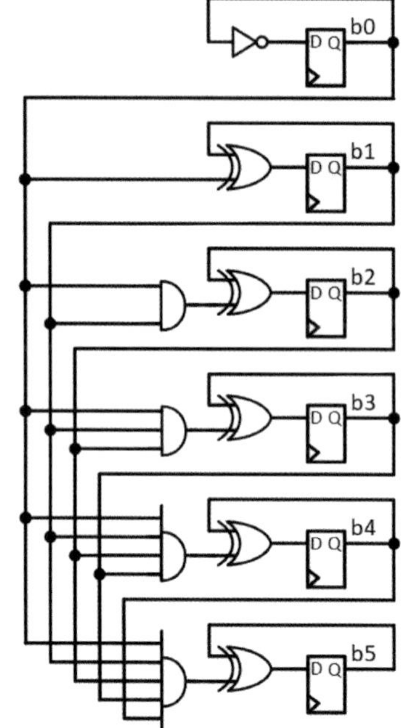

Since three inputs to an AND gate is the limit, we'll break the counter here:

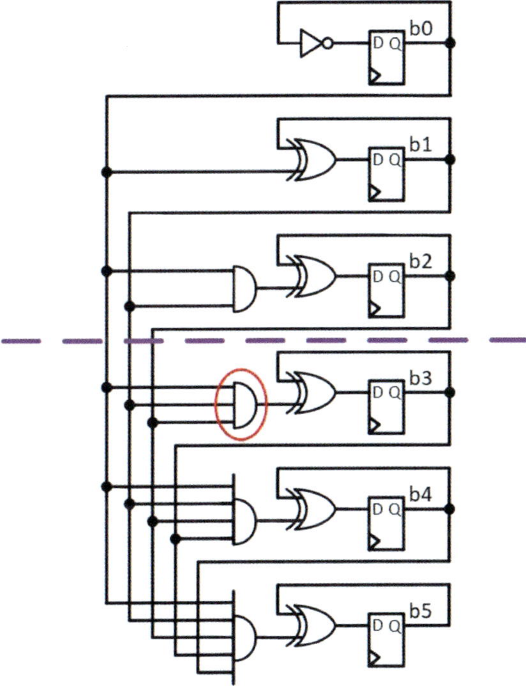

The circled AND gate decodes when the first three stages have reached their terminal value of all-ones and it's time to toggle the next stage (b3). However, notice that this three-input AND is actually included in the next stages:

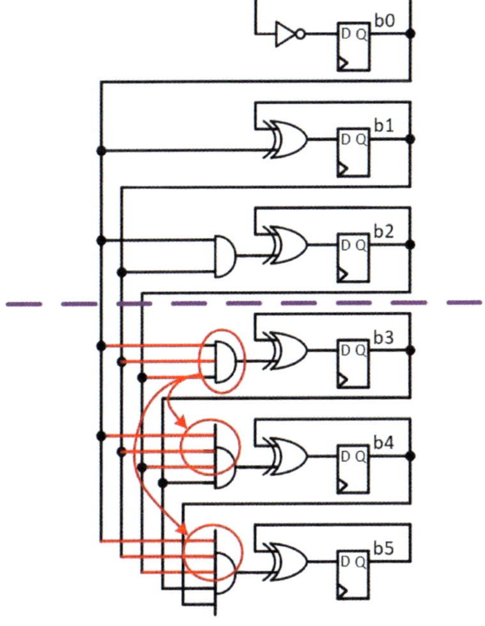

We'll insert our pipeline register here, and carry this three-input AND forward.

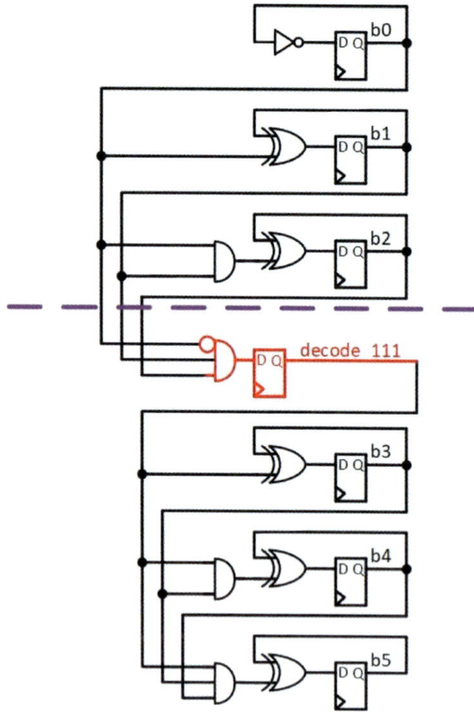

Notice that as we insert a pipeline delay, we change the decode value from "111" to "110", which is one count before the terminal value. This way, when the "110" decode is carried through the one-clock pipeline delay, the inputs to the next stages go active when the previous stages are "111", just how it worked before inserting the pipeline delay (thus, I named that signal "decode_111"). This sort of compensating adjustment is common when adding pipeline registers.

Here's a timing diagram:

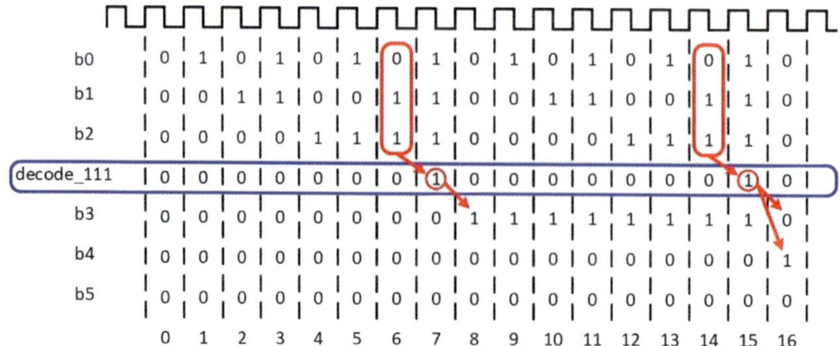

Index

About the Author

Blaine Readler graduated from Penn State with a degree in electrical engineering, and has been developing complex digital systems for over thirty-five years. He has worked with major corporations such as IBM, General Electric, and Northrup-Grumman, as well as small start-ups. He has been designing with FPGAs since their introduction thirty years ago. He has written two FPGA-related reference books, which have sold over ten thousand copies, and has taught introductory programming classes at community colleges. He holds a number of patents, including for the FakeTV home security product.

He also writes science fiction novels, which have very little to do with digital design.

He encourages you to visit him:
http://www.readler.com/

Printed in Great Britain
by Amazon

72651200R00165